SPRINGER TRACTS IN MODERN PHYSICS

Ergebnisse
der exakten Natur-
wissenschaften

Volume **73**

Editor: G. Höhler

Associate Editor: E. A. Niekisch

Editorial Board: S. Flügge J. Hamilton F. Hund
H. Lehmann G. Leibfried W. Paul

Springer-Verlag Berlin Heidelberg GmbH 1975

Manuscripts for publication should be addressed to:

G. Höhler, Institut für Theoretische Kernphysik der Universität, 75 Karlsruhe 1, Postfach 6380

Proofs and all correspondence concerning papers in the process of publication should be addressed to:

E. A. Niekisch, Institut für Grenzflächenforschung und Vakuumphysik der Kernforschungsanlage Jülich, 517 Jülich, Postfach 365

ISBN 978-3-662-15522-6   ISBN 978-3-540-37866-2 (eBook)
DOI 10.1007/978-3-540-37866-2

© by Springer-Verlag Berlin Heidelberg 1975
Originally published by Springer-Verlag Berlin Heidelberg New York in 1975
Softcover reprint of the hardcover 1st edition 1975

Library of Congress Cataloging in Publication Data. Haken, H.: Excitons at high density. (Springer tracts in modern physics; v. 73) Bibliography: p. Includes index. 1. Exciton theory. I. Nikitine, Serge, joint author. II. Bagaev, V. S. III. Title. IV. Series. QCL. S797 vol. 73 [QC176.8.E9] 539'. 08s [530.4'1] 74-22397

Excitons at High Density

Edited by H. HAKEN and S. NIKITINE

Contents

VII. Excitonic Polaritons at Higher Densities

I. Introduction

Survey

H. HAKEN and S. NIKITINE

Excitons at high density and polaritons offer a new field of research within the framework of semiconductor physics. The pioneering work of Nikitine, Gross and others has established that hydrogen-like excitons may be created in semiconductors. These excitons are made up of an electron of the conduction band and a hole of the valence band, usually described by effective masses, which are coupled together by the Coulomb interaction, possibly modified by polarization effects. At this stage a great variety of Rydberg constants and further generalizations beyond the hydrogen atom are found, because both the effective masses and the effective Coulomb interaction depend on the different crystals involved. Furthermore, in certain crystals tensor masses must be taken into account, and excitons may belong to different valleys of the conduction and valence band. Finally, the effective moments of the electron and the hole may differ from that of free electrons.

Detailed investigations of the different bound states of excitons are still in progress, but a new field has recently emerged thanks to the possibility of creating high concentrations of excitons. The usual way of producing such high concentrations is by shining laser light on the insulating crystals. In large classes of crystals excitons may form excitonic molecules ("biexcitons") in analogy to the hydrogen molecule. On the other hand, the effective interaction between excitons may also be repulsive, hence bose condensation of excitons may be possible. Bose condensation of biexcitons has also been considered and there are some experimental indications that such an effect exists.

Another field currently eliciting great interest is the formation of electron-hole droplets. Here the excitonic binding has broken up and some sort of metallic drop has been formed, bound together mainly by the Coulomb exchange interaction between the electrons and holes. A further possibility that has been discussed is the formation of polyexcitons. It is even suggested that excitons may form a crystal, presumably of the type of a quantum crystal, within the crystal. Thus excitons would form a new type of matter within matter. We are convinced that the various possibilities are by no means exhausted, because the mass ratio between the exciton and hole may vary over a wide range.

Many of the properties of excitons are investigated by optical means, so that the detailed study of the interactions between light and excitons at high intensity is of great importance. We have therefore included in this volume several articles dealing with the polariton concept. Polaritons consist of an exciton and a photon. At high concentrations of excitons or polaritons new effects of light propagation appear. Thus, one might expect the dispersion curve of the polariton to be changed and perhaps even to find self-induced transparency of excitons. Moreover, interactions between polaritons can give rise to a number of different scattering processes.

Excitons at high density may also produce stimulated emission of light; this process is of great practical interest because of the extremely high gain of such lasers. The present authors think that this field is worthy of particularly intensive development because many different processes may be competing with each other, and both temporal and spatial transient states may play an important role. It is hoped that the present book will stimulate further research work in this extremely fascinating field. The articles in this book (except those by Novikov and Rashba) were presented at an international symposium at Tonbach, Germany. We wish to thank the Deutsche Forschungsgemeinschaft and the Land Baden-Württemberg for financial support and Mrs. Funke for her efficiency in organizing this meeting.

Introduction to Exciton Spectroscopy *

S. NIKITINE

Contents

1. Introduction

This short paper should not be considered as a review of exciton spectroscopy but only as an introduction. For detailed information, the reader should consult review papers [1–4]. Some of these papers are, however, rather out-of-date. Exciton spectroscopy has developed very rapidly and, to the author's knowledge, no attempt has been made to summarize its present state. This introduction is not intended to fill this gap but to facilitate the approach to the reports that follow. It is limited to classic results: It is hoped that the non-specialist reader will find these remarks helpful.

2. Early Observations

Though some work was carried out on solid-state spectroscopy in the early twenties [5, 6], modern work was started some 22 years ago by Hayashi Mazakuzu [7], by Gross [8] and his group, and by the Strasbourg group [9]. Other groups joined this research later [10].

* This work was carried out within the frame of the agreement on French-German scientific cooperation between C.N.R.S., Paris, Université Louis Pasteur, Strasbourg, and Universität Stuttgart.

Semiconductors and insulating crystals, ionic or covalent, exhibit two types of spectra in the low-energy absorption edge:

a) This edge may exhibit one or several sharp absorption peaks. In some cases, they form series converging to a continuum. The series are sometimes hydrogen-like. The peaks have been shown to be due to the optical formation of excitons, an excited state of the crystal in which an electron and a hole are bound in a state resembling that of the hydrogen atom.

b) The edge may exhibit absorption steps. These steps have been recognized to be due to band-to-band transitions or to transitions to exciton states with the cooperation of phonons. They are known as indirect transitions.

The formation of excitons is essential for an understanding of this part of the spectrum. In high-energy absorption rather broad peaks are also observed. They might be ascribed to exciton transitions of higher energies, or to band-to-band transitions at points in the Brillouin zone where the joint density of states has a pronounced maximum.

Equidistant peaks apparently forming vibrational satellites of exciton peaks have been reported [11]. The separation of these peaks agrees well with the *LO* phonon of the crystal.

It has been shown by the Strasbourg group that the reflection spectra sometimes exhibit drastic anomalies [12]. Most of these anomalies correspond to the first and second exciton peaks of a series. Some exciton spectra are, however, weak and do not give rise to anomalies. The anomalies consist of a pronounced maximum and a sharp minimum of reflection, called excitonic residual ray and missing ray, respectively. Maxima of reflection can also be observed in the high-energy spectrum and can be ascribed to high-energy exciton transitions or to transitions at particular points in the Brillouin zone.

Some weak lines on the low-energy side of exciton peaks depend on the quality of the crystal and have been named "sensitive lines" by the Strasbourg group [13]. They have been ascribed to complexes formed by excitons captured by defects or impurities.

Exciton luminescence was first observed by Grillot [14] independently by the Strasbourg group [15], and by Archangelskaja et al. [16]. This luminescence exhibits narrow and broad lines and sometimes vibrational satellites.

3. Classification of Exciton Transitions

These observations have stimulated theoretical work and it has been shown by Elliott [17] and Haken [18] that several classes of exciton spectra can be described by the theory [3].

a) First-Class Spectra

This class corresponds to transitions according to the selection rules $k=0$, $\Delta k=0$, where k is the wave vector of excitons. In this class excitons in a spectroscopic S state are formed. A series of lines is expected:

$$v = v_\infty - R_{ex}/n^2 \quad \text{with} \quad n = 1, 2, 3 \dots$$

R_{ex} is an excitonic Rydberg constant, $R_{ex} = (R/\varepsilon^2)(\mu/m_0)$, where ε is a dielectric constant, m_0 the mass of the electron in the vacuum at rest, μ the effective mass of the exciton, and R the atomic Rydberg constant. Haken's correction [18] has to be used for the calculation of ε.

The oscillator strength per unit cell is of the form $f = \text{const}/\varepsilon^3 n^3$ and typical values are about 10^{-2} to 10^{-3}. For $v > v_\infty$ the theory predicts a continuous absorption, which corresponds to a band-to-band transition taking into account electron–hole interaction.

Some observed spectra are tentatively ascribed to this class. Though the quantitative agreement is in many cases not very good, the qualitative description and the orders of magnitude are well predicted by the theory. It has to be remembered that the theory is based on assumptions that are only roughly realistic in many cases. CuI, CuCl and CuBr are quoted as examples of such transitions [19]; one of these spectra is shown in Fig. 1. Only this class of spectra shows excitonic residual rays [12]: Fig. 2 shows one of the residual rays reported by the Strasbourg group [20].

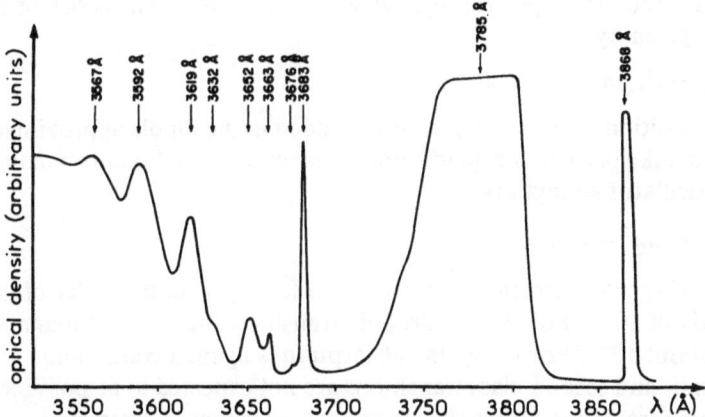

Fig. 1. Spectrum of CuCl (thin film) at 4.2 K showing a number of peaks. The spectrum is ascribed to the first-class type of exciton transition. The spectrum has a somewhat complicated character on account of the overlap of two series (the sharp and diffuse series) separated by the spin-orbit splitting of the valence band. Vibrational satellites can be seen on the high-energy side

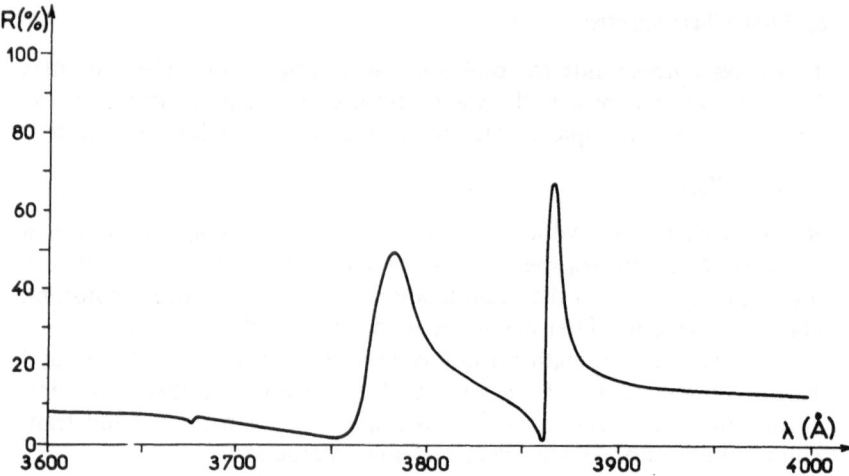

Fig. 2. Reflection spectrum of CuCl at 4.2 K showing strong anomalies, excitonic residual rays and missing rays for the first line of both series

b) Second-Class Transitions

Elliott [17] has shown that, when the matrix element of the transition moment is zero on account of the valence-band (V.B.) and the conduction-band (C.B.) wave functions having the same parity at the Γ point, another class of transitions is possible in the electric dipole approximation. The selection rules are $\Delta k = 0$, and $k \neq 0$ but small. In this case excitons in a spectroscopic P state are created. The series of lines is still given by

$$v = v_\infty - R_{ex}/n^2 \text{ but } n = 2, 3, 4\dots.$$

The transition to the $n = 1$ state is forbidden in the dipole approximation but can take place in the quadrupole or magnetic dipole approximations. The oscillator strength is

$$f = \text{const}(n^2 - 1)/\varepsilon^2 n^5.$$

Again Haken's corrections must be used for ε. f is of the order of magnitude of 10^{-6}. For the quadrupole transition the order of magnitude is of about 10^{-9}. For $v > v_\infty$ the absorption is again a continuum.

First- and second-class transitions are not expected to be possible for the same transition unless the crystal has no center of symmetry.

It has been shown by the Strasbourg group that the yellow and green series of Cu_2O are well described by the theory of second-class transitions. Indications in favor of such transitions have been reported for SnO_2, but transitions of this class are exceptional [21].

c) Indirect Transitions

When the extrema of V.B. and C.B. are not at $k=0$, or when the transition is forbidden for $k=0$, a phonon-assisted transition may take place. These transitions are frequent and give absorption curves exhibiting steps at equal distances on both sides of exciton states. Typical examples are Si and Ge. Cu_2O shows such a transition to the $n=1$ exciton state. Figure 3 illustrates the three classes of transitions.

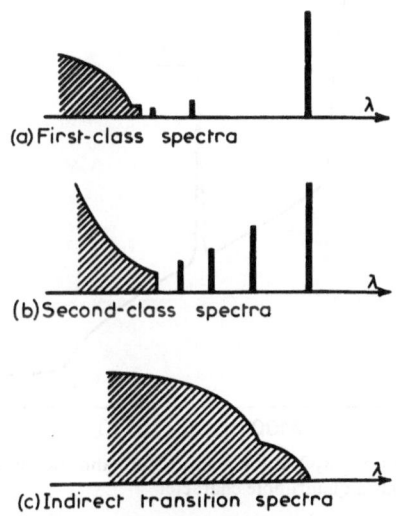

(a) First-class spectra

(b) Second-class spectra

(c) Indirect transition spectra

Fig. 3. Schematic representation of the three types of exciton spectra described in the text

d) High-Energy Spectra

Excitons transitions may appear not only on the fundamental edge but also at higher energies. In this case the transition involves higher C.B. or lower V.B. However, high-energy peaks are not necessarily due to excitons. They may arise from a special configuration of bands and, in particular, from high values of the joint density of states at particular points in the Brillouin zone.

4. The Spectrum of Cu_2O

The spectrum of Cu_2O is quite remarkable. On the low-energy side of the absorption lines two edges are observed at equal distances on both sides of a very sharp but weak line. The edges have been interpreted [3] as phonon(105 cm^{-1})-assisted transitions. The line is the $n=1$ forbidden quadrupole line of the yellow series [23, 24]. The yellow and green series are second-class transitions to the first C.B. [25, 26, 3]. In the blue and violet the absorption becomes very strong and two strong peaks are observed [27]; they are ascribed to first-class transitions to the next higher C.B. Some other peaks that have not yet been identified are observed at higher energies [28].

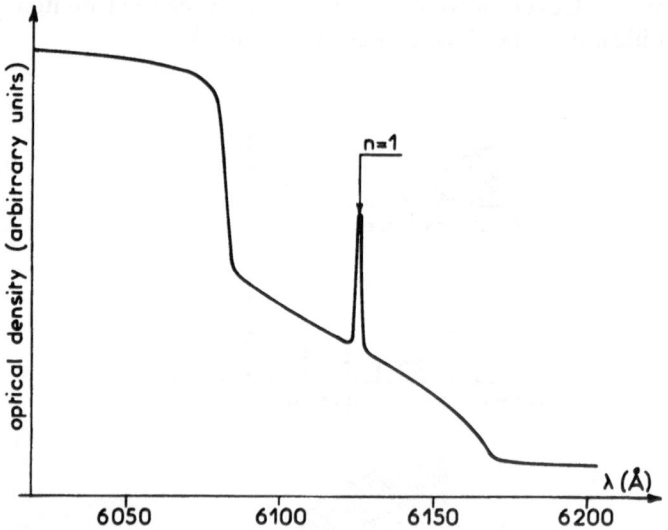

Fig. 4. The two red edges of Cu_2O at low temperature and the forbidden $n = 1$ line of the yellow series. This line has a quadrupole character

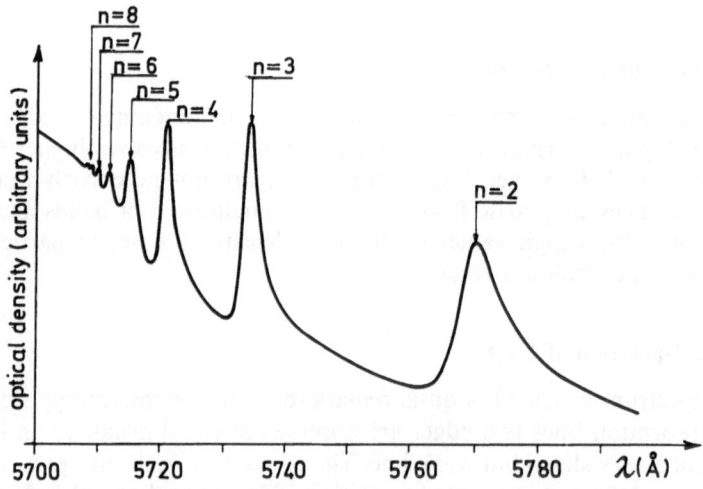

Fig. 5. The yellow series of Cu_2O at 4.2 K of an exceptionally good sample. This series, as well as the green series, is ascribed to second-class exciton transitions.

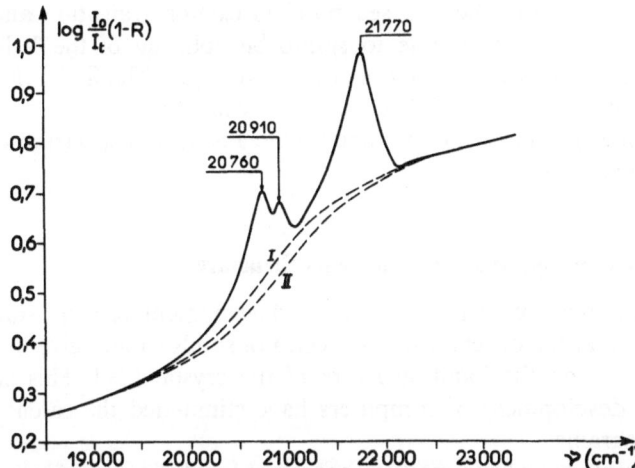

Fig. 6. Strong absorption peaks in the blue and violet part of the visible spectrum at 4.2 K. They are ascribed to the allowed first-class exciton spectrum of Cu_2O (transition to a higher conduction band)

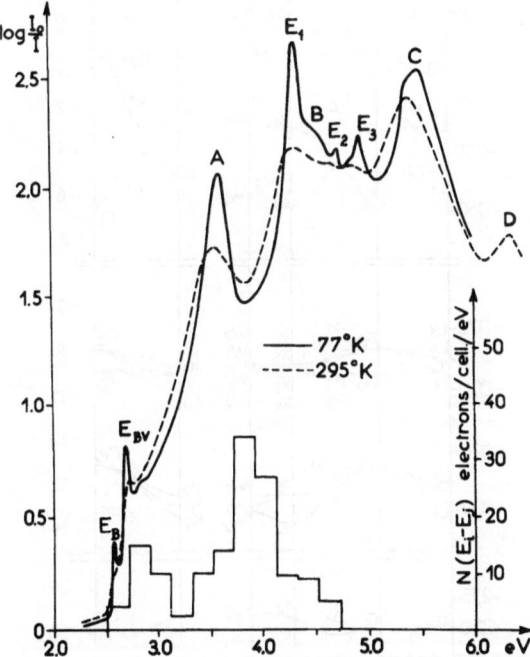

Fig. 7. The high energy spectrum of Cu_2O after Brahms. Representation of the density of states calculated by Dahl

The splitting of the two second-class exciton transition and of the two first-class peaks is due to spin-orbit splitting of the V.B. These bands are Γ_7^- and Γ_8^-, taking in account the spin. The first C.B. has the symmetry Γ_6^+, the second C.B. has Γ_{12}^+ symmetry.

Figures 4–7 illustrate the remarkable absorption spectrum of Cu_2O at low temperatures.

5. Solid-State Spectroscopy and Band Structure

It was pointed out very early in the development of solid-state spectroscopy that the observation of spectra of solids should give important information on the band structure of the crystal [29]. This situation and the development of computers have stimulated the calculation of band diagrams.

Attempts have been made to construct band diagrams from absorption spectra directly by the so-called method of pseudopotentials.

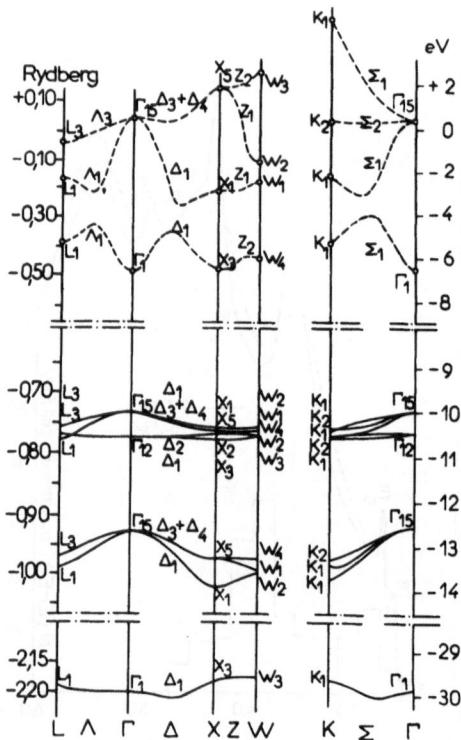

Fig. 8. Band-structure diagram of CuCl according to Song

However, attempts to calculate band structures by usual methods, in spite of the great amount of work involved, have also been successful. These calculations usually depend on one or several parameters, which are subsequently adapted to obtain the best fit with the observed spectrum. Band-structure calculations have been carried out for a great many substances during the last 10 years or so [30]. Figures 9 and 8 give the band structures of Cu_2O and CuCl [31, 32].

The reflection and absorption spectra of crystals in the visible and in the UV often present number of more or less sharp peaks, edges, shoulders and other singularities. Some of these peaks and edges can be assigned to a given type of transitions in the Brillouin zone, e.g. the exciton transitions on the low-energy edge of the spectrum (often at the

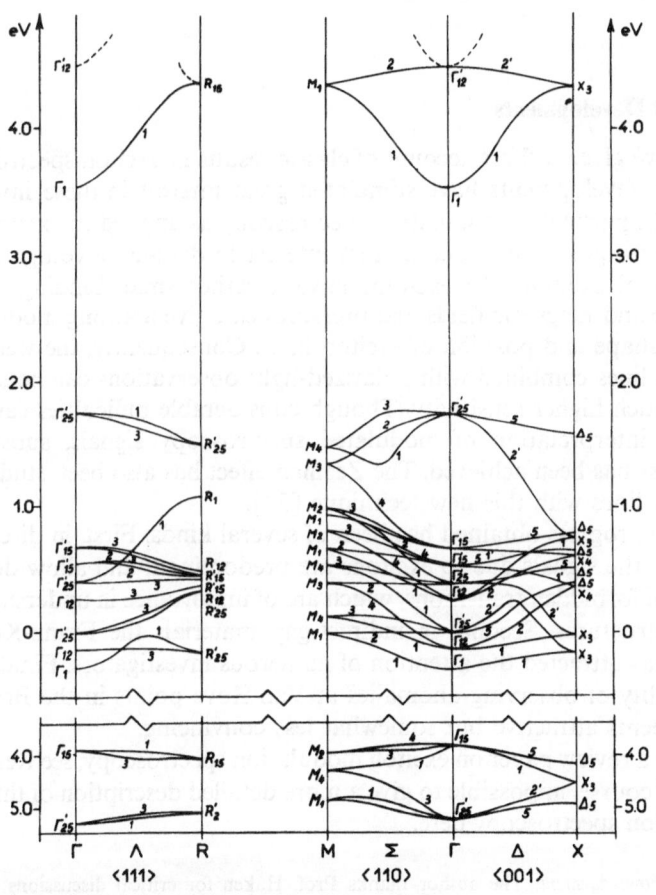

Fig. 9. Band-structure diagram of Cu_2O according to Dahl

Γ point of the zone). However, a major task of band-structure analysis is to assign in a reliable way the observed structures of optical spectra to critical points of a certain type and location in the Brillouin zone (van Hove points M_0, M_1, M_2, M_3) at which the joint density of states is responsible for the structures observed. In most cases this is very complicated and difficult. Any structure caused by a critical point is usually superimposed on absorption (or reflection) due to different transitions from noncritical areas of the Brillouin zone and may be practically impossible to identify or even to observe. More sensitive methods had to be found for such determinations, and such methods have been developed recently. Their application to exciton spectroscopy has been tried recently by the Strasbourg group and independently by other authors [33] and has proved very successful. This new development is discussed in the next section.

Recent Developments

We have given a short account of classic results in exciton spectroscopy. Recent developments have stimulated great interest in these investigations. In particular, modulation spectroscopy as applied to excitons has given new possibilities and a great interest to detailed investigations of spectra of excitons. As excitons have a rather small binding energy, electric and magnetic fields and pressures can give a strong modulation of the shape and position of exciton lines. Consequently, the weak forbidden lines combined with polarized-light observations can be studied with much higher sensitivity. Though considerable difficulties can arise in the interpretation of modulated spectroscopy signals, substantial progress has been achieved. The Zeeman effect has also been studied on exciton lines with this new technique [34].

The progress obtained has been of several kinds. First, in direct-gap crystals the signals due to excitons are predominant and allow detailed study of forbidden transitions, which are of importance in understanding band structures. Second, in indirect-gap materials the Franz-Keldysh effect has attracted the attention of numerous investigators. Finally, the possibility of observing anomalies at Van Hove points in the Brillouin zone seems attractive but somewhat less convincing.

For a review paper on exciton modulation spectroscopy, see Ref. [35]. It is, of course, impossible to give a more detailed description of this part of exciton spectroscopy here.

Acknowledgement. The author thanks Prof. Haken for critical discussions on this introduction.

References

The list of references given below does not claim to be complete.

1. Gross,E.F.: Nuovo Cimento Suppl. **3**, 672 (1956); — Advan. Phys. Sci. Moscow. **63**, 575 (1957).
2. Nikitine,S.: J. Chim. Phys. **55**, 621 (1958); — Phil. Mag. **4**, 1 (1959); — J. Phys. Chem. Solids **8**, 190 (1959).
3. Nikitine,S.: Progr. Semiconductors **6**, 235 (1962). See also, Optical Properties of Solids, p. 197. New York: Plenum Press 1969. In these two publications some bibliography can be found.
4. Knox,R.S.: Theory of Excitons, Solid State Phys. Suppl. **5** (1963); Academic Press. Dexter,D.L., Knox,R.S.: Excitons. New York: J. Wiley 1965.
5. Hilsch,R., Pohl,R.W.: Z. Phys. **48**, 384 (1928); **57**, 145 (1929); **59**, 812 (1930).
6. Obreimoff,I.W., Prikhot'ko: Phys. Z. Sovjetunion **9**, 48 (1936).
7. Hayashi,M., Katzuki,K.: J. Phys. Soc. Japan **5**, 380 (1950); **7**, 599 (1952). See also, Hayashi,M.: J. Fac. Sc. Hokkaido Univ. **4**, 107 (1952).
8. Gross,E.F., Karryev,N.A.: C.R. Acad. Sci. URSS **84**, 261 (1952); **84**, 471 (1952).
9. Nikitine,S., Perny,G., Sieskind,M.: J. Phys. Rad. **15**, 18 (1954); — C.R. Acad. Sci. Paris **238**, 67 (1954).
10. MacFarlane,G.G., McLeanT.P., QuarringtonJ.E., Roberts,V.: J. Phys. Chem. Solids **8**, 388 (1959). See also, Phys. Rev. **108**, 388 (1959). See also, Phys. Rev. **108**, 1377 (1957); — Proc. Phys. Soc. **71**, 863 (1958).
    Zwerdling,S., Lax,B., Roth,L.: J. Phys. Chem. Solids **8**, 397 (1959). See also, Phys. Rev. **108**, 1402 (1957); **109**, 2207 (1958).
11. Ringeissen,J., Lewonczuk,S., Coret,A., Nikitine,S.: Phys. Letters **22**, 571 (1966). See also, Ringeissen,J., Coret,A., Nikitine,S.: Intern. Conf. on Localized Excitations in Solids, Irvine, p. 298 (1968). New York: Plenum Press.
12. Nikitine,S., Reiss,R.: C.R. Acad Sci. Paris **242**, 238 (1956); **242**, 1003 (1956).
13. Nikitine,S.: Phil. Mag. **4**, 1 (1959). See also Ref. [3] for the very extensive literature on this subject; see in particular the review papers of:
    Hopfield,J.J.: 7ᵉ Congrès sur la Phys. des Semiconducteurs, 726 (1964), Paris.
    Broser,I.: J. Phys. Suppl. **28**, C-3, 95 (1967).
    and numerous papers of the Strasbourg Group, see Ref. [2 and 3] and
    Nikitine,S., Certier,M., Ringeissen,J., Merle,J.C.: C.R. Acad. Sci. Paris **262**, 198 (1966).
    Nikitine,S., Mˡˡᵉ Roth,D., Certier,M.: J. Phys. **27**, 329 (1966).
    Nikitine,S., Munschy,G., Ringeissen,J., Mˡˡᵉ Kirch,M.: J. Phys. Suppl. **28**, C-3, 116 (1967).
    Certier,M., Wecker,C., Nikitine,S.: J. Phys. Chem. Solids **30**, 2135 (1969).
14. Grillot,E.F., Bancie-Grillot,M., Pesteil,P., Zmerli: C.R. Acad. Sci. Paris **242**, 1794 (1956).
15. Nikitine,S., Perny,G.: J. Phys. Rad. **17**, 1017 (1956).
    Nikitine,S., Reiss,R.: C.R. Acad. Sci. Paris **244**, 2788 (1957); **245**, 52 (1957), and subsequent numerous publications of the Strasbourg Group.
16. Arkhangelskaia,V.A., Feofilov,P.P.: J. Phys. Rad. **17**, 824 (1956).
17. Elliott,R.J.: Phys. Rev. **108**, 1384 (1957).
    Elliott,R.J., Loudon,R.: J. Phys. Chem. Solids **8**, 382 (1959).
    See also, Overhauser,A.W.: Phys. Rev. **101**, 1702 (1956).
18. Haken,H.: Halbleiter Probleme II. Ed. W. Schottky. Berlin: Vieweg 1955; — Fortschr. Phys. **38**, 271 (1958); — J. Phys. Chem. Solids **8**, 166 (1959); — Z. Phys. **155**, 223 (1959).
19. Nikitine,S.: Progr. Semiconductors **6**, 235 (1962); — Opt. Properties Solids, p. 197. New York: Plenum Press 1969.
    Reiss,R.: Thèse. Strasbourg (1959): Cahiers de Phys. **13**, 129 (1959).
    Ringeissen,J., Nikitine,S.: J. Phys. Suppl. **28**, C-3, 48 (1967).

20. Ringeissen, J., Coret, A., Nikitine, S.: Intern. Conf. on Localized Excitations in Solids, Irvine, p. 298 (1968). New York: Plenum Press.
21. Nagasawa, M., Shwoya, S.: Phys. Letters **22**, 409 (1966).
22. Nikitine, S., Grun, J. B., Sieskind, M.: Proceedings of the Intern. Conf. on Semiconductors, Prague, 414 (1960). See also, Grun, J. B., Sieskind M., Nikitine, S.: J. Phys. Chem. Solids **19**, 189 (1961).
23. Gross, E. F.: International Conf. on Semiconductors, Prague, 407 (1960). Gross, E. F., Kapliansky, A. A.: C.R. Acad. Sci. URSS **132**, 98 (1960).
24. Grun, J. B.: Rev. Opt. **41**, 439 (1962). Nikitine, S., Grun, J. B., Certier, M., Deiss, J. L., Grosmann, M.: Proceedings Intern. Conf. Semiconductors, Exeter, 441 (1962). Nikitine, S., Grun, J. B., Certier, M.: Phys. Condens. Mater **1**, 214 (1963). Certier, M., Grun, J. B., Nikitine, S.: J. Phys. Rad. **25**, 361 (1964).
25. Nikitine, S., Grun, J. B., Sieskind, M.: J. Phys. Chem. Solids **17**, 292 (1961). See also, Grun, J. B., Nikitine, S.: J. Phys. **24**, 355 (1963).
26. Grun, J. B., Sieskind, M., Nikitine, S.: J. Phys. Rad. **22**, 176 (1961).
27. Grun, J. B., Sieskind, M., Nikitine, S.: J. Phys. Chem. Solids **21**, 119 (1961) Daunois, A., Deiss, J. L., Mayer, B.: J. Phys. Rad. **27**, 142 (1966).
28. Brahms, S., Nikitine, S.: Solid State Commun **3**, 209 (1965).
29. Nikitine, S.: J. Phys. Rad. **17**, 817 (1956).
30. See for example, Gautier, J.: J. Phys. Suppl. **28**, C3-3 (1967).
31. Dahl, J. P., Switendick, A. C.: J. Phys. Chem. Solids **27**, 931 (1966). See also, Brahms, S., Nikitine, S., Dahl, J. P.: Phys. Letters **22**, 31 (1966).
32. Song, K. S.: J. Phys. Suppl. **28**, C3-43 (1967).
33. Lange, H., Gutsche, E.: Phys. Stat. Sol. **32**, 293 (1969). Mohler, E.: Phys. Stat. Sol. **29**, K 55 (1968).
34. The bibliography on differential optical methods can be found for example on the paper of Seraphin, B. O.: J. Phys. Suppl. **8**, C3-73 (1967). See also, Cardona, E.: Intern. Conf. on Semiconductors, Moscow (1968). Cardona, M.: Modulation Spectroscopy. New York-London: Academic Press 1969.
35. Nikitine, S.: Congrès Européen de Spectroscopie Moléculaire, Liège (Belgique). Mémoires de la Société Royale des Sciences, Liège (1970). Deiss, J. L., Daunois, A.: Proc. of the First Intern. Conf. on Modulation Spectroscopy, Tucson (1972); — Surface Sci. **37**, 804 (1973).

Prof. Dr. S. Nikitine
Laboratoire de Spectroscopie
et d'Optique du Corps Solide
Université Louis Pasteur
5, rue de l'Université
F-67000 Strasbourg (France)

II. Biexcitons

Properties of Biexcitons[*]

S. NIKITINE

Contents

1. Introduction — High Concentration of Excitons

When the concentration of excitons becomes very high, a number of new effects, specific to this situation, can be observed. This situation can be induced by optical excitation with laser light or by electron beams or in *pn* junctions. These effects are of considerable interest and will be described here. The review will concern, however, only optical excitation of biexcitons. A more extended review has been given by the author elsewhere [1].

When the concentration of excitons obtained by optical excitation is sufficiently high, biexcitons can be formed. Biexcitons are molecules

---

[*] This work was carried out within the frame of the agreement on French-German scientific cooperation between C.N.R.S., Paris, Université Louis Pasteur, Strasbourg, and Universität Stuttgart.

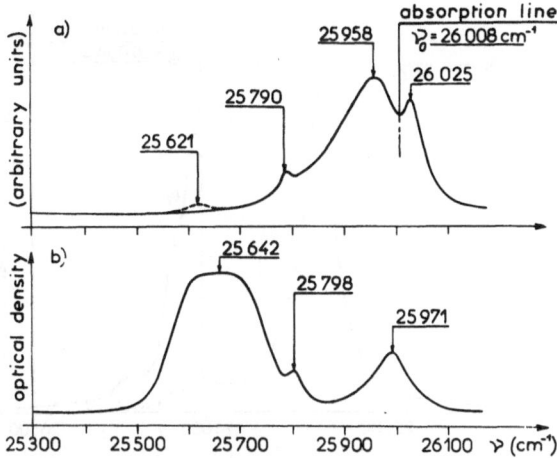

Fig. 1. One of the first spectra (1968) in which the luminescence of a biexciton recombination appears in CuCl at 77 K with laser light excitation

formed with two excitons in a similar way to H_2 [1]. They were predicted by Lampert and by Moskalenko [2] a long time ago and evidence for the existence of biexcitons has been found experimentally by the Strasbourg Group [3] and in other laboratories later [4, 5]. In an early publication, Haynes [6] was first to suggest the interpretation of his observation by a radiative decay of biexcitons formed in Si. These observations may be, however, interpreted in another way. Further on the mechanism of decay suggested by Haynes seems to be unprobable.

2. Experimental Situation for CuCl and CuBr

The experimental situation is as follows. The luminescence of CuCl and CuBr is now well known when the excitation is of a conventional type [7]. However, the luminescence excited by very intense UV laser radiation has been shown to have various rather remarkable properties [3]. At 77 K, the spectrum comprises mainly the resonance line v_0 and a new line v_B becoming very strong at high intensities of excitation. This line is not observed at low intensities of excitation light i. The difference of $v_0 - v_B = 350$ cm^{-1} for CuCl and 236 cm^{-1} for CuBr. The

---

[1] It is suggested to give the name of "biexcitons" to these molecules. The name often used of excitonic molecules is rather ambiguous. It is probable that molecules formed of more than two excitons may be observed. In this case, "excitonic molecule" could be used for the entire class of molecules, biexcitons, quadriexcitons, etc. ... as well as higher "polymers" of excitons, as suggested by Kittel.

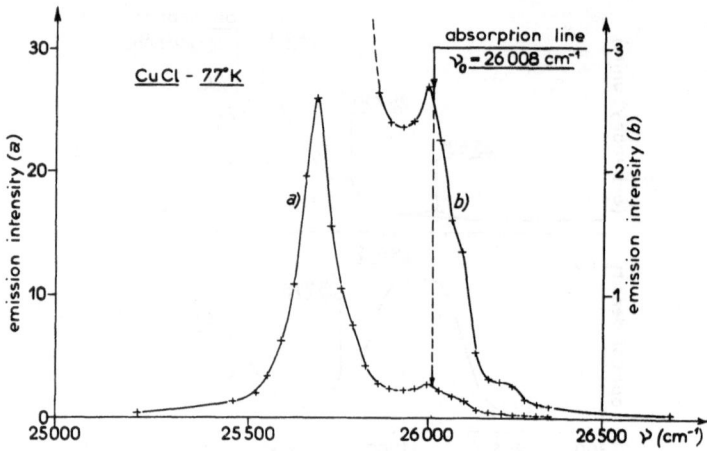

Fig. 2. A similar spectrum showing the great difference of intensities between v_B and v_0

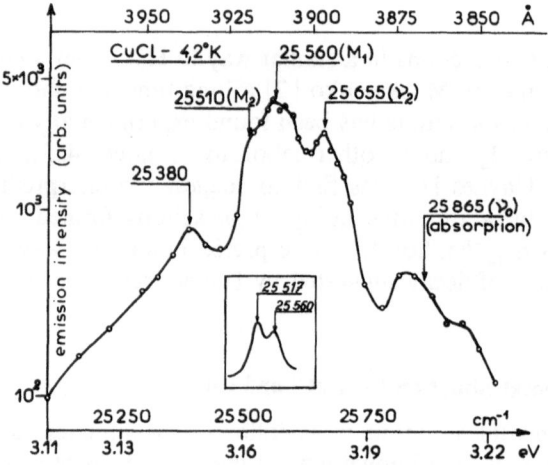

Fig. 3. The emission line v_B of CuCl excited by laser light at 4.2 K. This spectrum shows a structure M_1 and M_2. v_2 is an exciton acceptor and (ex-Lo) line

intensity of luminescence light I_B of line v_B can be represented by the relation

$$\log I_B = m_B \log i + ct.\tag{1}$$

In a wide range of intensities m_B is somewhat lower than 2 but tends progressively to $m_B = 1$ for very high intensities. At 4.2 K, similar experimental results are obtained. However, at this temperature other

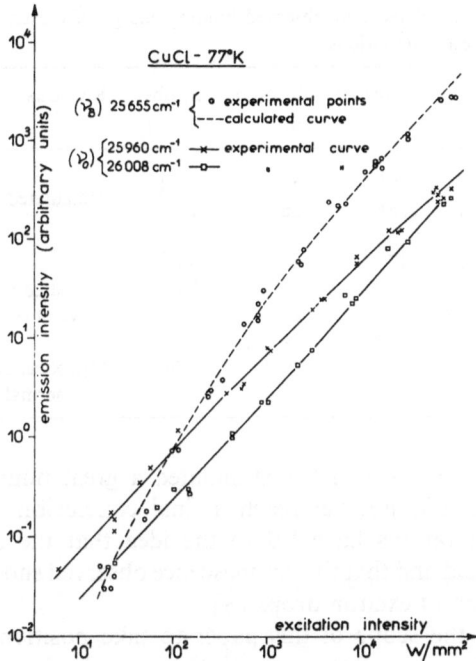

Fig. 4. Variation of I_B and I_0 taken for two different wave lengths as a function of i, the intensity of the incident radiation

lines overlap with the new lines mentioned above, and the variation of their intensity may influence I_B and I_0, the intensity of ν_0. It is assumed that this influence between ν_B and ν_0 is also present at 77 K and that it is responsible for some perturbations of the variation of both I_B and I_0 observed recently (Figs. 2–4).

I_0 is proportional to i for all intensities.

It was suggested that the line ν_B is due to a radiative recombination of biexcitons according to the process discussed later. The interpretation is based primarily on the comparison of observed and calculated values of the binding energy and on the kinetics of the process.

3. Experimental Situation for Other Substances

As stated above, Haynes has suggested that some of the peaks of luminescence observed in Si with even not too high excitation may be considered as stemming from a decay process involving the recombination of a biexciton. Similar experiments have been performed more recently for Si and Ge. As it will be seen below, this interpretation has

Table 1. Values of calculated and observed binding energy of biexcitons and numerical data involved in these calculations

	E_{ex} cm^{-1}	σ	E_B (calc.) cm^{-1}	E_B (obs.) cm^{-1}	$\dfrac{E_B \text{ obs.}}{E_B \text{ calc.}}$	Remarks
CuCl	1523	0.02	350*	350	1	*calculated from formula (20)
CuBr	873	0.01	235*	236	1	
ZnO	483	0.47	18.5			
CdS	226	0.17	17	43.5	3.1	
CdSe	122	0.3	7	24	3.4	evaluated from the corrected
Ge	34	0.665	1	28	28	curve of Hanamura
Si	64	0.447	2.4	44	18.3	
Cu$_2$O	1105	0.75	80**	99**	1.26	**process not consistent with the thermal variation

led to a controversy which has stimulated a great number of elegant experiments but did not yet reach a final conclusion. Objections to Haynes' interpretations have led to the idea that the excitons could condense to a fluid and that the luminescence observed should be ascribed to recombination of exciton drops [8].

It is out of the scope of this paper to take position in the above mentioned controversy. But a critical discussion is given below.

The experimental data concerning the experimental binding energy of biexcitons are given in Table 1, fourth column.

Experiments of different compounds have been suggested to give indications in favour of a biexciton interpretation. All these suggestions are based on strong arguments but they also contain shortcomings which have not yet been explained. In particular experiments on CdS [5], CdSe [5], Cu$_2$O [9] as well as the mentioned experiments on Si and Ge are discussed later. This discussion does not pretend to give a complete review of known data for all substances investigated.

It should also be mentioned that the study of the kinetics of the luminescence is mentioned in most of these papers, but the detailed description of the experiments and results is usually scarce. These experiments are, however, of some difficulty and the results, when they are sufficiently well established, are of importance.

4. Discussion of Different Processes Involving the Radiative Decay of Biexcitons

The usual procedure consists in exciting the crystal with UV laser light. The absorption of this UV excitation is usually in the continuum situated

Note added in proof: These new data on ZnO have been kindly communicated to the author by Dr. Klingshirn and Dr. R. Helbig, Erlangen, recently.

in the spectrum above the energy gap. This corresponds to the creation of hot free carriers.

At very high intensities of excitation in the continuum, a very large number of free carriers are created. As shown by Toyozawa, after thermalization, most of them combine at low temperatures to excitons. As the concentration of these excitons is very high, they may associate to form biexcitons.

a) One of the recombination processes of the biexciton (in CuCl and CuBr) may perhaps be as follows [10]: The biexciton radiates a photon $h\nu_B$ and a free exciton is left:

$$(\text{biex}) \to (\text{ex}) + h\nu_B . \tag{2}$$

The energy conservation now gives:

$$2(E_g - E_{ex}) - E_B + E_B^k = h\nu_B + (E_g - E_{ex}) + E_{ex}^k \tag{3}$$

which gives:

$$h\nu_B = E_g - E_{ex} - E_B + E_B^k - E_{ex}^k . \tag{3'}$$

E_g is the energy gap, E_{ex} the binding energy of the exciton, E_B the binding energy of the biexciton and E_B^k and E_{ex}^k are, respectively, the kinetic energies of the biexciton and of the resulting exciton. E_B^k is rather small at low temperatures, but E_{ex}^k is probably not small. Both account for the breadth of the line ν_B; the maximum of the line is, however, obtained in first approximation by neglecting both. This gives:

$$h\nu_B(\text{max}) = E_g - E_{ex} - E_B \tag{4}$$

as $E_g - E_{ex} = h\nu_0$ is well known from resonance emission:

$$E_B = h(\nu_0 - \nu_B) \tag{5}$$

The above relations can also be written in a shorter form introducing G_{ex} for the so called "exciton energy gap" $G_{ex} = E_g - E_{ex}$. This gives $h\nu_B(\text{max}) = G_{ex} - E_B$.

b) A second process of recombination has been suggested by Haynes: the biexciton radiates a photon $h\nu_B$ and a free hole and a free electron for direct transitions; one can write:

$$(\text{biex}) \to e + h + h\nu_B \tag{6}$$

where e and h are written for electron and hole. The energy balance now gives, neglecting again the kinetic energies:

$$h\nu_B = E_g - 2E_{ex} - E_B . \tag{7}$$

If such a process is considered:

$$E_B = h(v_0 - v_B) + E_{ex} \tag{8}$$

this value is larger by the amount E_{ex} than in process a). However, it appears from calculations of Kikuo Cho [11] that this process is much less probable than the process a). The same process can be written also for indirect transitions as imagined initially by Haynes. In this case, the process leads to the relation:

$$(biex) \rightarrow h v_B + e + h + \hbar \omega_{K_0} . \tag{9}$$

The last term being the phonon energy.

We will discuss these processes later with their advantages and short-comings.

c) Finally, a third process has been suggested by Gross and Krein-gold [9]. This process will be written in a somewhat modified form.

It is suggested that in the process a) the liberated exciton is in an excited state. This can then explain the observation of a reversed series of lines. The process is as follows:

$$(biex) \rightarrow h v_B + (ex)^+ \tag{10}$$

where $(ex)^+$ is written for an excited state of the exciton. The energy balance reads (writing G_{ex} (exciton gap) for $E_g - E_{ex}$)

$$2 G_{ex} - E_B = h v_B + G_{ex}(n) . \tag{11}$$

In this process, it is assumed that G_{ex} refers to the lowest state of the exciton and $G_{ex}(n)$ to the n-th excited state of the exciton. This gives again neglecting the kinetic energies:

$$h v_B = G_{ex} - E_B + G_{ex} - G_{ex}(n) . \tag{12}$$

The first two terms on the right hand side are identical to those of process a). The two others are:

$$G_{ex} - G_{ex}(n) = E_g - E_{ex}(1) - E_g + E_{ex}(n) = - E_{ex}(1) + E_{ex}(n)$$

where $E_{ex}(1)$ and $E_{ex}(n)$ are the binding energies of the exciton in the states $n = 1$ and in the n-th state. This gives now:

$$h v_B(n) = G_{ex} - E_B - R_{ex}(1) ch + R_{ex}(n) ch/n^2 . \tag{13}$$

Here, $R_{ex}(1)$ and $R_{ex}(n)$ are the Rydberg constants of the exciton for $n = 1$ and for the n-th state. It is well known that exciton levels are usually not exactly hydrogen-like on account of Haken's correction due to the choice of the dielectric constant, on account of what $R_{ex}(1) > R_{ex}(n)$.

This means that the $v_B(n)$ form a reversed series of lines, the limit of which is:

$$h v_B(\infty) = G_{ex} - E_B - E_{ex}(1)$$

which is identical with (7). Again for $n = 1$ we obtain the process a) and (4). The binding energy can be obtained from the formula:

$$E_B = h(v_0 - v_B(n)) - ch\, R_{ex}(1) + ch\, R_{ex}(n)/n^2 \qquad (13')$$

the first term on the right hand side is measured, $R(1)$ and $R(n)$ are usually known from the exciton spectrum. As it will be seen, no experimental evidence has been published in favour of such a process, which however seems to be quite possible.

No other processes[2] seem to have been suggested so far; the biexciton line v_B can, however, present fine structures of another origin which will be discussed later. It can be seen that in all the processes discussed in this section the binding energy of the biexciton can be evaluated from the experiment.

The experimental justification of the hypothesis of formation of biexcitons has so far been given in comparing a) the observed binding energy E_B(obs) with the calculated value E_B(calc) and b) the observed kinetics with the calculated one. It has to be mentioned that both present some difficulties. E_B(obs) can be evaluated with some accuracy but the calculated values E_B(calc) may well contain some inaccuracies. The comparison of the observed kinetics with the calculated one is also not quite simple but can be performed on a qualitative basis. The quantitative comparison is not possible so far as the different coefficients involved are not well known. With these restrictions, however, the comparison can be justified.

It is hoped that other arguments will be found in the near future in developing both experimental and theoretical research in this direction.

---

[2] *Remark.* An other process has been suggested to Hiroshi, Saito et al. [12], which is supposed to explain the observation of a new line $v_{B'}$. This process is as follows. It is supposed that the collision of two biexcitons is not elastic, giving as a result three excitons and a photon

$$(2 G_{ex} - E_B) + (2 G_{ex} - E_B) + E_B^K = h v_{B'} + 3 G_{ex} + E_{ex}^K$$

which gives

$$G_{ex} - 2 E_B + E_B^K - E_{ex}^K = h v_{B'} . \qquad (14)$$

The question arises whether this process in which $3 G_{ex}$ are liberated is realistic a biexciton being more stable then two excitons. The above process may however by plausible if $(E_B^K - E_{ex}^K)$ is sufficiently high.

5. Calculation of the Binding Energy of Biexciton E_B

a) The binding energy of biexcitons has been calculated as a function of E_{ex} and $\sigma = m_e^*/m_h^*$ by different authors [13–18]. It must be noted that these calculations are difficult. In the calculation by Hanamura [13], a variational method is used which may give somewhat too low values of E_B. The results of the calculations are given in form of a curve calculated by using a computer. This curve reproduced on Fig. 4 is easy to be used for $\sigma \leqq 1$ but the use of this curve is not very accurate at $\sigma \ll 1$ as the slope becomes very steep. Values deduced from this curve for some substances are given in Table 1. However, we have limited the use of the Hanamura's curve to values of $\sigma > 0.1$. We have preferred to use a direct analytical formula which can be given in this approximation. It is out of the scope of this review to give the details of Hanamura's calculations. It has to be mentioned, however, that the calculations of Hanamura do not take into account the polarisability of the lattice which may bring some corrections which can be substantial in some cases. An attempt to take care of these corrections has been made by Elkomoss and Stébé [19] for exciton complexes and by Büttner et al. recently, for biexcitons [17].

b) For $\sigma < 0.1$ an approximate calculation can be carried out in an analytical form.

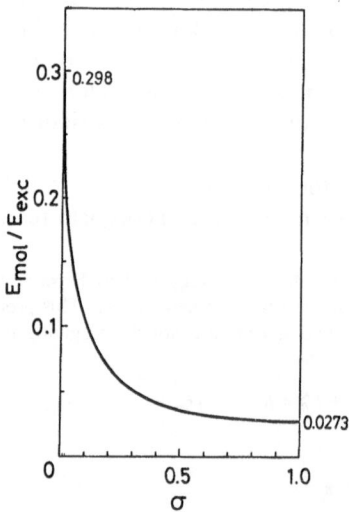

Fig. 5. The binding energy of the biexciton as a function of $\sigma = m_e^*/m_h^*$ according to the calculations of Hanamura

A quite good approximate value can be obtained from simplified calculations. The theory concerns small values of m_e^*/m_h^*: it was given by the author and Grun and Myzyrowicz some time ago [18a]; an improved version of this very simple theory is as follows [18c], taking advantage of the evident analogy of this problem with the theory of H_2.

Consider the depth of the potential energy well of H_2 and of the biexciton: D_{H_2} and D_B. The problem of the biexcitons is quite similar to that of the hydrogen molecule if $m_e^*/m_h^* \ll 1$, but the energy scale is changed in the ratio of E_{ex}/E_H where E_H is the binding energy of the hydrogen atom. So one can assume in a very good approximation:

$$D_B = D_{H_2}(E_{ex}/E_H). \tag{15}$$

In order to obtain from the above relation the binding energy of the biexciton, one has to take into account the zero-point vibration energy E_B^{vo} and the correction of non-adiabatic movement of the holes E_B^{ad}; this assumes that the Born-Oppenheimer approximation is justified. This gives then

$$E_B = D_B - E_B^{vo} - E_B^{ad} . \tag{16}$$

For H_2 the zero-point vibration energy is rather small in comparison to the depth of the potential well and the non-adiabatic correction is negligible.

For biexcitons, however, the zero-point vibration energy can be become quite large in comparison to D_B in some substances. This would give a limitation for the use of this method which can be written:

$$D_B \gg (E_B^{vo} + E_B^{ad}). \tag{17}$$

However, if this is not the case, the Born-Oppenheimer approximation collapses and the approximate theory given here is not applicable.

E_B^{vo} can be calculated from the relation which is readily found

$$E_B^{vo} = (E_{ex}/E_H)\,(m_e^*/m_h^*)^{1/2}\,(M_p/m_0)^{1/2}\,E_{H_2}^{vo} \tag{18}$$

where M_p/m_0 is the ratio of the mass of a proton to the mass of the electron at rest and $E_{H_2}^{vo}$ is the zero-point vibration energy of H_2.

The nonadiabatic correction is small. It should, however, not always be neglected. The calculation of this correction has been carried out by Kazamanyan [20].

The complete formula is now:

$$E_B = (E_{ex}/E_H)\,[D_{H_2} - \sigma^{1/2}\{M_p/m_0\}^{1/2}\,E_{H_2}^{v} - 10^5\,\sigma] \tag{19}$$

using the values of known constants this can be written as follows:

$$E_B = E_x 9.10\,10^{-2}\,[4.03 - 9.4\,\sigma^{1/2} - 10.1\,\sigma]\,\text{cm}^{-1}. \tag{20}$$

This formula has some analogy to the formula given independently by Kazamanyan (and also by Wehner). It has an advantage with respect to both as for the former the use of the experimental value of E_{ex} avoids the ambiguity on the use of a suitable dielectric constant and to some extent it takes into account the polarisability of the lattice. With respect to the second theory on top of the above advantage, the use of the correction for non-adiabaticity seems to be important as in some cases it is not negligible.

It has, however, the strict limitations of applicability to values of $\sigma \ll 1$, to say $\sigma < 0.1$.

The above formula has been used for the calculation of E_B for CuCl and CuBr, reported in Table 1.

c) *The question of anisotropic masses.* It was suggested some time ago that in the case of anisotropic masses of hole or electron the values of masses most favorable for the stability of biexcitons could be chosen when calculating E_B.

In a recent preprint, however, Akimoto [21] has kindly informed the author that a more rigorous treatment shows that the masses in this case should be between the harmonic $\bar{m}_H$ and geometric $\bar{m}_g$ mean values.

$$\bar{m}_g = (m_\perp^2 m_\parallel)^{1/3}$$

$$\bar{m}_H = 3[m_\perp m_\parallel / (2m_\parallel + m_\perp)]$$

where $m_\parallel$ and $m_\perp$ are masses in the parallel or perpendicular directions to the axes of symmetry. As these masses are usually not very different only one of them has been used ($\bar{m}_H$).

6. Comparison of Calculated and Observed Values of the Binding Energy E_B

Values of the binding energy E_B(calc) predicted by the calculation are given in Table 1 together with the corresponding observed data E_B(obs). The values of the other values are evaluated from Hanamura's curve[3]. Experimental data of E_{ex} are taken from the experiment in both cases. The comparison between both values is discussed in this section.

a) CuCl and CuBr

The surprising fit between the observed and calculated values of E_B should not be overestimated. The values of σ for both substances are estimated from the spectra with a moderate accuracy. As the value of E_B

---

[3] The values of the binding energy deduced from [14] are somewhat higher than those deduced from [13].

is very sensitive to the variation of σ in this range this may produce substantial differences if a better determination gives different values of σ. Furtheron, a calculation by Song of the masses from band structure considerations gives values of the same order but somewhat higher. Furtheron the formula (20) takes only partly into account the polarisability of the lattice in using an experimental value of E_{ex}.

A similar restriction to this comparison can be made for other substances as well; it seems justified to consider this fit as definitely exceptionnally satisfactory.

It will be seen that the study of kinetics gives also a plausible fit with predicted properties of biexcitons. It can be considered that substantial arguments in favour of an interpretation of v_B as corresponding to a radiative recombination of biexcitons, are available.

b) CdS and CdSe

For CdS and CdSe the values of E_B observed and calculated are in a ratio of somewhat more than 3. This might possibly be an argument against the assignment of v_B to a radiative recombination of biexcitons. The question should be answered whether another interpretation or an improvement of the existing theories are required.

Conversely, it seems questionable to postulate such high binding energies for biexcitons in spite of theoretical plausible calculations giving much lower binding energies.

c) Ge and Si

According to Table 1 the discrepancy between the experimental and theoretical values of E_B is very large for these substances. It has to be stated that in the calculation we have made, it was suggested on ground of the calculations of Cho that the process involved is:

$$(\text{biex}) \to h v_B + (\text{ex}) + \hbar \omega_K$$

and not the one of Haynes. The difference $h \Delta v$ is smaller by E_{ex} with respect to Haynes mechanism. Though this is a rather small correction, a considerable difficulty exists for the assignment of the process to a biexciton radiative recombination unless one postulates a high binding energy of biexcitons in spite of the theory.

As otherwise an amount of other arguments are rather in favour of biexciton interpretation (kinetics!) it seems to be prudent to postpone a final judgement.

It should, however, be kept in mind that the theory may not be perfect and that some new aspects could improve the situation (polarization of the lattice). The problem of possibility of formation of polyexcitons should also be examined.

d) Cu_2O

For this substance the binding energy has been evaluated from Hanamura's curve assuming one of the processes examined in Section 4. It can be seen that the binding energy deduced from the experiment and from Hanamura's theory are in rather good agreement.

Unfortunately the Rydberg constant obtained in the assumed process does not fit with the experimental value of Gross and Kreingold. If the binding energy is decuded from higher terms of the observed series another value namely $150 \, \text{cm}^{-1}$ is obtained which disagrees with the theoretical value.

The greatest criticism which can be made to a biexciton interpretation is, however, the temperature dependence of the spectrum which is observed at 2 K but disappears at 4.2 K. This obviously is in contradiction to the rather high value of the binding energy. Unless the formation of biexcitons in Cu_2O could be proved to be in competition with another process (or complex) stable at higher temperature, it appears from this analysis that this assumption is questionable. This statement is also reinforced by the low power source used by the authors which would not give rise to abnormally high "biexciton temperatures".

An alternative treatment by Pavinsky does not seem to improve the situation, as regards the interpretation in this problem.

In several other substances luminescence peaks have tentatively been assigned to biexciton radiative recombination. It is out of the scope of this paper to give a complete review of these experiments.

7. Exciton Molecules. Polyexcitons

The possibility of formation of higher polymers of excitonic molecules has been often suggested. Recently Hy-Yin-Wang and Kittel [22] have suggested that in multivalley conduction band semiconductors like Ge and Si polyexcitons may be formed. They have shown that if N is the number of equivalent "band edges" it is possible to place $2N$ electrons on the same molecular orbital without violating the Pauli exclusion principle. This result can have important consequences with regard to the stability of polyexciton molecules.

In Ge$\rightarrow 2N = 8$ and Si$\rightarrow 2N = 12$; the calculations have been carried out on the basis that $m_h \gg m_e$ and under the assumption that the holes are localised. A realistic comparison with the experiment can therefore not be expected so far. The suggestion gives, however, an indication for a possible direction of theoretical investigations.

The result of these calculations is as follows (we write here $(exc)_2$ for the biexciton, $(exc)_8$ for the octo-exciton, $(exc)_{12}$ for the dodeca-exciton).

The total binding energy in effective Rydbergs (exciton Rydbergs) are:

$(ex)_2$	$(ex)_8$	$(ex)_{12}$
0.2548	5.986	11.963

the binding energy per exciton is:

for	$(ex)_2$	$(ex)_8$	$(ex)_{12}$
	0.1274	0.748	0.997 .

It has also been possible to calculate the distances between excitons in excitonic Bohr radius units

$(ex)_2$	$(ex)_8$	$(ex)_{12}$
1.39	0.93	0.79 .

The only conclusion that can be drawn from this very schematic but stimulating idea is that the polyexcitons may be considerably more stable in some substances than free excitons and free biexcitons and therefore the possibility of the formation of such polyexcitons should be kept in mind.

As seen above the observed binding energy as obtained from the luminescence · peak ascribed to the biexciton is for some substances much higher than the calculated binding energy of a biexciton. In the light of the above schematic considerations one could question whether they could not be ascribed to polyexcitons. The kinetics of the process is, however, not yet available.

In the present state this idea cannot be considered else as a possibility which deserves consideration.

8. Line Width of v_B

We have seen that Eq. (2) contains E_B^K, the kinetic energies of the biexciton, and E_{ex}^K, the kinetic energy of the exciton. These terms have been neglected for simplicity in our analysis of the binding energy. They are responsible of the line width of v_B.

Calculations have been performed by Hanamura [23] and Shionoya, and also Cho [11] and Ostertag [24]. The calculations are of some complexity and a temperature of the biexciton gas different with respect to the temperature of the lattice must be considered. Furthermore, as shown by Cho, it is important in some cases to take into account the dependence of the transition matrix element on the wave vector k. The shape of the line depends on various factors.

The kinetic energy and the collisions of biexcitons with excitons and biexcitons must be taken into account. This gives, according to Hanamura, a line shape obtained by the convolution of a lorentzian function and a Boltzman function.

Leaving aside normalization coefficients, this expression is:

$$I(hv) \propto \int_0^\infty \Gamma \varepsilon^{1/2} [(G_{ex} - E_B - hv - \varepsilon)^2 + \Gamma^2]^{-1} \exp(-\varepsilon/kT) \, d\varepsilon. \tag{21}$$

Γ is the Lorentzian line width which is a function of the number of biexcitons per cm^3. $\varepsilon = G_{ex} - E_B - hv'$, v' is the integration variable and v is a variable of the position in the spectrum.

On account of the Boltzman distribution function the maximum of the line is at:

$$hv_{max} = G_{ex} - E_B - kT/2. \tag{22}$$

$kT/2 = 1.7 \, cm^{-1}$ for $T = 4.2 \, K$ and is not very important, but $kT/2 = 26.5 \, cm^{-1}$ for $T = 77 \, K$ which is of some importance and cannot be neglected. It has to be remembered that in this theory T can be higher than the temperature of the lattice. The above expression tends to be at low temperatures:

$$I(hv) \propto \Gamma [(hv)^2 + \Gamma^2]^{-1} \quad \text{for} \quad \Gamma \gg kT \tag{23}$$

and

$$I(hv) \propto \varepsilon^{1/2} \exp(-\varepsilon/kT) \quad \text{for} \quad \Gamma \ll kT.$$

Γ depends on n_{biex}. The applicability of one or the other formula depends not only on the temperature but also on the excitation. At low temperatures, the correction due to Cho is negligible.

In fact a more elaborate calculation is necessary for the intermediate cases and is in progress by Ostertag and Cho and will be published. elsewhere.

The fitting of the experimental curve with the theory leads to an apparent fit for

$$E_B = 323 \, cm^{-1}, \quad T = 16.7 \, K, \quad \Gamma \ldots 58 \, cm^{-1}.$$

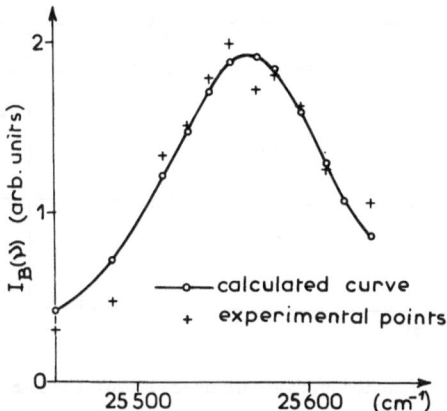

Fig. 6. The shape of a biexciton line in CuCl according to Cho and Ostertag taken from the thesis of Lévy

Though these figures seem reasonable, the fitting has the serious defect that the fine structure of the line has not been taken into account.

For higher temperatures

$$I(h\nu) \propto \int_0^\infty \Gamma \left[(G_{ex} - E_B - h\nu - \varepsilon)^2 + \Gamma^2 \right]^{-1} (1 + \varepsilon/E_B^K)^{-4} \varepsilon^{1/2}$$

$$\cdot \exp(-\varepsilon/kT)\, d\varepsilon. \tag{24}$$

The second term is an analytical expression of Cho's correction.

Though at higher temperature the fine structure is no more stable, the fitting of this curve does, however, not lead to plausible results. This point has been treated elsewhere by Ostertag and Cho.

It should finally be emphasized that except for CuCl and CuBr the line width is often of the order of the experimental binding energy and much larger than the theoretical binding energy. The physical meaning of this situation seems to indicate a very unstable state.

9. Spin Fine Structure of the Biexciton Line

Though the biexciton lines are usually rather broad the existence of a fine structure of the line could be seen or suspected [3, 4]. This structure can have several origins:

1. The fine structure of the exciton spectrum can cause a fine structure of the line ν_B.

2. The biexciton may also have rotational energy levels in analogy to a diatomic molecule.

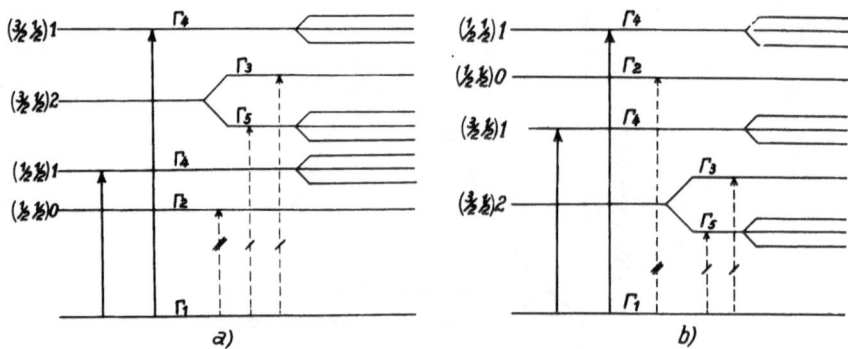

Fig. 7. Fine structure of exciton levels according to [25, 26]. The case a refers to CuCl and b to CuBr

The first possibility will be treated in this section; the second is treated in the next section.

We will limit this analysis to the simple case of CuCl.

The fine structure of CuCl has been given a long time ago by Nikitine, Ringeissen and Certier on ground of jj coupling [25] considerations and rediscussed on ground of group theory with Ringeissen and Sennett [26]. The fine structure is given in Fig. 7. The group theoretical notations are those given by Curie [27].

The allowed transitions are those of Γ_1 to Γ_4. This is a transition from the ground state[4]

$$^1S_0(\Gamma_1) \to {}^3S_1(\Gamma_4) \to \nu_0 .$$

The $^3S_1(\Gamma_4)(\frac{1}{2}\frac{1}{2})1$ state is a triplet orthoexciton state. The $^1S_0(\Gamma_2)(\frac{1}{2}\frac{1}{2})0$ (excited) state is a paraexciton or singlet state. The transition from the ground state to this state is forbidden in dipole and quadrupole approximations.

$$S_0(\Gamma_1) \to {}^1S_0(\Gamma_2) \to \nu_1 .$$

It has, however, been observed in presumably unperfect crystals and is also reinforced in a magnetic field but does not show any Zeeman splitting.

It is known that in CuCl the para-ortho separation is of $50\,\mathrm{cm}^{-1}$. The ν_1 emission line disappears at higher temperature. However, at low temperatures the Γ_2 state could have an appreciable population as presumably the lifetime of ν_1 is rather long.

---

[4] The analogy to atomic spectroscopy is useful, but must not be taken with a too strict meaning.

The state $^3S_1(\Gamma_4)(\frac{1}{2}\frac{3}{2})1$ of excitons in CuCl lies some $500\,\text{cm}^{-1}$ higher and the state $^5S_2$ splits into Γ_3 and Γ_5 but has not been observed and lies probably within the line width of the so-called diffuse line corresponding to the above configuration (Fig. 7a). For CuBr, the situation is reversed as shown on Fig. 7b.

The fine structure of the biexciton luminescence line is a problem of some complexity and a theory in general terms should be developed. A first approach is, however, given in an important paper by Bassani et al. [28].

With respect to the experimental situation we are going to discuss the case of CuCl only, which is probably the simplest, though already of some complexity. For this case $\sigma \ll 1$, however, it has to be remembered that the binding energy of the biexciton is of the order of spin orbit splitting of the valence bands (350 and $500\,\text{cm}^{-1}$) and therefore doubts can be expressed whether the spin coupling may no more be a good parameter. On account of this fact and of still scarce experimental and theoretical information this discussion is bound to be in a frame of a rather oversimplified qualitative picture.

With this restriction let us consider in a simplest first hypothesis that the ground state of biexcitons is not split. In this case as suggested in several papers [3d] and developed by Lévy [3h] it is assumed that the exciton formed in the radiative recombination process can be either the ortho (Γ_4) or the para (Γ_2) exciton.

$$\text{biex} \rightarrow h\nu_B + \begin{cases} \text{ex}(\Gamma_4) \\ \text{ex}(\Gamma_2). \end{cases} \tag{25}$$

This leads to the energy equations

$$G_B = h\nu_{B_1} + G_{\text{ex}}(\Gamma_4) \tag{26}$$

and

$$G_B = h\nu_{B_2} + G_{\text{ex}}(\Gamma_2) \tag{27}$$

here

$$G_B = 2(E_g - G_{\text{ex}}) - E_B. \tag{28}$$

In this process the line ν_B is split into two components, $\delta\nu$ apart:

$$\delta\nu = G_{\text{ex}}(\Gamma_4) - G_{\text{ex}}(\Gamma_2) \tag{29}$$

$\delta \nu$ is the orthopara splitting of the exciton lines. It is, of course, not known whether both processes [25] are equally probable.

These oversimplified considerations contain, however, two assumptions which are not supported by a theoretical proof, or, probably, in a better approximation no more correct.

a) First, it is not obvious that in the processes to be considered

$$
\mathrm{ex}(\Gamma_4) + \mathrm{ex}(\Gamma_4) \to \mathrm{bi}_{\mathrm{ex}}; \; \mathrm{ex}(\Gamma_2) + \mathrm{ex}(\Gamma_4) \to \mathrm{bi}_{\mathrm{ex}}
$$
$$
\mathrm{ex}(\Gamma_2) + \mathrm{ex}(\Gamma_2) \to \mathrm{bi}_{\mathrm{ex}}
$$

(30)

the same and unique biexciton state is obtained. If so, this would mean that in each of these processes the binding energy could be different, so that the ground state of the biexciton would not be split. But this assumption is not yet proved to be correct as no theory of formation of biexcitons from different excitons has been worked out.

An opposite extreme hypothesis could be that the binding energy for all these combinations is the same. Then the ground state of biexcitons should be split into three components and the resultant structure of ν_B is quadruplet (a sextuplet reducing to a quadruplet)[5].

Though it seems very imprudent to argue for these possibilities before a calculation is performed it is of some use to mention these possibilities.

b) Furthermore the assumption in [25] is certainly not quite correct. Let us again take the case of CuCl with $\sigma \ll 1$, which is in a plausible approximation similar to H_2. In this case the ground state of the biexciton is, by analogy to H_2, certainly split into two states, a para and an orthobiexciton. Let us remind of this difference.

It has been recognized that in H_2 the ground state is influenced by the spins of the nuclei. In ortho H_2, these nuclear spins are parallel; in the para variety they are anti parallel. These two possibilities are certainly also available in the biexciton problem. For $\sigma \ll 1$, the analogy is certainly pronounced. The "nuclear" spins are now the spins of the holes ($\frac{1}{2}$ and $\frac{3}{2}$ for CuCl). The coupling energy has not been calculated, but it can be noticed that though the moment of the holes is much larger than for protons, the distances between both holes are much longer in the biexciton. As the coupling energy is presumably proportional (like in H_2) to the momentum and $\bar{r}^{-3}$, $\bar{r}$ being the mean distance of electrons and holes, the para-ortho splitting of the biexciton is likely to be rather small but probably not negligible.

Bassani et al. [28] have given rather large splitting for $\sigma < 1$.

---

[5] These considerations have been developed at the Symposium in Tonbach. The present form of this section is somewhat more general.

Now the essential question is the symmetry of the wave functions with respect to the holes. So we are going to write the wave functions $\phi_{el}(S)$, $\phi_{rot}(S)$, $\phi_\Sigma(S)$ for the wave functions of the electrons, of the rotation of the biexciton, and the nuclear spin function. S indicates symmetric and A the antisymmetric character. The vibration which in the case of biexciton will reduce to the zero point vibrations is always symmetric and will not be considered[6]. The wave functions of the para and ortho biexciton must be antisymmetric and so it will be assuming the validity of Born Oppenheimer approximation, say:

$$\psi_B(\text{para}) = \phi_{el}(S)\,\phi_{rot}(S)\,\phi_\Sigma(A) \tag{31}$$

and

$$\psi_B(\text{ortho}) = \phi_{el}(S)\,\phi_{rot}(A)\,\phi_\Sigma(S). \tag{32}$$

It can be shown that $\phi_\Sigma(S)$ is threefold degenerate while $\phi_\Sigma(A)$ is single and therefore one should expect, like in H_2, 3 times more ortho biexcitons if an equilibrium is obtained.

Now it is shown for para and ortho hydrogen that the para terms can combine with even rotational levels and $J=0$ only, and ortho terms with odd rotational levels only. Furtheron the transition between para and ortho terms is strictly forbidden. So a biexciton created in an ortho state and in an odd rotational level should be stable and not recombine at low temperatures with the lower para $J=0$ state. Now the difference of energy between these two states is just the rotational energy if the coupling energy of the spins of the holes in the biexciton is very small. It is the rotational energy plus the para-ortho energy if this energy is not negligible. In both cases [25] is still to be taken into account. This discussion is, however, correct only if the thermal equilibrium of biexcitons in a crystal is similar to that of a H_2 gas. This is questionable in particular because of the small life time of biexcitons.

In conclusion it is seen that on ground of analogy of biexcitons in CuCl and H_2 molecules it can be expected that the ground state of biexcitons in this material is probably split in at least two terms and therefore the structure of the recombination line ν_B may be rather complicated. Further theoretical work and experimental results should bring more informations and help a quantitative interpretations.

It has to be remembered that this discussion is based on spin considerations. It is not sure that spins are good parameters in this case owing the the comparable magnitudes of binding energy of biexcitons in CuCl and the spin-orbit splitting of the valence band.

The interaction with the lattice has also not been taken in account.

---

[6] The higher vibrational terms are of the order of the binding energy of the biexciton.

10. Rotational Levels of Biexcitons

In the fine structure of the biexciton line v_B rotational levels can be expected.

Some considerations on these levels are given in this section. For simplification purposes, the case of CuCl and CuBr will be considered because on one hand, the exciton binding energy is large, on the other hand, the ratio σ is small. Furtheron, in these two crystals a structure has been observed.

It is supposed that $m_h^* \gg m_e^*$. The rotational energy of the biexciton is now:

$$E_B^r = (\hbar^2/2\mu_B R_B^2)\, j(j+1). \tag{33}$$

Here, μ_B is the effective mass of the biexciton and R_B the distance between the two holes. μ_B is in our case just equal to $m_h^*/2$ and R_B has been calculated by Hanamura. This formula can, however, be readily written in a more handy form:

$$E_B^r = E_{H_2}^r (M_p/m_0)(E_{ex}/E_H)\, \sigma\, j(j+1). \tag{34}$$

Here, E_H^r is the rotational energy of the hydrogen molecule H_2, $M_p/m_0 = 1840$. E_{ex}/E_H is the ratio of the exciton and hydrogen atom binding energies.

For $j = 0$ $E_B^r = 0$.

For $j = 1$ $E_B^r(1) = 2 E_B^{ro}$.

For $j = 2$ $E_B^r(2) = 6 E_B^{ro}$ etc. ...

with

$$E_B^{ro} = \hbar^2/2\mu_B R^2 = E_{H_2}^r (M_p/m_0)(E_{ex}/E_H)\, \sigma.$$

The structure should present lines diverging to higher energies. However, on account of the low binding energies for biexcitons this formula should be perturbed by centrifugal forces. The polarisation effects should also have an importance. The first line of the structure should lie now over the $j = 0$ line for CuCl and CuBr to the amount of:

CuCl   CuBr

$E_B^r(1) \simeq 61 \text{ cm}^{-1}$ $E_B^r(1) \simeq 17.5 \text{ cm}^{-1}$.

It has to be noted that formula (34) depends directly on σ. If a structure is observed which can be reasonably assigned to a rotational spectrum, formula (34) gives the possibility of a direct determination of σ, all other terms entering into (34) being known constants or measurable quantities.

It has to be noted that (34) does not depend explicitly on R_B, the interhole distance of the biexciton. This is, however, a result of an approximation in which R_B is taken to be $1.4 a_{ex}$. If this is not valid a correction should be made. This calculation may contain some ambiguities.

11. Kinetics of Formation of Biexcitons

In order to explain the behaviour of the kinetics it was found necessary by Knox and the author [29] to separate the excitons formed into two parts: the "optical" (density n_0) and "thermal" (density n_1). The biexcitons (density n_2) are assumed to be formed essentially from the very much larger pool of thermal excitons. Both classes of excitons are formed from the recombination of free carriers with rates $A_0 i$ (optical) and $A_1 i$ (thermal); $A_0 i < A_1 i$. The kinetic equations are readily written and the solution is as follows. Including only these processes which appear to be important the kinetic equations are:

$$dn_0/dt = A_0 i - n_0/\tau - k n_0 + k' n_1 ,$$

$$dn_1/dt = A_1 i + k n_0 - k' n_1 - k'' n_1 - B n^2 + C n_2 , \tag{35}$$

$$dn_2/dt = \tfrac{1}{2} B n_1^2 - C n_2 - k''' n_2 .$$

Here τ_0 is the life time of excitons, k and k' are the rates of conversion of optical to thermal excitons and of the reverse. k'' is the rate of radiationless dissociation of excitons. B is the bimolecular collision coefficient, C a recombination rate of biexcitons and k''' the rate of radiationless decay of biexcitons. The rates k', k''' are supposed to be negligible.

As the laser pulse lasts considerably longer then the life time of excitons it seems reasonable to solve the above equations under the hypothesis of a steady state. This gives then:

$$n_0 = A_0 \tau i/(1 + \tau k); \qquad n_1 = (k''/B) (\sqrt{i/i_0 + 1} - 1);$$

$$n_2 = [(k'')^2/2 B C] (\sqrt{i/i_0 + 1} - 1)^2 . \tag{36}$$

As $I_0 \propto n_0$, this gives the linear dependence of I_0 versus i.

As $I_B \propto n_2$, this can be shown to give for $i \ll i_0$, $I_B \propto i^2$ and for $i \gg i_0$, $I_B \propto i$.

Here:

$$i_0 = (k''^2/2B) (1 + k\tau) [A_1 + k\tau(A_0 + A_1)]^{-1} . \tag{37}$$

Though the experimental data give a m_B which is somewhat lower than 2 for $i \ll i_0$ but tends to 1 for $i \gg i_0$ for CuCl and CuBr, the agreement with the theory seems to be good. As stated above, the discrepancies are very likely to be due to overlapping of the lines v_B and v_0 at 77 K, and to overlapping of these two lines with other lines v_1 and v_2 at 4.2 K.

Therefore it seems that the binding energy of biexcitons and the kinetics of the observed luminescence being in reasonable agreement with the observations, a strong argument in favour of formation of biexcitons is given in this investigation for CuCl and CuBr.

It has also been shown that biexcitons can be formed when the excitation is obtained by the double-photon absorption technique.

Though several authors claim that the intensity in their experiments $I_B \propto i^2$ these delicate experiments have, however, not been described in details except in the case of Ge where an important investigation has been made (see the paper by Rogatchev [8]). This seems to be an argument in favor of a biexciton interpretation. It has, however, to be mentioned that such a variation may also appear in other processes. It is out of the scope of this paper to take a position in the controversy on this emission of Ge and Si.

Conclusions

As discussed in this paper the arguments in favor of the existence of biexcitons are very strong for CuCl and CuBr. In these two substances as well the kinetics as the value of the binding energy is in a reasonable agreement with the theory.

It is probable that biexcitons can be formed also in other substances. Some strong arguments seem to exist in the case of CdS, CdSe and also in Ge and Si. However, in these cases some important shortcomings are also to be noted: the calculated binding energy is several times smaller than the value obtained from the experiment. Furtheron the width of the biexciton recombination line is very large in comparison to the theoretical and also to the experimental binding energy.

It is hoped that new experiments and theoretical considerations will clarify these difficulties.

In the cases where the arguments in favor of biexcitons are sufficiently established it is of interest to study the fine structure (if any) of the biexciton line. It is shown that in this study new data of some importance can be obtained.

Finally, the physics of biexcitons, their possible Bose condensation and temperature effects can reasonably be hoped to make important progress in the near future.

References

1. Nikitine, S.: Cooperative Phenomena. Ed. by Haken and M. Wagner. Berlin-Heidelberg-New York: Springer 1973.
   Nikitine, S., Haken, H.: Confer. on Luminescence, Leningrad. Luminescence of crystals, molecules and solutions, p. 4. New York: Plenum Publ. Corp. 1972.
2. Lampert, M. A.: Phys. Rev. Letters 1, 450 (1958).
   Moskalenko, S. A.: J. Opt. Spectroscop. 5, 147 (1958).
3. a) Mysyrowicz, A., Grun, J. B., Levy, R., Bivas, A., Nikitine, S.: Phys. Letters 26 A, 615 (1968).
   b) Nikitine, S., Mysyrowicz, A., Grun, J. B.: Helv. Phys. Acta 41, 1058 (1968).
   c) Mysyrowicz, A.: Thesis, Strasbourg (1968).
   d) Bivas, A., Levy, R., Nikitine, S., Grun, J. B.: J. Phys. 31, 227 (1970).
   e) Bivas, A.: Thesis (3ᵉ Cycle), Strasbourg (1969).
   f) Knox, R., Nikitine, S., Mysyrowicz, A.: Opt. Commun. 1969, 19.
   g) Grun, J. B., Nikitine, S., Bivas, A., Levy, R.: J. Luminescence 1, 2, 241 (1970).
   h) See also Levy, R.: Thesis, Strasbourg (1973).
4. Goto, T., Souma, H., Ueta, M.: J. Luminesc. 1, 231 (1970).
   Souma, H., Goto, T., Ueta, M.: J. Phys. Sox. Japan 29, 697 (1970).
5. Saito, H., Shionoya, S.: Intern. Confer. on Luminescence, Leningrad (1972).
   Shionoya, S., Saito, H., Hanamura, E., Akimoto, O.: Sol. State Commun. 12, 223 (1973).
6. Haynes, J. R.: Phys. Rev. Letters 17, 860 (1966).
7. Nikitine, S., Ringeissen, J., Certier, M.: Acta Phys. Polon. 26, 745 (1964).
   Reiss, R.: Thesis (Strasbourg). Cahiers de Phys. 13, 129 (1959).
   Lewonczak, S., Ringeissen, J., Nikitine, S.: J. Phys. 32, 841 (1971)
8. See in this book the papers by A. A. Rogatcheff, V. S. Bagaieff and C. Benoit a la Guillaume.
9. Gross, E. F., Kreingold, F. J.: Zh. E.T.F. Letters 12, 68 (1970).
   Lvov, O. I., Pavinskii, P. P.: Zh. E.T.F. Letters 14, 253 (1971).
10. Mysyrowicz, A., Grun, J. B., Levy, R., Bivas, A., Nikitine, S.: Phys. Letters 26 A, 615 (1968).
    Nikitine, S., Mysyrowicz, A., Grun, J. B.: Helv. Phys. Acta 41, 1058 (1968).
11. Cho, K.: Opt. Commun. 8, 412 (1973).
12. Saito, H., Shionoya, S., Hanamura, E.: Solid State Commun. 12, 227 (1973).
13. Akimoto, O., Hanamura, E.: Solid State Commun. 10, 253 (1969).
14. Brinkman, W. F., Rice, T. H., Bell, B.: Phys. Rev. 8, 1570 (1973).
15. Wehner, R. K.: Solid State Commun. 9, 457 (1969).
16. Adamovski, J., Bednarek, S., Suffczinski, M.: Solid State Commun. 9, 2037 (1971).
17. Büttner, H., Pollmann, J.: Sol. State Commun. 12, 1105 (1973).
18. a) Nikitine, S., Grun, J. B., Mysyrowicz, A.: Helv. Phys. Acta 41, 1058 (1968).
    b) Nikitine, S.: Nonlinear optics Colloquium, Titisee (1972).
    c) Nikitine, S., Haken, H.: Confer. on Luminescence, Leningrad Luminescence of crystals, molecules and solutions, p. 4. New York: Plenum Publ. Corp. 1972.
19. Elkomoss, S. G., Stebe, B.: Nuovo Cimento 10 B, 255 (1972).
    Elkomoss, S. G.: Phys. Rev. B 6, 3913 (1972).
    Also short communication at the Symposium on Excitons at High Density and Polaritons, Tonbach, 1973.
20. Kazamanyan, Z. A.: Sov. Phys. 1, 341 (1967).
21. Akimoto, O.: In press. Technical Report of ISSP A N 592 (may 1973).
22. Shy-Yih-Wang, J., Kittel, C.: Phys. Letters 42, 189 (1972).

23. Hanamura, E.: Confer. on Luminescence, Leningrad (1972).
    Shionoya, S., Saito, H., Hanamura, E., Akimoto, O.: Solid State Commun. **12**, 223 (1973).
24. Ostertag, E.: Short communication to the Symposium on Excitons at High Density and Polaritons, Tonbach, 1973. A part of this work of Ostertag is given with permission of this author in the thesis of R. Lévy, Strasbourg (1973).
25. Nikitine, S., Ringeissen, J., Certier, M.: Acta Phys. Pol. **26**, 745 (1964).
26. Nikitine, S., Ringeissen, J., Sennett, J.: Confer. on Phys. of Semiconductors, Paris "Radiative Recombination", p. 274 (1964).
27. Curie, D.: Champ cristallin et luminescence. Paris: Gauthier Villars 1968.
28. Bassani, F., Forney, J. J., Quattropani, A.: Preprint, private communication.
29. Knox, R. S., Nikitine, S., Mysyrowicz, A.: Opt. Commun. **1969**, 19.

Prof. Dr. S. Nikitine
Université Louis Pasteur
Laboratoire de Spectroscopie et d'Optique du Corps Solide
5, rue de l'Université
F-67000 Strasbourg (France)

Biexcitons – Bose Condensation and Optical Response

EIICHI HANAMURA

Contents

1. Introduction

Recently theoretical physicists have paid considerable interest to open systems in contact with different kinds of reservoires or coherent driving forces [1, 2]. These open systems may be in non-stationary or stationary states far from thermal equilibrium. The system of exciton and biexciton gases as well as the radiation field emitted from these gases in heavily excited semiconductors offers a good example of these open systems.

Since Akimoto and Hanamura [3] pointed out that the biexciton is stable for any value of the electron to hole mass ratio, not only the existence of biexcitons [4] was confirmed in CdS [5] and CdSe [6] but also the Bose condensation [7] of the biexciton gas was observed. On the other hand, laser action due to radiative annihilation of biexcitons was found in CuCl and CuBr [8]. Thus the important role [9, 10] of biexcitons in heavily excited semiconductors has been recognized: they are the final states at low temperatures and below the Mott transition concentration, and act also as the radiative centers of these heavily excited semiconductors.

By the recent development of laser and pulse techniques, it became possible to create the quasi-equilibrium state of excitons at any high density in crystals by laser light with the desired intensity and pulse duration longer than pico-second. The excitations induced in crystals by e.g. the pico-second laser are stabilized into the thermal equilibrium of excitons in the time order of pico-seconds with the help of giant trapping effects [11] by emitting and absorbing phonons, and they are forming biexcitons in a time which is determined by the kinetics of this system as shown in Section 2. This state may be considered as a quasi-stationary state in the sense that it has a life time of micro-seconds in the case of indirect excitons and of the order of nano-seconds in the case of direct allowed excitons before they are annihilated by emitting the radiation corresponding to the interband transition (case 1). On the other hand, under the illumination of laser light with a duration longer then the life time of excitons, a stationary state of exciton and biexciton gases is generated in some cases (case 2) and laser action due to biexcitons is realized in other cases (case 3). In the cases 1 and 2, we want to consider the exciton and biexciton gases as the *relevant system* and the phonons emitted during the binding process of two excitons into biexcitons and the photons emitted in the radiative annihilation of biexcitons as the *irrelevant systems*. Reversely, in the case 3 the *relevant system* is the radiation field which is created by the radiative annihilation of biexcitons. The justification of these considerations is discussed in Sections 2 and 6, respectively.

The second characteristic feature of the system of exciton and biexciton gases is the large quantum effect caused by Bose-particles [9]: Wannier excitons and biexcitons have boson-like characters [12–14] as far as the number of excited electron-hole pairs remains very small compared to the total number of valence electrons. By making the best use of this boson-like characters, Hanamura [12] has developed a theory to treat excitons and biexcitons as bose-particles and to take into account as the effective interactions the residual interaction as well as those due to the Pauli principle acting between the component particles belonging to different complexes. Furthermore the quantum effect works more effectively in this system than in the quantum fluid of He which also obeys the Bose-Einstein statistics, because the ratio of the exciton or biexciton translation mass to the effective electron masses by two or three orders smaller than in the case of He. As a result, we predicted the Bose-condensation of biexcitons [9] and calculated the characteristic emission spectra [10] for this system, which were also experimentally observed [7]. This will be discussed in Sections 3 and 4.

According to the discussions in Section 2, the open system of excitons and biexcitons created by the radiation field can be considered to be stationary or quasi-stationary. This system shows a large variety of phenomena depending upon the degree of excitation, the excitation frequency and the duration of excitation pulses. Thus we find a biexciton gas interacting by elastic [5] and inelastic [6] collisions and Bose-condensation of biexcitons [7]. Because of the stationary state, we may apply a second weak external field [2] as a probe to this open system. In Section 4, the linear response theory is extended to detect the electronic state of exciton and biexciton gases and we show how the existence of biexcitons [5] and the Bose-condensation of biexcitons [7] have been concluded from the comparison of the observed emission spectra with this theoretical result.

Beyond the linear response theory, we can also study the coherent interaction between the biexciton gas and the radiation field. In Section 5 we represent the giant two-photon absorption due to biexcitons [15] as the first example beyond the linear response and indicate the possibility of using this property for a two-photon absorption spectroscopy to check the existence of biexcitons. As an example of the coherent interaction, we discuss laser action of biexcitons in Section 6. We discuss in Section 7, as the third example beyond the linear response, whether the result of self-induced transparency due to two-photon transition [16] can be applied to the self-induced transparency of biexcitons.

2. The Open System of Exciton and Biexciton Gases

We will discuss in this section under what conditions it is justified to treat the population dynamics of exciton and biexciton gases as the relevant open system. Then the rate equations derived here, will be solved.

The system of exciton and biexciton gases as well as of phonons and photons emitted by these gases is described according to the boson description [12–14] which was developed by Hanamura, as follows:

$$H/\hbar = \Sigma \omega_M b_M^+ b_M + \Sigma \omega_X b_X^+ b_X + \Sigma \omega_0 a_0^+ a_0 + \Sigma \omega_p a_p^+ a_p$$

$$+ \Sigma (F a_p^+ b_M^+ b_X b_{X'} + \text{h.c.}) + \Sigma (G_1 a_0^+ b_X^+ b_M + \text{h.c.}) \tag{1.1}$$

$$+ \Sigma (G_2 a_0^+ b_X + \text{h.c.}) + \Sigma (G_3 a_0^+ b_X^+ b_{X'}^+ b_{X''}^+ b_M b_{M'} + \text{h.c.}),$$

where b_M, b_X, a_0, and a_p are the boson annihilation operators of biexciton, exciton, photon and phonon, respectively. The F-term describes the binding process of two single excitons X and X' into the biexciton M with the emission of a phonon p. The G_1-term describes the radiative annihilation of a biexciton leaving the other electron-hole pair as free exciton X. The G_2-term corresponds to the radiative annihilation of an exciton or a bound exciton and G_3-term describes the so-called P_M-process [6] in which two biexcitons collide, one of these four involved excitons is radiatively annihilated and the other three excitons remain in the crystal. The terms which describe the collisional effects among excitons and biexcitons are omitted from this Hamiltonian because they have no effects on the population dynamics of $\Sigma_X b_X^+ b_X$ and $\Sigma_M b_M^+ b_M$, which is pursued in this section. The total system of this Hamiltonian should be still considered to be an open system. The effect of the pumping field corresponding to exciton creation is taken into account by the negative damping $-\delta_{pX}$ and the fluctuation forces f_{pX}. The photons emitted by the radiative annihilation of biexcitons escape from the crystal with the decay constant γ_0 and the fluctuating force f_0. The phonons emitted in the F process also decay into the irrelevant branches with the decay constant γ_p and the fluctuating force f_p. We specify the damping constant γ_0 as $\gamma_0 = c/l$, where l is the linear dimension of the crystal and c is the light velocity. The momentum dependence of the matrix elements F and G is neglected for simplicity, which is justified for low temperature and not so high density of excitons and biexcitons. It is pointed out here that the external pumping of biexcitons is also possible by the giant two-photon absorption [15] due to biexcitons which will be discussed in Section 5.

In order to derive the rate equations from the microscopic point of view, at first we derive the Langevin equations for a_0^+, a_p^+ and b_X^+ and the Heisenberg equations for b_M^+ by using the Hamiltonian (1.1) and the consideration as an open system, as follows:

$$(\mathrm{d}/\mathrm{d}t - i\omega_0 + \gamma_0)\, a_0^+ = iG_1^* \Sigma b_M^+ b_X + iG_2^* \Sigma b_X^+$$
$$+ iG_3^* \Sigma b_{M'}^+ b_M^+ b_{X''} b_{X'} b_X + f_0, \tag{1.2}$$

$$(\mathrm{d}/\mathrm{d}t - i\omega_p + \gamma_p)\, a_p^+ = iF^* \Sigma b_{X'}^+ b_X^+ b_M + f_p, \tag{1.3}$$

$$(\mathrm{d}/\mathrm{d}t - i\omega_X - \delta_{pX})\, b_X^+ = 2iF a_p^+ b_M^+ b_{X'} + iG_1^* \Sigma b_M^+ a_0 + iG_2 \Sigma a_0^+$$
$$+ 2iG_3^* \Sigma b_{M'}^+ b_M^+ b_{X''} b_{X'} a_0 + f_{pX}, \tag{1.4}$$

$$(\mathrm{d}/\mathrm{d}t - i\omega_M)\, b_M^+ = iF^* \Sigma b_{X'}^+ b_X a_p + iG_1 \Sigma a_0^+ b_X^+$$
$$+ 2iG_3 \Sigma a_0^+ b_X^+ b_{X'}^+ b_{X''}^+ b_{M'}, \tag{1.5}$$

where we confined ourselves to exciton pumping for simplicity. Not only because the F and G_1 terms give the dominant contribution to the population dynamics but because the effect of the G_2 and G_3 terms can be easily put together into the final result, we keep only the F and G_1 terms in the process of calculations. From Eqs. (1.4) and (1.5) we derive the following equations for the time development for the number operators of excitons at the state X and biexcitons at the state M:

$$(b_M^+ b_M)' = i(F^* \Sigma b_X^+ {}_{X'}^+ a_p b_M - F \Sigma b_M^+ a_p^+ b_{X'} b_X)$$
$$+ i(G \Sigma a_0^+ b_X^+ b_M - G^* \Sigma b_M^+ b_X a_0), \tag{1.6}$$

$$(b_X^+ b_X)' = 2\delta_{pX}(b_X^+ b_X) + f_{pX} b_X + b_X^+ f_{pX}^*$$
$$+ 2i(F \Sigma b_M^+ a_p^+ b_{X'} b_X - F^* \Sigma b_X^+ b_{X'}^+ a_p b_M) \tag{1.7}$$
$$+ i(G^* \Sigma b_M^+ b_X a_0 - G \Sigma a_0^+ b_X^+ b_M).$$

Because the right hand sides of Eqs. (1.6) an (1.7) are represented in terms of $a_0^+ b_X^+ b_M$ and $b_M^+ a_p^+ b_{X'} b_X$, we write down the equations for these product operators by using Eqs. (1.2) to (1.5) as follows:

$$(a_0^+ b_X^+ b_M)' = \{i(\omega_0 + \omega_X - \omega_M) - \gamma_0 + \delta_{pX}\}(a_0^+ b_X^+ b_M)$$
$$+ iG^* \Sigma b_{M'}^+ b_{X'} b_X^+ b_M + iG^* \Sigma a_0^+ b_{M'}^+ a_{0'} b_M - iG^* \Sigma a_0^+ b_X^+ b_{X'} a_0$$
$$+ 2iF \Sigma a_0^+ a_{p'}^+ b_{M'}^+ b_{X'} b_M - iF \Sigma a_0^+ b_X^+ a_{p'}^+ b_{X'} b_{X''} \tag{1.8}$$
$$+ f_0 b_X^+ b_M + a_0^+ f_{pX} b_M,$$

$$(b_M^+ a_p^+ b_{X'} b_X)' = \{i(\omega_M + \omega_p - \omega_X - \omega_{X'}) - \gamma_p + \delta_{pX} + \delta_{pX'}\} b_M^+ a_p^+ b_{X'} b_X$$
$$+ iF^* \Sigma b_{X''}^+ b_{X'''}^+ a_{p'} a_p^+ b_{X'} b_X + iF^* \Sigma b_M^+ b_{X''}^+ b_{X'''}^+ b_{M'} b_{X'} b_X$$
$$- 2iF^* \Sigma b_M^+ a_p^+ \{b_{X''}^+ b_X b_{M'} a_{p'} + b_{X'} b_{X''}^+ b_{M'} a_{p'}\} + b_M^+ f_p b_{X'} b_X \tag{1.9}$$
$$+ b_M^+ a_p^+ (f_{pX'} b_X + b_{X'} f_{pX}) + iG \Sigma a_{0'}^+ b_{X''}^+ a_p^+ b_{X'} b_X$$
$$- iG \Sigma b_M^+ a_p^+ a_{0'}^+ b_{M'} (b_X + b_{X'}).$$

As to the irrelevant system of photons and phonons, we assume [17] that (i) the relaxation times $(2\gamma_0)^{-1}$ and $(2\gamma_p)^{-1}$ of photons and phonons are very short compared to the time constants of excitons and biexcitons which are to be determined and (ii) that the irrelevant system is sufficiently large and so weakly coupled to the relevant system of excitons and biexcitons that the thermal equilibrium of the irrelevant system is never disturbed appreciably. According to the 1st assumption of the Markovian process for the relevant system, we can integrate Eqs. (1.8) and (1.9) easily and the expressions for $(a_0^+ b_X^+ b_M)$ and $(b_M^+ a_p^+ b_{X'} b_X)$ coincide with the solutions which are obtained by neglecting the time derivative on the left hand sides in comparison with the first terms of Eqs. (1.8) and (1.9), respectively.

We insert these expressions of $(a_0^+ b_X^+ b_M)$ and $(b_M^+ a_p^+ b_{X'} b_X)$ into Eqs. (1.6) and (1.7), take the expectation value with respect to the phonon and photon coordinates with the assumption (ii) of thermal equilibrium, and pick up the diagonal component of these equations with respect to the numbers of excitons and biexcitons. The resulting rate equations are as follows

$$\dot{N}_M = -\Sigma \frac{2\gamma_0 |G|^2}{\gamma_0^2 + (\omega_0 + \omega_X - \omega_M)^2} \{N_M(1 + N_0 + N_X) - N_0 N_X\}$$

$$+ \Sigma \frac{2\gamma_p |F|^2}{\gamma_p^2 + (\omega_M + \omega_p - \omega_X - \omega_{X'})^2} \tag{1.10}$$

$$\cdot \{N_X N_{X'}(1 + N_p + N_M) - 2N_p N_M(1 + N_X + N_{X'})\} ,$$

$$\dot{N}_X = 2\delta_{pX}(N_X + 1)$$

$$+ \Sigma \frac{2\gamma_0 |G|^2}{\gamma_0^2 + (\omega_0 + \omega_X - \omega_M)^2} \{N_M(1 + N_0 + N_X) - N_0 N_X\}$$

$$- 2\Sigma \frac{2\gamma_p |F|^2}{\gamma_p^2 + (\omega_M + \omega_p - \omega_X - \omega_{X'})^2} \tag{1.11}$$

$$\cdot \{N_X N_{X'}(1 + N_p + N_M) - 2N_p N_M(1 + N_X + N_{X'})\} ,$$

where we applied the technique of laser theory [18] and used the relation $\langle f_{pX} b_X + b_X^+ f_{pX}^* \rangle_R = 2\delta_{pX}$. Excitons and biexcitons are considered to be in stationary states within a time of the order of pico-seconds just after the pumping through mutual collisions and interactions with phonons, and the kinetic energies obey the Boltzmann distribution with effective temperature T. Therefore we may consider the distributions on the specified momentum state: N_X and N_M as well as N_0 and N_p to be much smaller than 1, and we keep the lowest order terms in these number operators in Eqs. (1.10) and (1.11). We also replace

$$2\gamma_0 |G|^2 / \{\gamma_0^2 + (\omega_0 + \omega_X - \omega_M)^2\} \tag{1.12}$$

and

$$2\gamma_p |F|^2 V / \{\gamma_p^2 + (\omega_M + \omega_p - \omega_X - \omega_{X'})^2\}$$

by the average value α and β, respectively. This is justified at least when $\hbar\gamma_0$ and $\hbar\gamma_p$ are much larger than kT because $\omega_0 + \omega_X - \omega_M$ and $\omega_M + \omega_p - \omega_X - \omega_{X'}$ are considered to be of the order of $kT/\hbar$. This is the case if T is lower than 10 K and l is less than 0.1 mm. Then by summing up Eqs. (1.10) and (1.11) over the states M and X and dividing them by the crystal volume $V = l^3$, we get the simplified expressions for the exciton concentration $x = \Sigma_X N_X / V$ and the biexciton concentration $y = \Sigma_M N_M / V$:

$$\dot{y} = -\alpha y + \beta x^2 , \tag{1.13}$$

$$\dot{x} = P + \alpha y - 2\beta x^2 , \tag{1.14}$$

where $P = \Sigma_X 2\delta_{pX}(N_X + 1)/V$. When we apply the same approximations to the G_2 and G_3 terms and introduce the corresponding parameters

$$\kappa \equiv 2\gamma_0 |G_2|^2 / \{\gamma_0^2 + (\omega_0 - \omega_X)^2\}$$

and

$$\gamma \equiv 2\gamma_0 |G_3|^2 V / \{\gamma_0^2 + (\omega_0 + \omega_X + \omega_{X'} - \omega_{X''} - \omega_{M'} - \omega_{M''})^2\},$$

we obtain the following rate equations [19]:

$$\dot{y} = \beta x^2 - \alpha y - 2\gamma y^2, \tag{1.15}$$

$$\dot{x} = \begin{pmatrix} 0 \\ P \end{pmatrix} + \alpha y + 3\gamma y^2 - \kappa x - 2\beta x^2. \tag{1.16}$$

Here it should be noted that the term with N_X in the expression of P corresponds to the coherent excitation of excitons and that the experimental situation of incoherent excitation of excitons may be described by $P^* = \Sigma_X 2\delta_{pX}/V$. The case with P in the first term of Eq. (1.16) corresponds to the steady pumping (case 2) and the case of 0 corresponds

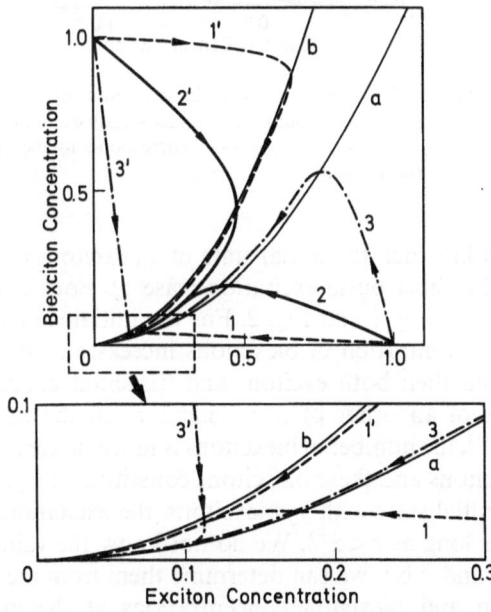

Fig. 1. The population dynamics of single excitons (x/x_0) and biexcitons (y/y_0) for the initial states with x_0 excitons or y_0 biexcitons under short pulse excitation (case 1). Between the normalization factors, there is the following relation: $\alpha y_0 = \beta x_0^2$ and (1.1′), (2.2′), and (3.3′) correspond to $\alpha/\beta x_0 = 0.1$, 1, and 10, respectively, for $\gamma = \kappa = 0$. On the curves a and b, x and y have a maximum value.

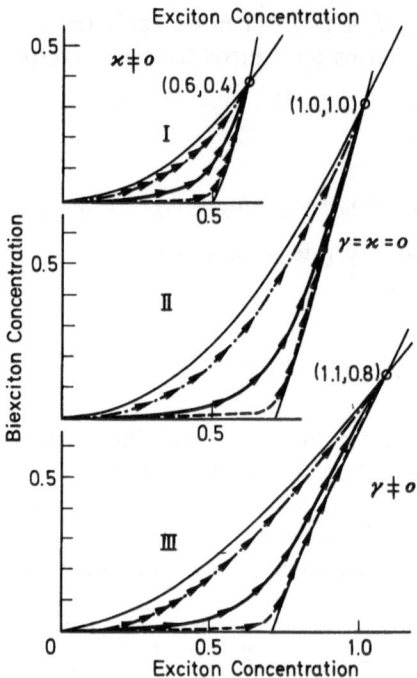

Fig. 2. Shows how the steady state of the normalized exciton and biexciton concentrations $x/\sqrt{P/\beta}$ and $y/(P/\alpha)$ is brought about in time under steady excitation (case 2). I: $\gamma = 0$, $\kappa \neq 0$, II: $\gamma = \kappa = 0$, III: $\gamma \neq 0$, $\kappa = 0$. $-\cdot-$, $—$, $----$ correspond to the various excitation power $\alpha/\sqrt{\beta P} = 10, 1, 0.1$, respectively.

to the situation in which the initial state of x_0 excitons or y_0 biexcitons is created by the short pulse excitation (case 1). Both cases have been solved as shown in Fig. 1 and Fig. 2. For the initial state of x_0 excitons in case 1, the concentration of biexcitons increases as $\beta x_0^2 t$ and shows a maximum, and then both exciton- and biexciton concentrations decrease in times of an order of α^{-1}. Therefore in the time interval of $(\beta x_0)^{-1} < t < \alpha^{-1}$, the number of biexcitons is much larger $(y/x \simeq \beta x_0 t \gg 1)$ than that of excitons and these biexcitons constitute the quasi-stationary state. For the initial state with y_0 biexcitons, the excitations exist almost as biexcitons as long as $t < \alpha^{-1}$. We do not know the values of the parameters α, β, γ, and κ but we can determine them from the time dependence of exciton and biexciton concentrations at the initial stage of population dynamics as shown in Table 1 for two initial conditions $(x_0, 0)$ and $(0, y_0)$. These values will be checked by comparison of experimental data with Fig. 1. Under steady excitation (case 2), while the exciton concentration increases in time and the biexciton concentration

Table 1. Short Pulse Excitation

$t = 0$	$x_0, \; y = 0$	$x = 0, \; y_0$
Initial stage	$\begin{cases} \dfrac{x - x_0}{x_0} = -(\kappa + 2\beta x_0)\, t \\[2mm] y = \beta x_0^2 t \end{cases}$	$\begin{cases} x = \alpha y_0 \left(1 + \dfrac{3\gamma y_0}{\alpha} \right) t \\[2mm] \dfrac{y - y_0}{y_0} = -(\alpha + 2\gamma y_0)\, t \end{cases}$

Steady Excitation

Initial stage	$x = Pt,$	$y = \dfrac{1}{3}\, \beta P^2 t^3$
stationary state	$P < \alpha^2/\gamma$	$P > \kappa^2/\beta$
	$\begin{cases} x = \dfrac{1}{2\beta}\, (\sqrt{\kappa^2 + 4\beta P} - \kappa), \\[2mm] y = \dfrac{1}{2\alpha\beta}\, (2\beta P + \kappa^2 - \kappa\sqrt{4\beta P + \kappa^2}), \end{cases}$	$\begin{cases} x^2 = \dfrac{1}{2\beta\gamma}\, (\alpha^2 + 4\gamma P - \alpha\sqrt{\alpha^2 + 4\gamma P}) \\[2mm] y = \dfrac{1}{2\gamma}\, (\sqrt{\alpha^2 + 4\gamma P} - \alpha) \end{cases}$

increases gradually in proportion to the third power of time at the initial stage, the stationary state of exciton and biexciton gases is created in shorter times of the order of $(\beta P)^{-1/2}$ or $(\alpha\beta P)^{-1/3}$ as the excitation power P increases. The stationary concentrations are given as follows; for such an excitation power as $\gamma P/\alpha^2 \ll 1$,

$$x_s = (2\beta)^{-1} (\sqrt{\kappa^2 + 4\beta P} - \kappa), \quad y_s = (2\alpha\beta)^{-1} (\kappa^2 + 2\beta P - \kappa\sqrt{\kappa^2 + 4\beta P}),$$

and for such an excitation power as $\kappa \ll \sqrt{\beta P}$,

$$x_s^2 = (2\beta\gamma)^{-1} (\alpha^2 + 4\gamma P - \alpha\sqrt{\alpha^2 + 4\gamma P}), \quad y_s = (2\gamma)^{-1} (\sqrt{\alpha^2 + 4\gamma P} - \alpha).$$

For any case the biexciton concentration at stationary state is much higher than that of excitons as long as $P \gg \alpha(\kappa + \alpha)/\beta$, or $P \gg \alpha^2(\beta - \gamma)/(\beta - 2\gamma)^2$. From the experimental results [20], we can estimate that $\alpha^2/\gamma \gg \kappa^2/\beta$ for CuCl and CuBr and $\alpha^2/\gamma \lesssim \kappa^2/\beta$ for CdS [5]. In the former case (CuCl), we expect that $y_s \propto P^2$ for $P < \kappa^2/2\beta$, $y_s \propto P$ for $\kappa^2/2\beta < P < \alpha^2/2\gamma$ and $y_s \propto P^{1/2}$ for $P > \alpha^2/2\gamma$. In the latter case (CdS), we expect the direct transition from P^2 dependence to $P^{1/2}$ dependence as observed in CdS. By using these 4 relations, we can also determine four parameters α, β, γ, and κ. As a conclusion of this section, the stationary state of biexciton gas is brought about under the steady excitation and we may also consider the excited state under the short pulse laser to be in the quasi-stationary state of biexcitons if we observe the system in the time interval $(\beta x_0)^{-1} < t < \alpha^{-1}$ by the short time slit.

3. Bose-Condensation of Biexcitons

The biexciton is considered to be the basic unit at low temperatures and at low concentrations below the Mott transition because of the large quantum effect discussed in Section 1. The biexciton gas is not liquefied nor solidified because the repulsive interaction due to the Pauli effect overcomes the exchange attraction and its light translational mass prevents the formation of more complicated complexes of charged particles due to van der Waals attraction. Furthermore, as the result of the discussion of Section 2, a great majority of excitations exist as the stationary or quasi-stationary states of the biexciton gas. On the other hand, the biexciton with any translational momentum is radiative according to the process in which an electron-hole pair in the biexciton is annihilated through light emission and the other pair is left as a single exciton absorbing the remaining momentum. In this sense, the biexcitons play the role of the radiative centers of many exciton systems. Basing upon these characters of quantum fluid and radiative centers, Hanamura [9, 10] pointed out the advantage of biexcitons for the observation of Bose-condensation and presented their characteristic emission spectrum at 0 K. Very recently evidence of the Bose condensation was found in the emission spectrum of CdSe [7] under pico-second laser pulse excitation in agreement with these predictions with respect to the position of the sharp line and the characteristic spectrum as well as the temperature and concentration dependence of these phenomena. Making the best use of the character of biexcitons as radiative centers, we can get fruitful information on the state of this quantum fluid. Therefore we discuss the state of Bose-condensed biexcitons in this section and the emission spectrum characteristic of Bose-condensed biexcitons at finite temperature in the next section.

Because we may consider the heavily excited state as an assembly of biexcitons according to the results of Section 2, we can rewrite the Hamiltonian of the quasi-equilibrium of the heavily excited state in terms of the ideal boson operator [12–14] $C^+\begin{pmatrix} PP' \\ pp' \end{pmatrix}$ corresponding to $a^+_{CP\uparrow} a^+_{CP'\downarrow} a_{\omega p'\downarrow} a_{vp\uparrow}$, by expanding the Hamiltonian in terms of C^+ and C as in the case of many exciton systems [12]. The harmonic part of this Hamiltonian is diagonalized under the following transformation in terms of the wave function f_μ [3] for the internal motion μ of biexciton:

$$C^+\begin{pmatrix} PP' \\ pp' \end{pmatrix} = \sum_{\mu K} 1/\sqrt{V} \delta_{K, P+P'-p-p'} f_\mu \begin{pmatrix} PP' \\ pp' \end{pmatrix} C^+_{\mu K},$$

and it describes the independent motion of biexcitons, where $C^+_{\mu K}$ and $C_{\mu K}$ are the creation and annihilation operators of biexcitons with the internal level μ and the translational momentum K. As far as we are concerned with a system at low temperature and low concentration, only the lowest internal level need be kept and our model Hamiltonian is described as follows:

$$H_0 = \sum_K E_m(K) C_K^+ C_K + \frac{1}{2} \sum_{K,K',q} W(q;K,K') C_{K+q}^+ C_{K'-q}^+ C_{K'} C_K, \qquad (3.1)$$

where $E_m(K) = 2E_{1s} - E_m^b + K^2/4M$ (E_{1s} is the excitation energy of an exciton, E_m^b is the molecular binding energy and $M = m_e + m_h$) and the effective interaction $W(q;K,K')$ is composed of not only the dynamical interaction but also the kinematical interaction coming from the Pauli effect acting between the component particles belonging to different biexcitons. Although the evaluation of $W(q;K,K')$ is very hard in contrast to the case of single excitons, we need only the value $W_0 = W(0;0,0)$ as long as the discussion is confined to low temperatures $kT \ll E_{ex}^b M/\mu$ and low concentrations $Na_0^3 \ll \mu/2M$, where $\mu = m_e m_h/(m_e + m_h)$. $E_{ex}^b = e^4 \mu/2\varepsilon_0^2 \hbar^2$ is the exciton binding energy and $a_0 = \hbar^2 \varepsilon_0/e^2 \mu$ is the exciton Bohr radius. The effect of multiple scattering between biexcitons is taken into account by replacing W_0 by $W = 4\pi f_0 a_0^2 E_{ex}^b \mu/M$ in terms of the scattering amplitude f_0 of two biexcitons in vacuum. f_0 is considered here to be a parameter of the order of $13a_0 M/3\mu$ twice the scattering amplitude between two single excitons [14].

In terms of the thermal Green's function of biexciton in the Bogoliubov approximation:

$$G_m(i\varepsilon_n, p) = u_p^2 [i\varepsilon_n - E(p)]^{-1} - v_p^2 [i\varepsilon_n + E(p)]^{-1} \qquad (3.2)$$

where

$$E(p)^2 = (p^2/4M)(2N_0 W + p^2/4M), \quad v_p^2 = [p^2/4M + N_0 W - E(p)]/2E(p)$$

and $u_p^2 = 1 + v_p^2$ and the energy is measured here with respect to the chemical potential $\mu_m = E_m(0) + NW$, the temperature dependence of the concentration of biexcitons condensed to the $K = 0$ state, is determined by solving self-consistently the following equation:

$$N' = N - N_0 = -\beta^{-1} \sum_{p,n} G_m(i\varepsilon_n, p), \qquad (3.3)$$

$$= (2\pi^2)^{-1} \int_0^\infty p^2 \, dp \, \{(u_p^2 + v_p^2)[\exp E(p)\beta - 1]^{-1} + v_p^2\} \qquad (3.4)$$

where $\beta = 1/kT$. From this expression, we can evaluate the N and T dependences of N_0/N. The N dependence of N_0/N at $0\,\mathrm{K}$ is given as the solution of

$$1 - N_0/N = (4f_0/a_0)^{3/2} \tfrac{1}{3} \pi^{-1/2} (N_0/N)^{3/2} (Na_0^3)^{1/2}. \qquad (3.5)$$

The concentration dependence of the critical temperature in this approximation coincides with that of ideal bosons and is given as

$$kT_c = 2\pi(\mu/M) E_{ex}^b \{N a_0^3/\zeta(\tfrac{3}{2})\}^{2/3}, \tag{3.6}$$

where $\zeta(\tfrac{3}{2}) = 2.612$.

As the concentration of biexcitons increases, the critical temperature increases according to Eq. (3.6), but this concentration cannot exceed that corresponding to the Mott transition at which biexcitons become unstable [9] against the formation of metallic electron-hole gas or liquid drops. For the biexciton system with concentrations of 10^{16} cm^{-3} in CdSe, for example, 23 % of biexcitons are condensed into the $K=0$ state at 0 K. N_0/N decreases as the temperature rises and they show a phase transition from the condensed state to the normal state at 4.8 K.

4. Linear Response of this Open System

According to the discussion in Section 2, the open system of excitons and biexcitons supported by a strong radiation field is considered to be stationary or quasi-stationary.

We can therefore apply a second weak external field as a probe to this open system [2]. The linear response theory to the probe field is so extended as to confirm the existence of biexcitons and the Bose-condensation of biexcitons from the emission spectra of the heavily excited semiconductors.

4.1. Biexcitons in Normal State

At first, let us calculate the emission spectra from the normal system of biexcitons. Because almost all the excitations exist as biexcitons in the stationary state, the equilibrium state is described by the Hamiltonian (3.1). However, when we want to describe the final state in which the single exciton is created after the emission of biexcitons and a small number of biexcitons are dissociated into single excitons, we should resort to the following Hamiltonian:

$$H = H_0 + \sum_K E_{ex}(K) B_K^+ B_K + V_0 \sum C_{K+q}^+ B_{K'-q}^+ B_{K'} C_K. \tag{4.1}$$

Here $E_{ex}(K)$ is the exciton energy $(E_{ex}(K) = E_{1s} + K^2/2M)$, B_K is the annihilation operator of this exciton and V_0 is the effective interaction between an exciton and a biexciton. Then the emission spectra are expressed as follows:

$$I(\omega) = 2\pi \sum_{i,f} Z^{-1} e^{-\beta E_i} |\langle f|H'|i\rangle|^2 \delta(\omega + E_f - E_i) \tag{4.2}$$

where $H' = g_0 \Sigma_K B_K^+ C_K$ is the radiative part of the electron-radiation interaction and the matrix element g_0 is approximated by a constant. $I(\omega)$ is rewritten in terms of the thermal Green's function χ in the stationary state as follows:

$$I(\Omega) = -2g_0^2 n(\Omega) \operatorname{Im} \chi(i\omega_n)|_{i\omega_n = -\Omega - i\delta}, \qquad (4.3)$$

where

$$\Omega = \omega - (\mu_m - \mu_{ex}), \quad n(\Omega) = [\exp \beta\Omega - 1]^{-1},$$

$$\chi(i\omega_n) = \int_0^\beta \chi(\tau) \exp(i\omega_n\tau) \, d\tau,$$

$$\chi(\tau) = -\left\langle T_\tau \sum_K C_K^+(\tau) B_K(\tau) \sum_{K'} B_{K'}^+ C_{K'} \right\rangle$$

and

$$A(\tau) = \exp(\tau\tilde{H}) \, A \exp(-\tau\tilde{H})$$

with $\tilde{H} = H - \mu_m \sum_K C_K^+ C_K - \mu_{ex} \sum_K B_K^+ B_K$

Here we introduced the chemical potentials μ_m and μ_{ex} of biexcitons and excitons, respectively. We take into account the level shifts $\Sigma_m(\varepsilon, K)$ and $\Sigma_{ex}(\varepsilon, K)$ and the level broadenings Γ_m and Γ_{ex} for a biexciton state and an exciton state, respectively, due to their interactions with $N - 1$ biexcitons, in the lowest order approximation. In terms of these Green's functions, which describe the exciton and biexciton propagation in the biexciton medium, the emission spectrum is obtained as a function of the emission frequency ω as follows:

$$I(\Omega) = -2g_0^2 T \operatorname{Im} \sum_n \int (2\pi)^{-3} dK \, g_m(i\varepsilon_n, K) \, g_{ex}(i\varepsilon_n - i\omega_e, K)|_{i\omega_e = -\Omega - i\delta}$$

$$= (4g_0^2 M^{3/2}/\pi\hbar^3) \int_0^\infty [(\Omega - \varepsilon)^2 + \Gamma^2]^{-1} \Gamma n(\varepsilon) \sqrt{\varepsilon} \, d\varepsilon \qquad (4.4)$$

$$= 2\pi g_0^2 (2M/\pi\hbar^2 kT)^{3/2} \exp(-\Omega/kT) \cdot \sqrt{\Omega^2 + \Gamma^2}$$
$$\cdot [\sin(-\alpha + \Gamma/kT) E_i(\Omega) + \cos(\alpha - \Gamma/kT) E_r(\Omega)],$$

where

$$E_r(\Omega) - iE_i(\Omega) = \operatorname{erfc}(y - ix)$$

with

$$\binom{y}{x} \equiv [(\sqrt{\Omega^2 + \Gamma^2} \mp \Omega)/2kT]^{1/2}, \qquad (4.5)$$

$$\tan \alpha = (\sqrt{\Omega^2 + \Gamma^2} - \Omega)/\Gamma, \quad \Omega = \omega - (\mu_m - \mu_{ex}), \quad \Gamma = \Gamma_m + \Gamma_{ex}, \qquad (4.6)$$

$$\mu_m = E_m + NW_0 \quad \text{and} \quad \mu_{ex} = E_{1s} + NV_0.$$

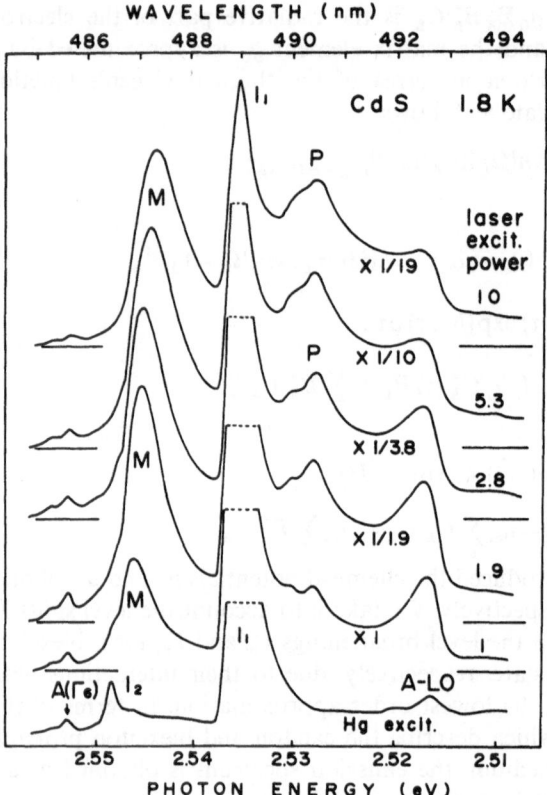

Fig. 3. Luminescence spectra of a CdS crystal at 1.8 K under 337.1 nm laser light excitation. The highest excitation level corresponds to about 40 kW/cm². The bottom shows the spectrum under mercury lamp (365.0 nm) excitation.

We approximated Γ_m and Γ_{ex} by the values at the bare kinetic energies of biexcitons and excitons [10]. We notice that this Γ is proportional to the biexciton concentration N in this approximation. Therefore at low concentrations, Γ is less than kT but this relation is reversed at high concentrations. Therefore for both limiting cases, we obtain the following expressions:

$$I(\omega) = \begin{cases} A'(\mu_m - \mu_{ex} - \omega)^{1/2} \exp -(\mu_m - \mu_{ex} - \omega)/kT, & \Gamma \ll kT, \\ A\Gamma[(\mu_m - \mu_{ex} - \omega)^2 + \Gamma^2]^{-1}, & \Gamma \gg kT. \end{cases} \quad (4.7)$$

This indicates that the spectral shape changes from the inverse Boltzmann distribution at low densities into the Lorentzian shape at high densities. Now we analyse the experimental data of CdS basing upon these theoretical results. In Fig. 3 are shown the luminescence spectra

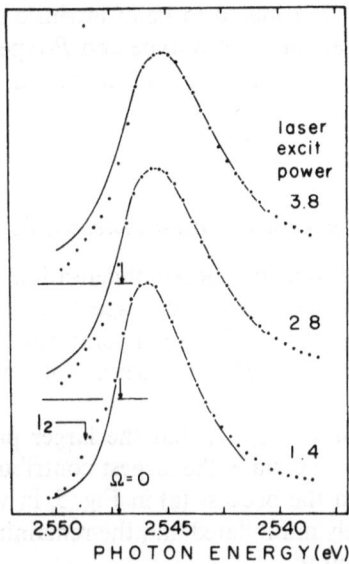

Fig. 4. The fitting of M line shape, shown in Fig. 3, to Eq. (4.4). The points indicate the experimental data, while the solid lines show the calculated curves.

of a CdS crystal under nitrogen laser illumination of various intensity levels. It is noted that under laser excitation, a new emission line designated as M is produced at $2.546 \sim 2.544$ eV. This M-line has the following four important features:

1. The asymmetric line shape of the M-line as observed fits the theoretical one fairly well with the effective temperature of $20\,\mathrm{K}$ as shown in Fig. 4. The inverse Boltzmann type at low excitation is modified into a Lorentzian form as the excitation power increases.

2. This comparison also gives us the zero point of kinetic energy ($\Omega = 0$) and the energy difference between $\Omega = 0$ and an exciton gives us the binding energy of the biexciton as $5.4\,\mathrm{meV}$ for CdS. This value is also fairly in agreement with the numerical calculation of molecular binding energy by Akimoto and Hanamura [3]: $4.2\,\mathrm{meV}$, which is obtained after the correction to variational calculation is taken into account.

3. The amplitude of the new M-line increases as the square of excitation intensity and this shows that two excitons are responsible for the M-line. This is in accordance with the result in Section 2 below the critical excitation power.

4. This *M*-line is considered to be of intrinsic nature because it has been observed independently of *N*-type and *P*-type samples. These four features support the inference that the *M*-line is the evidence for the existence of biexcitons in CdS.

4.2. Linear Optical Response of Bose-Condensed Biexcitons

In Section 3 we discussed the Bose-condensation of biexcitons and in this section we will discuss the emission spectrum characteristic of Bose-condensation of biexcitons at finite temperatures. Then we show in the next section how we succeeded in observing its evidence according to these predictions.

Taking into account the fact that the larger part of biexcitons are condensed into the $K = 0$ state, the largest contribution to the emission spectrum comes from the process (a) in Fig. 5, in which the condensed biexciton is radiatively annihilated and the remaining electron-hole pair propagates as an exciton.

This gives the sharp emission line at $\omega = E_{1s} - E_m^b + N(W - V_0)$, $(\Omega = 0)$, as follows:

$$I_a(\Omega) = 2\pi g_0^2 N_0 \{1 + n(\tilde{E}_{\mathrm{ex}}(0))\} \, \delta(\Omega + \tilde{E}'_{\mathrm{ex}}(0)). \tag{4.8}$$

The next largest contribution comes from the processes b, c and d in Fig. 5 as follows:

$$I_{bcd}(\Omega) = 2\pi g_0^2 \sum_K [A_K n(E(K)) \{1 + n(\tilde{E}_{\mathrm{ex}}(K))\} \, \delta(\Omega + E'_{\mathrm{ex}}(K) - E(K)) \tag{4.9}$$

$$+ B_K \{1 + n(E(K))\} \{1 + n(\tilde{E}_{\mathrm{ex}}(K))\} \, \delta(\Omega + E'_{\mathrm{ex}}(K) + E(K))],$$

where

$$A_K = \left\{ u_K - \frac{N_0 V_0(u_K - v_K)}{E(K) - E'_{\mathrm{ex}}(K)} \right\}^2, \quad \text{and} \quad B_K = \left\{ v_K - \frac{N_0 V_0(u_K - v_K)}{E(K) + E'_{\mathrm{ex}}(K)} \right\}^2.$$

$$E'_{\mathrm{ex}}(K) = K^2/2M \quad \text{and} \quad \tilde{E}_{\mathrm{ex}}(K) = E'_{\mathrm{ex}}(K) + \mu_{\mathrm{ex}} - \tfrac{1}{2}\mu_m.$$

$n(\tilde{E}_{\mathrm{ex}}(K))$ is evaluated by putting the chemical potential of an exciton μ_{ex} equal to $E_g - E_{\mathrm{ex}}^b + N V_0$. The first term of Eq. (4.9) gives the higher energy side band coming from the emission of biexcitons thermally excited over the phonon-like-dispersion. This band has a singularity at $\omega = \omega_0 + (2 - \sqrt{3}) N_0 W$, coming from a one-dimensional van Hove singularity of the joint density of states. The lower energy side band coming from the second term corresponds to the emission process in which the condensed biexcitons are excited into the K and $-K$ pair

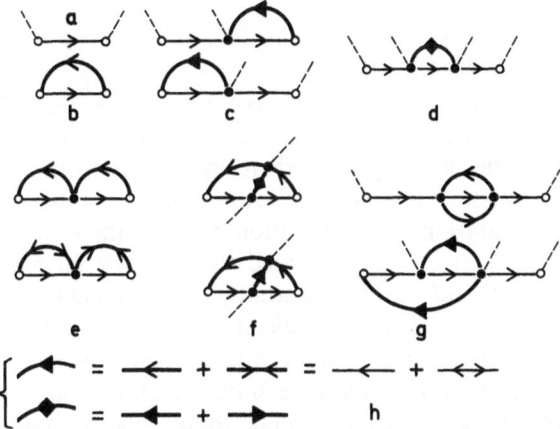

Fig. 5. The diagrams contributing to the emission spectrum. The dotted and thick solid lines describe the propagations of the Bose-condensed molecules and ones with the finite momentum, respectively. The thin solid line shows the exciton propagation created by the emission process denoted by the white circle. The black circle means the exciton-molecule or molecule-molecule scattering.

states of biexcitons, one of them is radiatively annihilated and the collective or individual mode of biexciton with the momentum K or $-K$ and an exciton remain in the crystal. As the temperature increases, the molecules thermally excited on the particle-like-dispersion curve also give the lower energy side band through the first term. This will grow into the well known emission spectrum characteristic of a Boltzmann distribution of biexcitons in the normal state, which was discussed in Section 4.1. In this case, however, other diagrams which were discarded here will become important. The contributions from the processes of Figs. 5e and 5f are shown to be by an order of $(N' V_0/N_0 W)$ smaller and the contribution of Fig. 5g is by an order of $(N' V_0^2/N_0 W^2)$ smaller than one of Figs. 5b, 5c, and 5d. As long as the discussion is confined to such a low concentration as $N' = N - N_0 \ll N_0$, we can neglect the contributions of Figs. 5e $\sim$ 5g. The value of V_0/W is estimated to be of the order of 0.5 in the crudest approximation but it will be much smaller due to the polarization effect of biexcitons around the exciton. At 0 K, the sum $I(\Omega)$ of $I_a(\Omega)$ and $I_{bcd}(\Omega)$ is evaluated analytically [10] as follows:

$$I(\Omega) = 2\pi g_0^2 N_0 \delta(\Omega) + (4g_0^2/3\pi a_0^3 E_{ex}^b)(M/2\mu)^{3/2} (N W_0/E_{ex}^b)^{1/2} g(\Omega/N W_0)$$

where (4.10)

$$g(y) = \{\tfrac{1}{3}(2+4y) - Z\}^{1/2} \{y + 2 - \tfrac{3}{2}Z\}/Z$$

and
$$Z = \{\tfrac{1}{9}(2+4y)^2 - \tfrac{4}{3}y^2\}^{1/2} .$$

As a conclusion, we can at first confirm the Bose-condensation of the biexcitons by the presence of the singular emission spectrum at the higher energy side, strongly dependent on the temperature, in addition to the sharp emission line and the lower energy side band due to collective mode excitation. The separation between the sharp line and the singularity, $(2-\sqrt{3})\,n_0\,W$, is estimated to be 1 meV for $n_0 = 10^{16}\ cm^{-3}$ and $W = (52\pi/3)\,E^b_{ex}\,a_0^3$ in CdSe. Secondly by analyzing the spectrum, we can check the dispersion of the collective mode and the consistency of the theory. It is not sure, however, whether the van Hove singularity survives after taking into account the finite momentum of emitted photon. From the experimental point of view, these detailed studies look too difficult at the present stage.

4.3. Experimental Results [7]

In order to realize the Bose-condensation of biexcitons, we must accumulate the biexcitons at a high concentration ($10^{16}\ cm^{-3}$) and keep the biexciton system at low temperatures below 4.8 K according to the phase diagram discussed in Section 3. In the emission spectra under nano-second pulse excitation of a nitrogen laser, however, we could not find any evidence of Bose-condensation. This absence of Bose-condensation is reasonable because the system of biexcitons has the effective temperature of 20 K for CdS and 8.4 K for CdSe although the crystal was put in liquid He. These effective temperatures of biexcitons were estimated from comparison of these spectra with theoretical ones in Section 4.1. The cause for the rise of exciton temperature, as observed, is attributed mainly to the radiative processes associated with inelastic collisions of biexcitons, in which hot single excitons with large kinetic energies are produced and the distribution in the momentum states of biexcitons become different from those in thermal equilibrium. Thus in the case of this nitrogen laser excitation, before the biexciton concentration reaches the threshold for Bose-condensation, the biexciton temperature is raised beyond the critical temperature, and as a result, Bose-condensation is not achieved. We have thought that Bose-condensation could be achieved, if a sufficient concentration of biexcitons is accumulated before the heating processes due to the inelastic collisions occur and that this could be brought about by using a pico-second laser pulse as the excitation source, thus making the generation rate of electron-hole pairs faster by three orders of magnitude. Fig. 6b shows one of the experimental results of emission spectra under pico-second laser pulse excitation. We point out how we conclude the Bose-condensation of biexcitons from these experimental data.

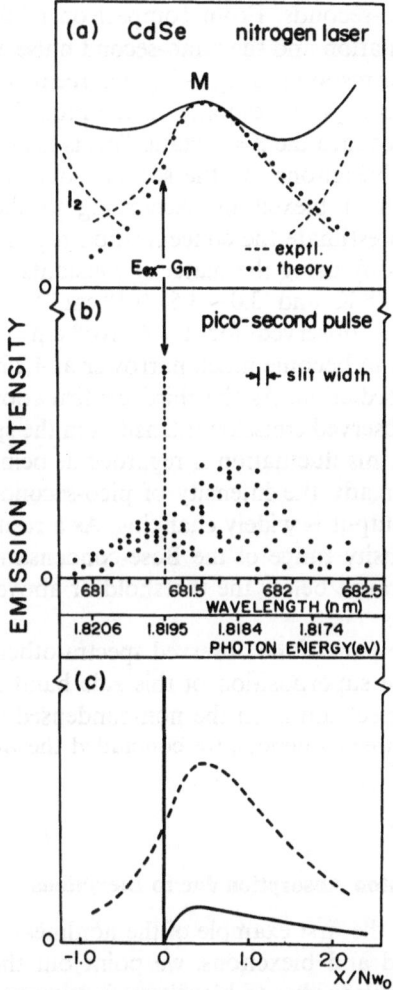

Fig. 6. (a) Spectrum of M line luminescence of a CdSe crystal at 4.2 K under nitrogen laser excitation. (b) Luminescence spectrum of a CdSe crystal at 1.8 K under pico-second light pulse excitation. (c) Theoretical luminescence spectrum for the Bose-condensed system of excitonic molecules at 0 K (solid line) from Eq. (4.10) and that for the non-condensed system at 8.2 K (dotted line) from Eq. (4.4). $X = E_{1s} - E_m^b - h\nu$.

The pico-second light pulses of neodymium glass laser were produced by means of mode-locking. The one-shot output of this mode-locked laser is a train composed of about 30 pico-second light pulses produced successively with a time interval of 7 ns. The width of one pulse was measured to be 7 ps and the maximum intensity of the pulse corresponds to the light density 2 GW/cm^2.

The electron-hole gas is created by 2-photon absorption of this pulse light and the electrons and holes are bound into biexcitons in a time order shorter than pico-seconds. From comparison of the data of the pico-second pulse excitation and the nano-second pulse excitation, this sharp line is found to correspond to $E_{ex} - E_m^b$ in agreement with the theoretical prediction. This energy corresponds to the radiative annihilation of bi-excitons condensed into the $K = 0$ state. This is the first evidence of Bose-condensation of biexcitons. As the second confirmation, we estimated the concentration of biexcitons. According to the phase-diagram in Section 3, we can estimate the concentration region for Bose-condensa-tion of biexcitons by using the material constants of CdSe as $0.84 \sim 3.5$ $\cdot 10^{16}$ cm^{-3} for 1.8 K and $3.0 \sim 3.5 \cdot 10^{16}$ cm^{-3} for 4.2 K. The sharp emission line was observed for $1 \sim 4 \cdot 10^{16}$ cm^{-3} at 1.8 K and this region was found to become much narrower at 4.2 K in agreement with the theoretical prediction. As the third confirmation, let us discuss the line shape. The observed emission intensities in the spectra fluctuate over a wide range but this fluctuation is regarded as being rather reasonable. As mentioned already the intensity of pico-second pulses in the train of the one-shot output is widely changing. As a result, only some pulses fall into the intensity range of the Bose-condensation and most of the pulses enter the range below the threshold or above the upper limit for the condensation.

Therefore, the part of the observed spectra other than the sharp line is regarded as the superposition of this side band from the condensed system and the spectrum from the non-condensed state.

From these three evidences, we concluded the Bose-condensation of biexcitons.

5. Giant Two-Photon Absorption due to Biexcitons

In this section, as the first example of the nonlinear interaction between the radiation field and biexcitons, we point out theoretically that the two-photon absorption due to biexcitons is extremely enhanced due to the same effect of giant oscillator strength as in bound excitons and the resonance effect, in contrast to the ordinary two-photon absorption. As a result, the existence of biexcitons will be confirmed also by the two-photon absorption spectroscopy.

We give the expression of the transition probability for the two-ω-photon excitation of an insulating crystal from its ground state $|g\rangle$ to the excited state $|e_m\rangle$ with a biexciton as follows;

$$W_m^{(2)} = (2\pi/\hbar)\,(e/m)^4\,(2\pi\hbar/\kappa\omega V)^2$$

$$\cdot N_\omega^2 \left| \langle e_m | \, \varepsilon \cdot \sum_j P_j \sum_i \frac{|i\rangle \langle i|}{E_{ig} - \hbar\omega} \, \varepsilon \cdot \sum_{j'} P_{j'} \, |g\rangle \right|^2 \cdot \delta(E_{eg} - 2\hbar\omega), \qquad (5.1)$$

where N_ω is the number of ω-photons in the incident beam, κ is a dielectric constant for frequency ω, V is the volume of the crystal, $P_j = \exp(i K_0 \cdot r_j) p_j$, and p_j is the momentum of the j-th electron. As the intermediate state $|i\rangle$, we only need take the one exciton state by paying attention to the resonance effect described below. Then we can give the transition probability of two-photon absorption due to biexcitons in terms of the exciton wave function for the internal motion $\phi_{1s}(r)$ and the Fourier K-component of the relative motion of two excitons in a biexciton $g(K)$ as follows:

$$W_m^{(2)} = (2\pi/\hbar)(e/m)^4 (2\pi\hbar N/\kappa\omega V)^2 \frac{(p_{cv} \cdot \varepsilon)^4}{(E_{1s} - \hbar\omega)^2} |\phi_{1s}(0)|^4 g^2(0) \qquad (5.2)$$
$$\cdot V\delta(2E_{1s} - E_m^b - 2\hbar\omega),$$

where E_m^b is the binding energy [3] of the biexciton, and $|\phi_{1s}(0) g(0)|^2 = 64(a_B/a_0)^3$. Here a_B is the average distance of excitons in the biexciton [3]. When this transition probability is compared with that of one photon absorption due to an exciton:

$$W_{ex}^{(1)} = (2\pi/\hbar)(e/m)^2 (2\pi\hbar N/\kappa\omega V)(p_{cv} \cdot \varepsilon)^2 V|\phi_{1s}(0)|^2 \delta(E_{1s} - \hbar\omega),$$

$W_m^{(2)}/W_{ex}^{(1)}$ is estimated to be of an order of $3 \cdot 10^{-15}(N/V)$, where $(p_{cv} \cdot \varepsilon)^2/m \simeq 3$ eV, $(a_B/a_0) \simeq 3$, $\hbar\omega \simeq 3$ eV and $E_{1s} - \hbar\omega = E_m^b/2 \simeq 20$ meV (CuCl) [4] or 3 meV (CdS) [5] are estimated, corresponding to the case of CuCl or CdS.

Therefore when we use a dye laser as the excitation source with tunable frequency above the photon density $N/V \simeq 10^{15}$ cm^{-3}, we can expect the two-photon absorption coefficient as strong as the single exciton absorption coefficient. Under this condition, it is guaranteed that the two-photon absorption coefficient due to biexcitons overcomes very much the absorption coefficient of an exciton at the exciton absorption tail $E_{1s} - E_m^b/2$. Furthermore in comparison with the case of an ordinary two-photon absorption due to a single electron excitation, we have two strong enhancement factors in our case; the first enhancement factor $64(a_B/a_0)^3$ comes from the fact that our case corresponds to the two-electron excitation and the second one comes from the resonance effect

$$\{(\tilde{E}_{ig} - \hbar\omega)/(E_{1s} - \hbar\omega)\}^2,$$

where $\tilde{E}_{ig}$ is the effective energy difference between the intermediate state and the state in the valence band. The first enhancement is explained as follows: in the process of transition from $|i\rangle$ to $|e_m\rangle$, we only need excite another exciton in the range within the molecular radius a_B around the first exciton in order to make a biexciton coherently, in contrast with the ordinary case where the electron excited to the intermediate state should interact again with the second photon. As a result, we have the factor $64\,(a_B/a_0)^3 \simeq 10^3$. The second enhancement factor of the resonance effect is estimated to be of an order of 10^4 at $\omega = \omega_{1s} - E_m^b/2\hbar$ if we assume $\tilde{E}_{ig} \simeq 5\,\text{eV}$ and $E_m^b = 40\,\text{meV}$ (CuCl). Because it is difficult to consider the other factors of the two-photon absorption process to work 10^6 times more unfavourably in the case of the biexciton, we can expect the two-photon absorption peak due to biexcitons to be observed dominantly. As a result, in two-photon spectroscopy, the biexciton will be confirmed as the absorption peak at $E_{1s} - E_m^b/2$ embedded in the rather weak background of the one-photon absorption tail of an exciton and ordinary two-photon absorption. This prediction was substantiated in CuCl by Gale and Mysyrowicz with the expected absorption intensity [20].

6. Laser Action of Biexcitons

We have discussed mainly the stationary states of biexcitons and excitons and their linear response to a second weak external field. As examples of the coherent interactions between the radiation field and biexcitons beyond the linear response, we discuss laser action of biexcitons in Section 6 and self-induced transparency in Section 7. Under the following 2 conditions, i.e. if the damping constant of the radiation field γ_0 due to the escape of the light through the end faces of the crystal, and the damping constant γ_p of phonons, are larger than the time constant of excitons and biexcitons, α and β, and if the system of phonons and photons may be considered to be still in the thermal equilibrium, we could treat the relevant system of biexcitons and excitons as a stationary state.

However, increasing the linear dimension of the crystal beyond 0.1 mm the decay constant of light due to the escape through the end faces of the crystal $\gamma_0 = c/l$ becomes small and it will be almost determined by the reabsorption process as $\gamma_0 = c\kappa_0$, where κ_0 is the reabsorption coefficient. In such a situation, the first assumption is not satisfied and as a result, the second assumption of thermal distribution of photons is also broken. In this case, we are rather justified in treating the emitted light as the relevant system beyond the linear response, because the radiation field has a feedback effect on the emission of biexcitons and the biexcitons rather obey this radiation field adiabatically.

It should be noted that the mutual collisons of excitons and biexcitons did not modify the population dynamics of excitons and biexcitons, but these frequent phase changes due to these collisions modify the coherent interaction between the radiation field and the biexcitons. Therefore we take into account these phase modulations by the damping constants (γ_X, γ_M) and their fluctuating forces (f_X, f_M) as follows;

$$(d/dt - i\omega_0 + \gamma_0) \, a_0^+ = iG^* \Sigma b_M^+ b_X + f_0 \,, \tag{6.1}$$

$$(d/dt - i\omega_X + \gamma_X - \delta_{pX}) \, b_X^+ = iG^* \Sigma b_M^+ a_0 + 2iF \Sigma a_p^+ b_M^+ b_X + f_X + f_{pX} \,, \tag{6.2}$$

$$(d/dt - i\omega_M + \gamma_M) \, b_M^+ = iG \Sigma a_0^+ b_X^+ + iF^* \Sigma b_X^+ b_X^+ \cdot a_p + f_M \,. \tag{6.3}$$

From these Langevin equations for the radiation field and for the exciton- and biexciton-creation operators, we get the coupled equations for the envelope of the radiation field $\tilde{a}_0^+$ and the exciton- and biexciton-concentrations X and Y as follows:

$$\ddot{\tilde{a}}_0^+ + (\gamma_0 + \gamma_M + \gamma_X - i\Delta) \, \dot{\tilde{a}}_0^+ + \gamma_0 \{(\gamma_M + \gamma_X) - i\Delta\} \, \tilde{a}_0^+ \\ - |G|^2 V(Y - X) \, \tilde{a}_0^+ = \tilde{F}_{\text{tot}}^+(t) \,, \tag{6.4}$$

$$\dot{X} = P + \alpha Y(1 + \tilde{a}_0^2) - 2\beta X^2 - \alpha \tilde{a}_0^2 X \,, \tag{6.5}$$

$$\dot{Y} = -\alpha Y(1 + \tilde{a}_0^2) + \beta X^2 + \alpha \tilde{a}_0^2 X \,, \tag{6.6}$$

where $\Delta = \omega_M - \omega_X - \omega$ and $\tilde{a}_0^+ = e^{i\omega t} a_0^+$, and $\tilde{F}_{\text{tot}}^+(t)$ is the total fluctuating force.

Due to the presence of the coherent radiation field $\tilde{a}_0$, the population dynamics of excitons and biexcitons is also modified as shown by Eqs. (6.5) and (6.6), where the induced emission of biexcitons and the induced absorption due to single excitons are taken into account by $\alpha Y \tilde{a}_0^2$ and $\alpha X \tilde{a}_0^2$, respectively. From these equations, we can derive the threshold excitation power [18] for laser action by requiring a positive $\tilde{a}_0^2$ under the conditions of resonance and stationarity as follows:

$$P_{\text{th}} = [\tfrac{1}{2}\sqrt{\alpha^2/\beta} + \sqrt{\alpha^2/4\beta + (\gamma_M + \gamma_X)/V}]^2 \,, \tag{6.7}$$

where we used the simplification of $\alpha = 2G^2/\gamma_0$. The population inversion is inevitable for laser action and in our case as shown by the last term in Eq. (6.4), the population Y of biexcitons is required to exceed the exciton-population X in order to bring about the laser action. This so-called population inversion is realized when the excitation power is larger than $\alpha(\kappa + \alpha)/\beta$ or $\alpha^2(\beta - \gamma)/(\beta - 2\gamma)^2$ as discussed in Section 2. The result of Eq. (6.7) for laser action was derived under the neglect of κ and γ processes in §2, i.e. in the case where the biexcitons are formed most effectively. This neglect of κ and γ is justified in the region where the biexciton concentration is linearly proportional to the excitation power. In this situation, the population inversion is brought about when the excitation power exceeds α^2/β. As shown by Eq. (6.7), in order to bring about laser action due to biexcitons, in addition to this population inversion, the excitation power must overcome the loss, coming from the phase destruction of biexcitons and excitons by mutual collisions.

Now basing upon these results let us speculate about the reason why laser action was observed in CuCl and CuBr but was not in CdS and CdSe. The expression Eq. (6.7) is applicable by itself to the case of CuCl and CuBr where κ and γ processes are negligible and this threshold excitation power is estimated to be $10^{28} \sim 10^{29}/cm^3 \cdot$ sec according to this equation in agreement with the experimental results [8].

In the case of CdS, as shown by the direct change from square to square root dependence of biexciton concentration on the excitation power, both κ and γ processes work effectively in reducing the biexciton concentrations as discussed in the population dynamics in Section 2. This situation requires a higher excitation power for bringing about the so-called population inversion in CdS than in CuCl.

The second demerit of biexciton laser action in CdS comes from its large exciton Bohr radius of 30 Å in comparison with 7 Å in CuCl. Because of this large Bohr radius, the collisional broadening of biexcitons and excitons works more effectively in CdS and this requires a stronger excitation power for laser action in CdS. Unfortunately, the higher excitation brings about an enhancement of the γ process which converts the two biexcitons into three single excitons and also a larger collisional effect.

Therefore in CdS, the phonon-assisted exciton annihilation or the exciton–exciton inelastic collisional processes show laser action [21]. This is the first example of coherent interaction between the radiation field and biexcitons.

7. Self-Induced Transparency [16]

As the second example of the coherent interaction between biexcitons and the radiation field, we discuss the possibility of self-induced transparency due to biexcitons.

We treated self-induced transparency (S.I.T.) due to two-photon transition in the system of three-level atoms [16]. The lowest level a and the highest level c are assumed to have the opposite parity to the intermediate level b. Then the dipolar transition due to the radiation field is allowed only for the transitions between levels a and b and between b and c. The level spacing between b and c is not so close to that between a and b. Then for a rather strong radiation field with the frequency ω, this field will be attenuated due to this two-photon absorption process corresponding to the process of two-photon excitation from a to b via the b-level as the virtual intermediate state, when the resonance condition for the two-photon transition is satisfied. Here arises the question whether the self-induced transparency by this three-level atomic system is possible for such a coherent radiation field that the stimulated two-photon emission process as well as the two-photon absorption process are brought about simultaneously. In the discussion of S.I.T., we assume a very short pulse in comparison with the relaxation times of excitations. Therefore we are allowed to discuss the closed system described by the following Hamiltonian of the electrons and the radiation systems;

$$H = \sum_l (\varepsilon_c c_l^+ c_l + \varepsilon_b b_l^+ b_l) + ig \sum_l \{(c_l^+ b_l + \gamma b_l^+ a_l) A_l^+ - \text{h.c.}\} + H_R , \quad (7.1)$$

where the zero point of the energy scale was taken at the atomic ground state a and a_l, b_l and c_l are the annihilation operators of the electrons in the state a, b, and c of the l-th atom, with the energy ε_a, ε_b, and ε_c, respectively. Here A_l^+ and A_l^- are the positive and negative frequency parts of the quantized vector potential and g and γg are the electron-radiation interaction matrix elements corresponding to the electron transition from b to c and from a to b, respectively. By using Heisenberg equations of motion for the excitation operators of electrons because of the closed system and Maxwell's equation for the radiation field, and by taking fully into account the nonlinearity of the system, we derived the nonlinear differential equation for the radiation field under the following two assumptions;

1. The slowly varying envelope approximation

$$|\partial^2 \hat{A}^\pm/\partial t^2| \ll \omega_0 |\partial \hat{A}^\pm/\partial t| , \quad |\partial(\widehat{b^+ a})/\partial t| \ll |(\varepsilon_b - \omega_0)(\widehat{b^+ a})|$$

and

$$|\partial(\widehat{c^+ b})/\partial t| \ll |(\varepsilon_c - \varepsilon_b - \omega_0)(\widehat{c^+ b})| ,$$

where

$$(b_l^+ a_l) = \exp(i\omega_0 t - ikl)(\widehat{b^+ a})_l , \ (c_e^+ b_e) = \exp(i\omega_0 t - ikl)(\widehat{c^+ b})_l$$

and

$$A^\pm(x, t) = \exp(\mp i\omega_0 t \pm ikx) \hat{A}^\pm(x, t)$$

and we can skip the subscript l because of the local character of the Hamiltonian.

2. The distortionless propagation of waves: the slowly varying envelope amplitude of the field as well as the corresponding atomic variables depend on space and time only in the combination of $\tau = t - x/v$ where the velocity v of the pulse envelope is a parameter still to be determined. Then the envelope of the radiation field $\hat{A}^\pm = A \exp(\pm i\phi)$ is

given by the solution of the following second order nonlinear differential equation:

$$A\,d^2 A/d\tau^2 - (dA/d\tau)^2 = 2pA^4(C - qA^2),\tag{7.2}$$

where $C = c_b - c_c + \gamma^2(c_a - c_b)$ and c_a, c_b, c_c are the initial electron population of atoms before the radiation field pulse comes in. For the resonant case $\varepsilon_c = 2\omega_0$, p and q are given in terms of material constants $\alpha \equiv 2\pi\hbar^2 c^2 n/\omega_0$ and $\Delta = \varepsilon_b - \omega_0$, as follows:

$$p = \frac{(1+\gamma^2)\,\gamma^2\alpha g^4}{2(c/v - 1)\,\Delta^2} \quad \text{and} \quad q = \frac{1+\gamma^2}{2\alpha\gamma^2}\,(c/v - 1).\tag{7.3}$$

When we take the initial condition corresponding to the pulse propagation, Eq. (7.2) is integrated as follows:

$$A^2 = A_0^2[1 + 2pA_0^2(t - z/v)^2]^{-1},\tag{7.4}$$

where

$$A_0^2 = 2/q = 4\alpha\gamma^2(1+\gamma^2)^{-1}(c/v - 1)^{-1}$$

and the pulse width $\Delta\tau = \sqrt{q/4p}$. Therefore when we give the pulse-width $\Delta\tau$ or the maximum intensity A_0^2 externally, the pulse-velocity v is determinated, respectively, as follows:

$$v = c\Delta/[\Delta + 2\alpha\gamma^2 g^2\,\Delta\tau]$$

and

$$v = cA_0^2/[A_0^2 + 4\gamma^2(1+\gamma^2)^{-1}\alpha].$$

When we make the b and c levels of the three level atoms correspond to the single-exciton and biexciton states, respectively, we can expect the self-induced transparency due to biexcitons. In order to get a more solid confirmation and check this possibility qualitatively, we must solve the more realistic model of biexcitons. This is one of our future problems.

References

1. Haken, H.: Z. Physik **263**, 267 (1973).
2. Weidlich, W.: Z. Physik **248**, 234 (1971).
3. Akimoto, O., Hanamura, E.: J. Phys. Soc. Japan **33**, 1537 (1972); — Solid State Commun. **10**, 253 (1972) and **11** (1972), No. 9, p. xiii.
4. Mysyrowicz, A., Grun, J. B., Levy, R., Bivas, A., Nikitine, S.: Phys. Letters A **26**, 615 (1968).
   Souma, H., Goto, T., Ohta, T., Ueta, M.: J. Phys. Soc. Japan **29**, 697 (1970).
5. Shionoya, S., Saito, H., Hanamura, E., Akimoto, O.: Solid State Commun. **12**, 223 (1973).
6. Saito, H., Shionoya, S., Hanamura, E.: Solid State Commun. **12**, 227 (1973).

7. Kuroda, H., Shionoya, S., Saito, H., Hanamura, E.: J. Phys. Soc. Japan **35**, 534 (1973); — Solid State Commun. **12**, 533 (1973).
8. Souma, H., Koike, H., Suzuki, K., Ueta, M.: J. Phys. Soc. Japan **31**, 1285 (1971). Shaklee, K. L., Leheny, R. F., Nahory, R. E.: Phys. Rev. Letters **26**, 888 (1971).
9. Hanamura, E., Inoue, M.: Proc. 11th Intern. Conf. on Phys. of Semiconductors, p. 711. Warszawa 1972.
10. Hanamura, E.: Luminescence of Crystals, Molecules and Solutions, p. 121. Ed. by F. Williams, 1972.
11. Lax, M.: J. Phys. Chem. Solids **8**, 66 (1959).
12. Hanamura, E.: J. Phys. Soc. Japan **29**, 50 (1970).
13. Hanamura, E.: Proc. 10th Intern. Conf. on Phys. of Semiconductors, p. 487. Cambridge 1970.
14. Hanamura, E.: Solid State Commun. **11**, 485 (1972).
15. Hanamura, E.: Solid State Commun. **12**, 951 (1973).
16. Hanamura, E.: Opt. Commun. **9**, 396 (1973): J. Phys. Soc. Japan **37**, Dec. (1974).
17. Haake, F.: Springer Tracts in Modern Physics **66**, 98 (1973).
18. Haken, H.: Handbuch der Physik, Bd. XXV/2 C. Berlin-Heidelberg-New York: Springer 1970.
19. Compare with Knox, R. S., Nikitine, S., Mysyrowicz: Opt. Commun. **1**, 19 (1969).
20. Gale, G. M., Mysyrowicz, A.: Colloque sur les semiconducteurs a grand gap, Montpellier, Sept. 5.–8. 1973.
21. Haken, H., Nikitine, S.: Springer Tracts Mod. Phys. this volume.

Prof. Dr. E. Hanamura
Institute for Solid State Physics
The University of Tokyo
Roppongi, Minato-Ku
Tokyo, Japan

III. Electron-Hole Droplets

Properties of Electron-Hole Drops in Germanium Crystals

V. S. BAGAEV

Contents

1. Introduction

The task we have set was to investigate the conditions of exciton condensation in germanium and to study the physical properties of electron-hole drops (EHD) formed under condensation [1]. One can expect that the properties of such an excited crystal state as EHD are most pronounced in the presence of a strong external excitation [2]. When carrying out our experiments, we have paid particular attention just to this circumstance.

We shall analyse in detail three groups of experiments which allow us to detect some properties of EHD in Ge and to obtain quantitative estimates of equilibrium carrier concentration n_0 in EHD, mean concentration of nonequilibrium carriers excited in a crystal $\bar{n}$, the number of carriers N in a drop, its radius a and the number N_d of EHD.

These experiments are as follows: Investigation of the influence of uniaxial homogeneous and inhomogeneous deformation on EHD [3–6]; studies in strong magnetic fields [7–9]; investigation of light scattering by EHD [10–13]. The effect of unaxial deformation and strong magnetic field on EHD was studied by the method of photo-luminescence.

2. Uniform Deformation Effects

The radiation spectrum of EHD and free excitons under a uniaxial deformation of a sample was examined in detail. An excitation source of nonequilibrium carriers was a He–Ne laser generating light with

$\lambda = 1.15\,\mu$. Measurements were carried out on germanium samples with a residual impurity concentration $\lesssim 10^{12}\,\mathrm{cm}^{-3}$. The samples were oriented in the directions [111], [110], and [100] to an accuracy of $2°$. The experiment was divided into two parts. The first part was to determine the influence of a uniaxial homogeneous deformation on the EHD radiation spectrum.

It should be noted that the homogeneity must be high since even small inhomogeneities in the deformation led to considerable changes in the experimental results. In the second part the influence on EHD of a uniaxial inhomogeneous deformation with a certain gradient along the sample was studied. The energetic position of the EHD radiation line maximum (the most intense EHD radiation line with the emission of a LA phonon will be designated as C), the spectral position of the radiation line maximum of free excitons E_{fe}, the shape of line C, the behavior of its intensity and of that for free excitons were investigated as a function of the deformation P at different temperatures for the case of a uniaxial homogeneous deformation. Measurements were carried out for cubic samples with a side $\sim 2\,\mathrm{mm}$. Particular attention was paid to the fact that friction between the sample and a holder compressing it should be minimal.

We have already reported on the kind of the dependence of $C^{\max}(P)$ and $E_{fe}^{\max}(P)$ shown in Fig. 1 [5, 14]. It has also been discussed by Benoît à la Guillaume et al. [15].

For the direction of the stress along [110] $E_{fe}^{\max}$ is shifted linearly within the whole pressure interval; it is connected with narrowing of the forbidden gap. For the dependence $C^{\max}(P)$ two regions are characteristic. Within the pressure interval from 0 to 250 kg/cm^2 (P_{cr}) the peak

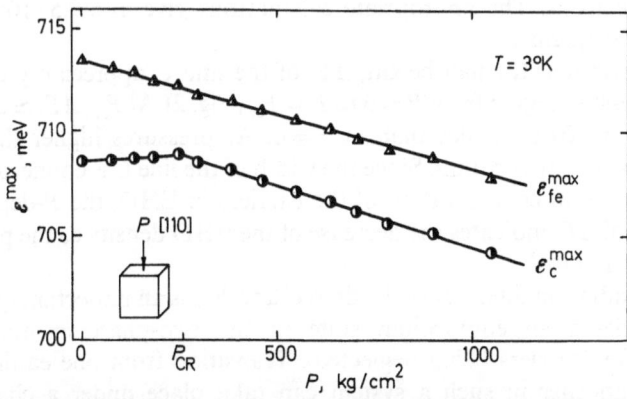

Fig. 1. Plots of $E_c^{\max}(P)$ and $E_{fe}^{\max}(P)$ for the case of uniform deformation

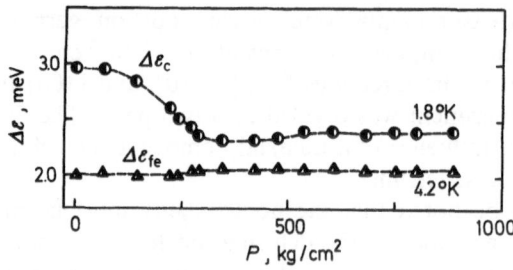

Fig. 2. Plots of the half-widths $\Delta E_c(P)$ and $\Delta E_{fe}(P)$

of the line C is practically not shifted. At pressures higher than P_{cr}, C^{max} is going in parallel with the shift of E_{fe}^{max} and, therefore, with the width of the forbidden gap. In the case when the deformation is directed along [111] and [100] the character of the pressure dependences $C^{max}(P)$ and $E_{fe}^{max}(P)$ is analogous with the only exception that the value of P_{cr} is different (~ 320 kg/cm^2 and ~ 700 kg/cm^2).

This fact has a simple physical explanation. It is well known that under a uniaxial compression of germanium samples the density of states is lowered in allowed bands. This results in the increase of kinetic energy of carriers in EHD or an internal pressure in the drop. The necessity to maintain equilibrium of attraction and repulsion forces leads to the increase of an equilibrium volume of the drop and, therefore, to the decrease of an average particle density in EHD. From the data on the shift $C^{max}(P)$ one can determine the density of carriers in EHD at $P = P_{cr}$. In fact, the break in the curve $C^{max}(P)$ corresponds e.g. for $P \parallel [111]$ to the fact that the magnitude of a relative shift of the minima in the conduction band is approximately equal to the Fermi energy of the electrons. The appropriate calculations give $n_0 \simeq 9.5 \cdot 10^{16}$ cm^{-3} at $P = 320$ kg/cm^2.

The width at the half-height, ΔE_c of the line C appreciably changes in the pressure region from $P = 0$ to $P = P_{cr}$ (Fig. 2). At P_{cr}, $\Delta E_c \simeq 2.4$ meV which is by 20% smaller than at $P = 0$. At pressures higher than P_{cr}, ΔE_c is practically constant. Since the width of the line C is connected with the equilibrium concentration of the carriers in EHD, the P-dependent decrease of ΔE_c indicates the decrease of the EHD density as the pressure varies from 0 to P_{cr}.

The radiation intensity of the lines C and E_{fe} is an important quantity characterizing an equilibrium state in the two-phase exciton-EHD system (free carriers being neglected). Transition from one equilibrium state to another in such a system can take place under a change of temperature or binding energies φ of carriers in EHD. Figure 3 represents

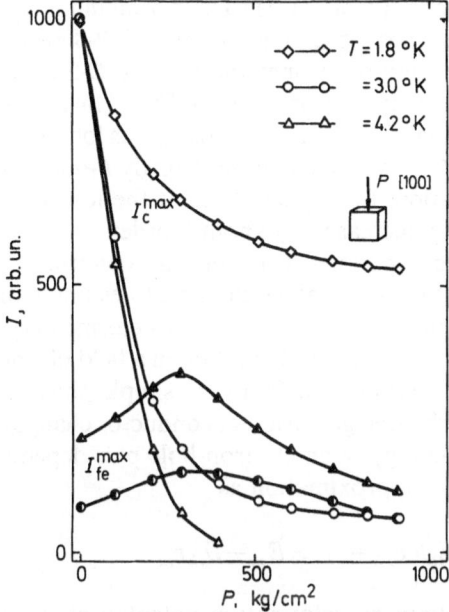

Fig. 3. Plots of line intensities I_c^{max} and I_{fe}^{max} vs. pressure at different temperatures

the pressure-dependent intensities I_c^{max} and I_{fe}^{max} at three different temperatures. It is seen from the figure that the intensity I_c^{max} drops as the pressure increases. The most sharp decrease is observed in the pressure interval from 0 to P_{cr}, whereas the intensity I_{fe}^{max} is slowly growing in this pressure interval and starts decreasing at $P > P_{cr}$. The decrease of the intensity I_c^{max} with the pressure increase from 0 to P_{cr} is clear if one takes into account that the binding energy φ and the drop equilibrium density n_0 decrease. EHD is partially vaporized increasing the exciton gas density. Evaporation is strongly temperature-dependent. At temperatures 3 K and 4.2 K (Fig. 3) the intensity I_c^{max} becomes 20–30 times lower as the pressure decreases which is connected with the decrease of an overall number of particles in EHD under pressure and constant temperature. At $T = 1.8$ K the intensity I_c^{max} is only two times lower since the temperature is not high enough for evaporation even at a lowered value of the binding energy φ.

3. Uniaxial Inhomogeneous Deformation Effects

The very interesting property of EHD detected up to now is, perhaps, to accelerate up to high velocities in external inhomogeneous fields.

Extremely high mobility of EHD in a field of an inhomogeneous deformation was noticed when observing an unusual *P*-dependence of intensity and energetic position of the maximum of the *C*-line under an evidently inhomogeneous uniaxial compression [3]. Thus in thin samples or those of a pyramidal shape a decrease of P_{cr} was found where at a pressure higher than P_{cr}, C^{max} was much more strongly *P*-dependent than under a uniform deformation. In this case even at the lowest possible temperatures the intensity decreased by about 2 orders.

The following experiments in question confirm the validity of the previous explanation and make it possible to estimate the mean velocity of EHD motion in a non-homogeneously deformed crystal.

Our assumption on the EHD motion in a field of a nonhomogeneous deformation was based on the following simple physical considerations. The width of a forbidden gap in a semiconductor changes under deformation. Hence, the energy of an electron-hole pair depends on a deformation ε_{ik} in the first approximation as

$$E_g(\varepsilon_{ik}) = E_{g_0} + \partial E_g/\partial \varepsilon_{1k} + \cdots \simeq E_{g_0} - D_{ik}\varepsilon_{ik} \qquad (1)$$

where D_{ik} is a summary deformation potential of the electron and the hole. Therefore, in a field of an inhomogeneous deformation ε_{ik} r) the force

$$F = V n_0 D_{ik} \operatorname{grad} \varepsilon_{ik} \qquad (2)$$

acts upon EHD of a volume *V*. Thus under the action of the force *F*, EHD must move to the region of maximum deformation at a velocity *v*

$$v = (D_{ik}\tau_p/M) \operatorname{grad} \varepsilon_{ik}. \qquad (3)$$

Here *M* is the mass of an electron-hole pair, τ_p is the time of relaxation.

For observing EHD motion in a field of inhomogeneous deformation, the sample shown in Fig. 4 has in our opinion the most convenient geometry. The area of the cross-section of such a sample depends on the square of the height, the maximum deformation being in the center of the sample. The pressure distribution along the height of the sample and the forces acting on the drops are shown schematically in Fig. 4. Such a distribution of the deformation gradient permits to obtain the slowing down motion of EHD produced, for example, at the end of the sample in the direction towards the centre, as into a trap, which they cannot escape. The height of the samples was 8–10 mm and the ratio of the areas of their bottoms to the central cross-section area was ~1.5. Two series of experiments were carried out: when the generation region

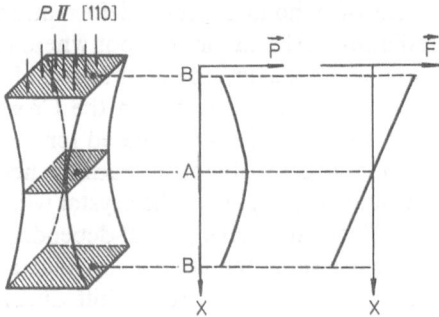

Fig. 4. Shape of sample with specified deformation gradient; approximate distribution of the pressure P over the height of the sample; distribution of the force F acting on the particles

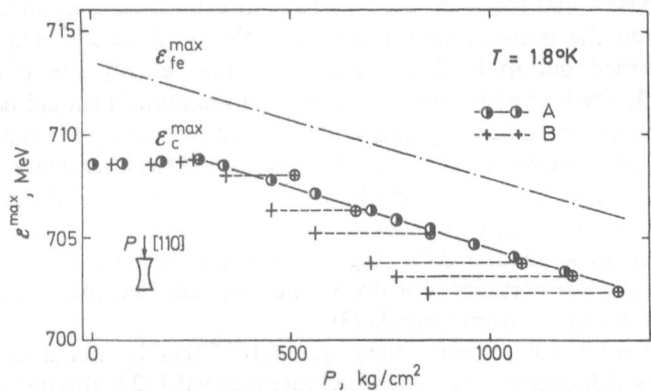

Fig. 5. Plots of $E_c^{max}(P)$; ◑-Section A illuminated; +-Section B illuminated; ⊕-Section B illuminated and pressure recalculated for Section A

of nonequilibrium carriers (a focused laser beam) was in the central part of a sample (Cross-section A) and when it was near the end of the sample (Cross-section B). The P-dependence of the energetic position of C^{max} for both these cases is presented in Fig. 5. Knowing the P-dependence of C^{max} in the pressure region exceeding P_{cr} for a homogeneous deformation, we can to a high accuracy determine the actual pressure in the region from which this sample radiates by the spectral position of C^{max} in the case of an inhomogeneous deformation. When the carriers are excited in the Cross-section A, the P-dependence of C^{max} (P is taken for the area of the Cross-section A) does practically not differ from the case of a homogeneous deformation (Fig. 5). Thus at the moment of radiation, EHD are in the spot where they were produced since this is a region with a maximum pressure and almost zero deformation gradient. If the drops are produced in the Cross-section B, the P-dependence of C^{max} is stronger (P is calculated for the area of the Cross-section B). EHD in this case obviously radiate not from the Cross-section B but from some other part of the crystal where the value of P is essentially higher. When analyzing the P-dependence of the spectral position of C^{max}, we arrive at the conclusion that at any P excluding those smaller than or close to P_{cr}, EHD shift during their lifetime τ_0 into the Cross-section A where the pressure is maximum (Fig. 5). The distance travelled by EHD in this experiment was ~ 4–4.5 mm. The EHD lifetime τ_0 measured by the decrease of photoluminescence was ~ 20 μsec. In the case of an inhomogeneous deformation, however, it was 3–4 times less and for pressures $P > P_{cr}$ it was 5–6 μsec [15]. Thus EHD pass 4 mm for $t < 5$ μsec. We have directly measured under pulse excitation the transit time for these samples at $T = 1.6$ K. The transit time turned out to be 2 μsec and, thus, the velocity $v \simeq 10^5$ cm/sec. This velocity is close to that of sound in germanium. It should be noted that an electron-hole drop as a united system bound by internal forces must start to intensively generate phonons as soon as it reaches a velocity close to that of sound. This mechanism will apparently limit the further increase of its velocity.

From experimental data (Fig. 6) we can estimate the force acting upon a pair of particles in a drop. Then we can find the value of the relaxation time τ_p from formula (3).

If $F = 3.2 \cdot 10^{15}$ erg/cm, then $\tau_p \simeq 6.10^{-8}$ sec. It should be noted that τ_p is T-dependent. In the temperature interval 4–2 K this dependence is close to linear. Hence, the velocity of EHD motion must depend not only on the size of the deformation gradient but on the temperature as well. This is confirmed by changes in the P-dependence of C^{max} (Fig. 6) resulting from a change in the temperature. It is easily seen that with the temperature increase larger pressure gradients are required for EHD to pass from B to A for the time $t \leqq \tau_0$.

Note that in the whole pressure interval under investigation (up to 1200 kg/cm^2) at $T = 5$ K and $T = 4.2$ K (at low excitation level, when radiation of the line C is absent) no deviation of the P-dependence of E_{fe}^{max} from a linear one was registered. When in the spectrum both the lines C and E_{fe} are registered simultaneously, the dependence $E_{fe}^{max}(P)$ is shifted (Fig. 6a and b). This is connected with an exciton "tail" of EHD, i.e. with the drop evaporation during its motion.

The P-dependence of the C-line width also gives evidence for the motion of EHD in a field of an inhomogeneous deformation (Fig. 7).

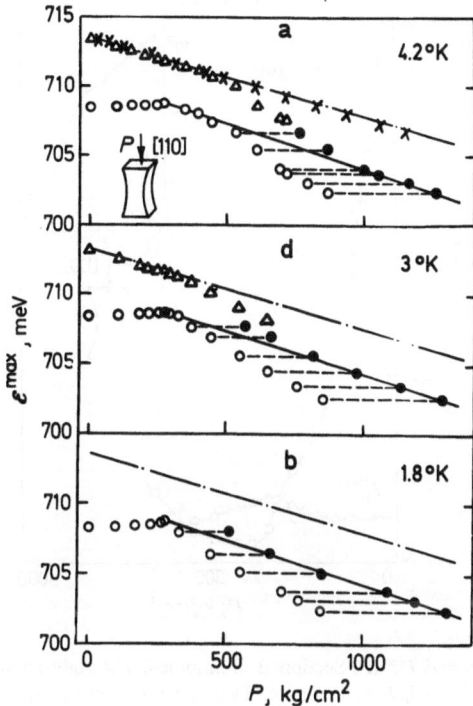

Fig. 6. Plots of $E_c^{max}(P) - $ ○ and $E_{fe}^{max}(P) - $ △, Section B illuminated at three different temperatures. Position of $E_{fe}^{max}(P) - $ × at $T = 4.2$ K, in the absence of emission in the C-line (low level of excitation). $E_c^{max}(P) - $ with P recalculated for the Section A

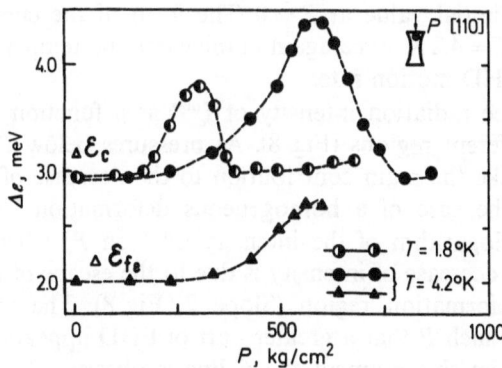

Fig. 7. Plots of half-widths $\Delta E_c(P)$ and $\Delta E_{fe}(P)$ for the case of nonuniform deformation and Section B illuminated

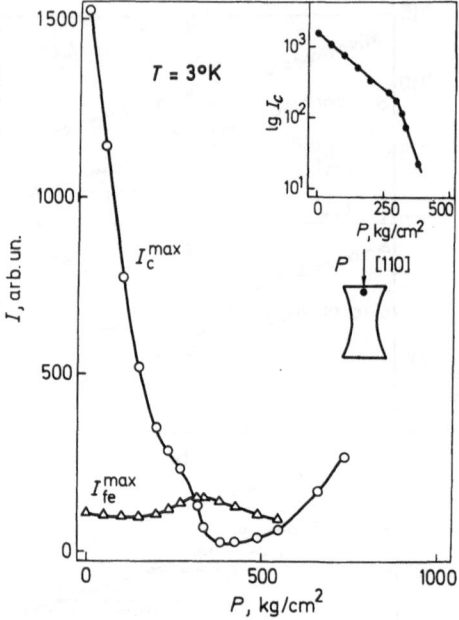

Fig. 8. Plots $I_c^{max}(P)$ and $I_{fe}^{max}(P)$ Section B illuminated. The right-hand corner shows the same relation, but with $\lg I_c$ as ordinate. Two different sections of the slope are seen: 1 – at $0 < P < P_{crit.}$ and 2 – at $P > P_{crit.}$ $P_{cr} \sim 300$ kg/cm²

At $T = 1.8$ K the C-line is considerably widened when the pressures are of order P_{cr}. Under further increase of the pressure, when a larger part of EHD occurs in the Cross-section A, the width of the line diminishes to about its initial value at $P = 0$. The form of the curves $\Delta E_c(P)$ at $T = 1.8$ and $T = 4.2$ K once again emphasizes the temperature dependence of the EHD motion rate.

Finally, the radiation intensity of I_c^{max} as a function of P at 4.2 K has three different regions (Fig. 8). At pressures below P_{cr}, EHD evaporations make the main contribution to the decrease of intensity, the same as in the case of a homogeneous deformation [Slope 1, Fig. 8 showing the logarithm of the intensity C^{max} on P] whereas at higher pressures the decrease of intensity is due to the escape of EHD into the maximum deformation region (Slope 2, Fig. 8). The intensity stops decreasing at such P that a greater part of EHD appears in the Cross-section A. From this moment the C-line is observed to narrow again which may account for an increase of the signal since in Fig. 8 the amplitude of the C-line and not its integral is plotted.

4. EHD in High Magnetic Fields

Another strong perturbation is an external magnetic field. It is well known that a strong magnetic field changes the exciton binding energy and a state density of the valence band and the conduction band. The presence of a strong magnetic field must influence the equilibrium state in the exciton gas – electron-hole liquid system. Taking into account that carriers in EHD are degenerate, such effects as the Shubnikow-de-Haas and de-Haas-van Alphen effects may be expected to appear in the magnetic field.

Photoluminescence of EHD and free excitons was investigated at temperatures 1.5–4.2 K in a constant magnetic field up to 100 kG. To increase the luminescence output, the germanium samples were made in Weierstraß's sphere forms. The direction of the magnetic field coincided

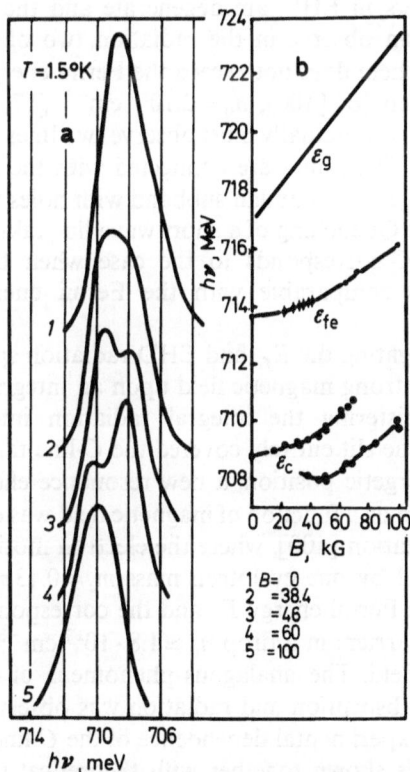

Fig. 9. (a) Spectra of radiation of Ge in high magnetic field; (b) Plots of energy position of E_c^{max} and E_{fe}^{max} vs. magnetic field for the Case B $\parallel$ [100]

with that of excitation and with one of the main three crystallographic directions in which the samples were oriented. Figure 9 presents the germanium recombination radiation spectra at $T = 1.5\,\text{K}$ in their dependence on the magnetic field for $B \parallel [100]$. It is seen from this figure that within the interval of the magnetic fields 40–70 kG two EHD radiating lines are observed which shift to the short wave side with the growth of B. The magnitude of splitting (Fig. 9b) is growing linearly with the magnetic field and at $B = 70$ kG it reaches 2.2 meV. A qualitatively similar picture of photoluminescence occurs for the other two crystallographic directions [111] and [100]. Figure 9 also presents our and also by Benoît à la Guillaume et al. [16] published experimental data on the B-dependent position of the radiation line E_{fe}. It is seen that the binding energy of the carriers in EHD increases with the increase of the magnetic field.

We shall try to explain the change in the EHD radiation spectrum in a magnetic field (Fig. 9a). Taking into account the fact that non-equilibrium carriers in EHD are degenerate and the magnetic field is quantizing, one can observe in the radiation two or more lines if the distance between them does not exceed the Fermi energy E_f. In the case under consideration ($B \parallel [100]$, $n_0 = 2.10^{17}\,\text{cm}^{-3}$ [17, 18]) in the EHD radiation spectrum one actually must observe two lines within a magnetic field of 40–70 kG. These lines are connected with the recombination of the electrons filling a zero Landau subband with holes that fill two lower Landau subbands. Quenching of a short wave line taking place at a field intensity $B > 70$ kG corresponds to the case when the magnitude of splitting becomes comparable with the Fermi energy for holes at $B \sim 70$ kG.

Besides investigating the E_{fe} and EHD radiation spectra, we studied the influence of a strong magnetic field upon an integral intensity of the C-line. While registering the integral radiation intensity when the spectral width of the slit entirely covered the C-line taking into account the shift of its energetic position, a new resonance effect of the C-lines intensity oscillation as a function of magnetic field was observed (Fig. 10) [8]. For the orientation [100], where the electron motion in all the four valleys is described by one cyclotron mass $m_c = 0.13\,m$, one succeeded in determining the Fermi energy E_f and the corresponding equilibrium concentration of carriers in a drop $n_0 \simeq 1.8 \cdot 10^{17}\,\text{cm}^{-3}$ from the period in the reciprocal field. The analagous phenomena of oscillation in the long wave length absorption and radiation was observed in [19].

In Fig. 10 the experimental dependence of the C-line intensity on the magnetic field H is shown together with theoretical values of Landau levels of electrons and holes. We have explained the observable minima of the dependence $J_c(H)$ by the electron Fermi energy oscillations due to

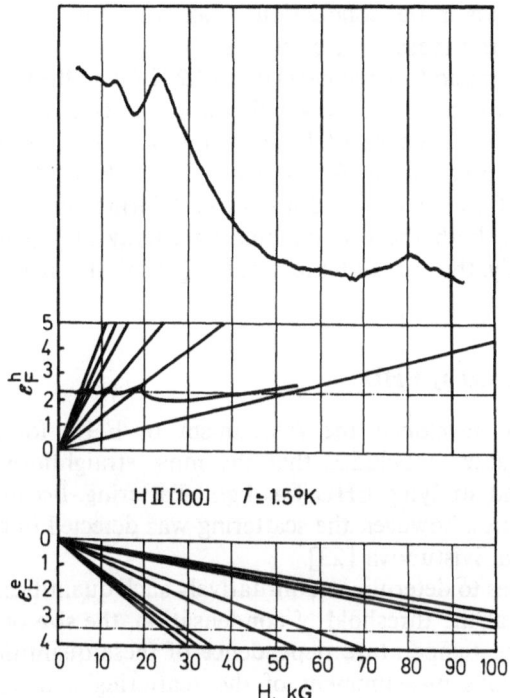

Fig. 10. Oscillations of the integrated radiation intensity in the C-line vs. the magnetic field for the Case $B \parallel [100]$. The lower part of the figure shows the Landau energy quantization diagram in the conduction band and in the valence band (in accordance with [20]), and also the oscillations of E_f^e at 2.10^{17} cm^{-3}; which explains the minima of the radiation intensity. E_f^e (at $H = 0$) = 2.3 meV

Landau quantization of a degenerate plasma in EHD [20]. In fact, when a Fermi level goes through a Landau level, the Fermi energy increases which must lead to EHD evaporation, since the work function from drops into the exciton gas becomes less. However, theoretical calculations of the carrier interaction energy in EHD in the presence of a magnetic field carried out by Keldysh and Silin showed that the mechanism suggested here, although taking place, is not the main one [21]. It is seen from the calculations that when Fermi levels go through a Landau level, the value of an equilibrium density of EHD and that of a quantum efficiency change which can lead to a considerable shift in the integral intensity of EHD radiation.

The dependence $J_c(H)$ for the other two orientations [110] and [111] turns out to be more complicated due to the existence of two families of

Landau subbands in the conduction band, which are described by two different cyclotron masses for electrons.

When the temperature increases up to 4.2 K for all the three directions of B with respect to the main crystallographic directions, the oscillatory structure of the dependence $J_c(H)$ is smeared out. Note also that with the temperature increase up to 6–8 K, when at $B = 0$ the C-line is absent, the presence of a magnetic field leads to a "flaring" of the C-line, since as B increases, both the exciton binding energy and φ increase and, correspondingly, the critical temperature T_{cr} of EHD creation also rises.

5. Light Scattering by EHD

After Keldysh developed the mechanism of EHD formation [22], many investigators concluded that the most straightforward method of detecting and studying EHD was light scattering. Because of experimental difficulties, however, the scattering was detected only in 1971 by Pokrovskii and Svistunova [23].

Our task was to determine quantitatively such quantities as the mean concentration at the threshold of condensation, the size of EHD, their density and the temperature dependence of these quantities with aid of a simultaneous measurement of the scattering and absorption of light by EHD and their radiation. In EHD with their large carrier density, the refractive index κ and dielectric permeability ε differ from those of a host crystal. This must cause light scattering by EHD. If the complex refractive index m of EHD in the crystal is small $(m - 1) \ll 1$ and the change in the optical path length due to travelling of light through a particle of radius a is also small, i.e. $(4\pi \varkappa_0 a/\lambda)(m - 1) \ll 1$, the scattering theory of Rayleigh-Gans is valid. For small angles of scattering the light flux $d\phi(\theta)$ can be represented by [10]

$$d\phi(\theta) = \frac{8\pi^3 a^6 n_0^2 N_d d(4e^4 \lambda_0^2 + m^{*2} c^4 \varkappa_0^2 S^2)}{9\lambda_0^2 \varkappa_0^2 m^{*2} c^4} \phi_0 Y^2(u) \sin\theta \, d\theta \tag{4}$$

where λ_0 is the wavelength of the scattered light in vacuum, d is the scattering length, S is the cross-section of the light absorption by free carriers, m^* is a reduced effective mass of an electron and a hole, N_d is the drop concentration, ϕ_0 is the incident light flux, θ is the angle of scattering in vacuum. In this expression the first term in brackets takes into account the contribution of the real part of the refractive index and the second term that of the imaginary part.

The angular intensity distribution of the scattered light is defined only by the size of the scattering particles. If the light absorption by a

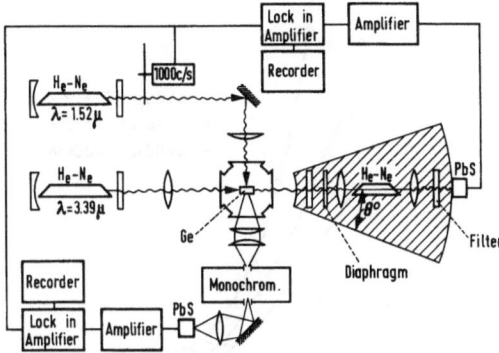

Fig. 11. Block diagrams of experimental arrangement for measurements of light scattering

crystal containing EHD is simultaneously measured together with the scattering, one can find both the mean concentration of nonequilibrium carriers and the concentration of carriers in a drop as well as the drop concentration N_d. In fact, the energy absorbed by a crystal per unit time:

$$W = \phi_0 \, S \bar{n} \, d$$

is determined by a total concentration $\bar{n}$ of absorbing particles. From this the quantity n_0 is determined. If at low temperatures one ignores the contribution of free carriers and excitons considering that almost all the nonequilibrium carriers are in EHD, then $\bar{n} = \frac{4}{3}\pi a^3 n_0 N_d$. Then, having measured the size of the drop a, one can find from (4) the quantity n_0 and then N_d.

The scheme of the experiment is presented in Fig. 11. The source of excitation was a He–Ne laser of a power of ~ 10 mW working at the wavelength $1.52\,\mu$. Since the absorption coefficient of such a radiation in germanium is comparatively small at low temperatures, the excitation would be practically uniform along the sample. The exciting radiation was modulated with frequency of $1000\,\mathrm{sec}^{-1}$ and focused on a wide crystal side with the aid of a cylindric lens into a narrow line parallel to a laser beam with a wavelength of $3.39\,\mu$. The mean concentration of nonequilibrium carriers in that part of the crystal where the scattered beam passed, estimated by the absorption of radiation with the wavelength $3.39\,\mu$, was $3.10^{14}\,\mathrm{cm}^{-3}$ at a temperature ~ 2 K.

The scattering of He–Ne laser radiation at the wavelength $3.39\,\mu$ was investigated. The scattered light going out of a crystal was amplified by an optical quantum amplifier on a He–Ne mixture and registered

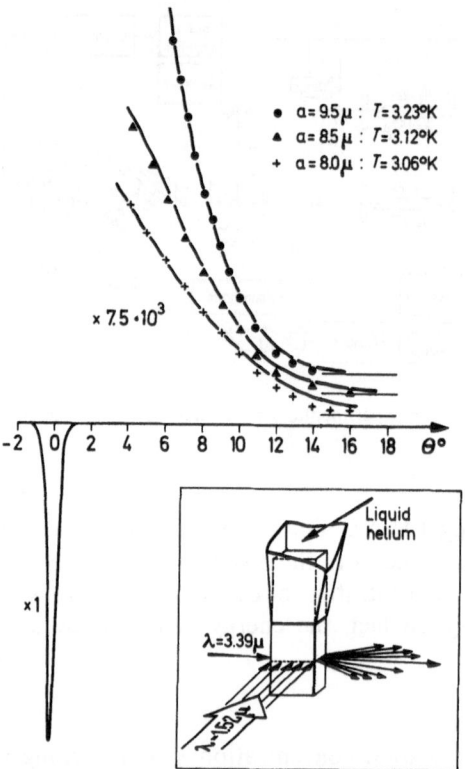

Fig. 12. Angular distribution of scattered light intensity at three different temperatures, the lower part shows the absorption signal at $P = 3.12$ K and sketch of sample illumination

by a cooled PbS photodetector. A set of diaphragms fixed before a quantum amplifier cut out a light beam scattered at a certain angle and defined the angular resolution of the system. A PbS detector together with the quantum amplifier were moved by a synchronous motor in a horizontal plane along the arc of a circle in the center of which a sample was under investigation. With such a registration method the angular distribution of the scattered light is determined only by the radius of scattering particles.

The installation was sensitive enough to follow the change in EHD size up to a condensation temperature close to which the integral scattering signal diminishes sharply. The measurements were carried out on polished germanium samples with a residual impurity concentration from 10^{11} cm^{-3} up to 10^{13} cm^{-3} (for different samples).

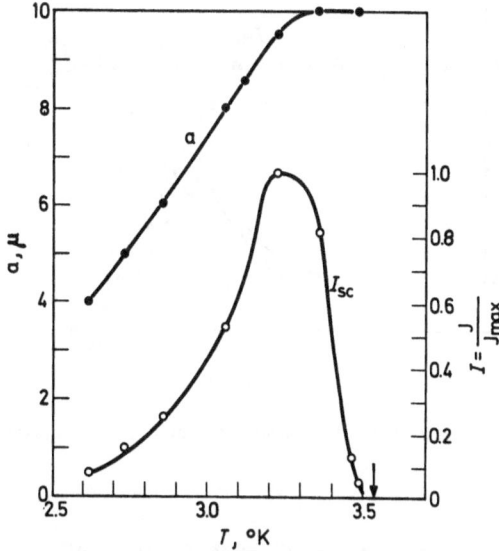

Fig. 13. Plots of EHD radius and scattered light intensity vs. temperature, in the angle of 8°

The samples had the dimensions $15 \times 5 \times 2$ mm and were fixed in a helium volume of a cryostat as is shown in Fig. 12. This figure also presents the mutual disposition of exciting and scattered radiation. The angular intensity distribution of light with $\lambda = 3.39 \mu$ scattered by EHD at three different temperatures is given as well. The points on these curves show the calculated values. The measurements were carried out with an angular resolution of 30'. An absorption signal of laser radiation with $\lambda = 3.39 \mu$ whose form represents the intensity distribution in a laser beam and whose phase is opposite to the scattering signal phase is shown to the left of the figure.

From the aforementioned dependences it is seen that the average radius of drops increases with temperature. It is also seen from the figure that the good agreement of calculation points with the experimentally observed dependence ceases in the region of rather large scattering angles. This fact is apparently due to the presence in the sample of EHD of radii smaller than the average one.

The dependence of the EHD radius a on the temperature and intensity of the light scattered by drops at the angle 8° (in vacuum) is presented in Fig. 13. The EHD radius is monotonously growing with temperature. It reaches the dimension 10μ at a temperature close to that of condensation (3.5 K). At first the scattering signal increases as the temperature

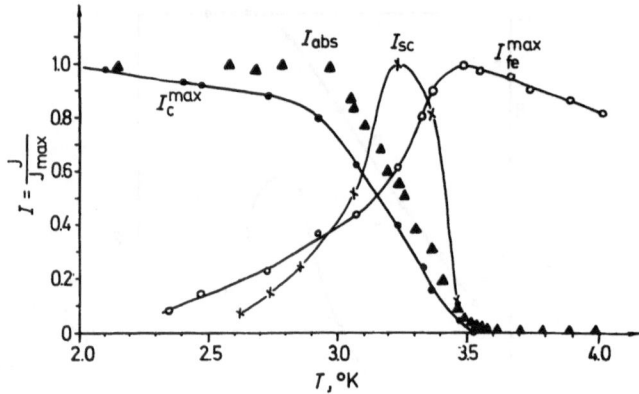

Fig. 14. Plots of scattered light, absorption and radiation vs. temperature.●-I_c^{max}, ○-I_{fe}^{max}, ▲-$I_{abs.}$, ×-$I_{scat.}$

rises, which is due to the growth of the EHD radius, and then, having passed the maximum, it begins to decrease since as the temperature rises the EHD concentration diminishes.

It is easily seen from Fig. 14 that at the temperature of 3.5 K, when the radiation line observed together with the scattering vanishes, the scattering signal vanishes simultaneously and the absorption signal decreases greatly. In this case a considerable increase in the radiation intensity of free excitons is observed.

The increase of the EHD radius with the temperature increase may be connected with the decrease of the number of "working" condensation centers, i.e. condensation takes place only on "deep" centers. The drop concentration being estimated from measurements of absorption and scattering at $T = 3.8$ K was about 100 times smaller than at $T = 2.6$ K. (At low temperatures the values of drop concentration N_d and drop density n_0 found from these experiments are $N_d \cong 2.10^6$ cm^{-3}, $n_0 \simeq 1.7 \cdot 10^{17}$ cm^{-3}.)

At present the role of surface tension forces in the exciton condensation is not yet clear. However, if these forces are supposed to be rather strong, the temperature dependence of the EHD radius could be explained on the basis of the wellknown considerations about the formation of nucleation centers of "critical" dimensions. In this case condensation would not take place in the centers but occurs due to the exciton gas density fluctuations.

In conclusion we will stress two more results which were obtained while investigating the interaction between EHD and radiation of a CO_2 laser. When a rather wide beam ($d \simeq 2.5$ mm) of radiation with

$\lambda = 10.6\,\mu$ passed through the condensation region (the diameter of the exciting light beam was $\sim 200\,\mu$) the picture of diffraction on a cloud of nonequilibrium carriers was observed. The diffraction picture vanished with the increase of temperature simultaneously with the disappearance of the C-line. The radius L of the localization region ~ 0.6 mm was determined by the angular position of the first diffraction maximum.

These observations show that under usual conditions EHD are extremely immobile. The second result is connected with the heating of EHD by the field of a light wave of CO_2 laser. When the radiation of CO_2 laser (power up to 60 mW) passes through the excitation region, a considerable diminishing of the C-line in the spectrum of photo-luminescence is observed which is connected with EHD evaporation. The intensity of free exciton radiation increases.

References

1. Keldysh,L.V.: Proceedings of the Ninth International Conference on Semiconductor Physics, Moscow, 1969, p. 1384.
2. Keldysh,L.V.: Eksitony v poluprovodnikakh (Excitons in semiconductors). Nauka 5 (1971).
3. Bagaev,V.S., Galkina,T.I., Gogolin,O.V., Keldysh,L.V.: ZhETF, Pis. Red. 10, 309 (1969); JETP Letters 10, 195 (1969).
4. Bagaev,V.S., Galkina,T.I., Gogolin,O.V.: Kratkie soobshch. po fizike. FIAN N 2, 42 (1970).
5. Bagaev,V.S., Galkina,T.I., Gogolin,O.V.: In: Eksitony v poluprovodnikakh (Excitons in semiconductors). Nauka 19 (1971).
6. Alekseev,A.S., Bagaev,V.S., Galkina,T.I.: ZhETF 63, 9, 1020 (1972).
7. Alekseev,A.S., Bagaev,V.S., Galkina,T.I., Gogolin,O.V., Penin,N.A., Semenov,A.N., Stopachinskii,V.B.: ZhETF, Pis. Red. 12, N 5, 203 (1970).
8. Bagaev,V.S., Galkina,T.I., Penin,N.A., Stopachinskii,V.B., Churaeva,M.N.: ZhETF, Pis. Red. 16, N 3, 120 (1972).
9. Alekseev,A.S., Bagaev,V.S., Galkina,T.I., Gogolin,O.V., Penin,N.A., Semenov,A.N., Stopachinskii,V.B.: Trudi FIAN, Moscow, Nauka 67, 109 (1973).
10. Sybeldin,N.N., Bagaev,V.S., Tsvetkov,V.A., Penin,N.A.: FTT 15, 1, 174 (1973).
11. Bagaev,V.S., Zamkovets,N.V., Penin,N.A., Sybeldin,N.N., Tsvetkov,V.A.: Thesis II. Intern. Conf. "Lasers and their application". Dresden, DDR, 1973, k 107.
12. Bagaev,V.S., Penin,N.A., Sybeldin,N.N., Tsvetkov,V.A.: FTT 15, 11, 3269 (1973).
13. Bagaev,V.S., Zamkovets,N.V., Penin,N.A., Sybeldin,N.N., Tsvetkov,V.A.: PTE, N 2 (1974) (in print).
14. Bagaev,V.S., Galkina,T.I., Gogolin,O.V.: Proc. X. Intern. Conf. on Phys. of Semicond., Cambridge, Massachusets, 500, 1970.
15. Benoît à la Guillaume,C., Voos,M.: Phys. Rev. B5, N 8, 3079 (1972).
16. Benoît à la Guillaume,C., Parodi,O.: J. Electr. and Cont. 6, 4 (1959).
17. Pokrovskii,Ja., Kaminskii,A., Svistunova,K.: Proc. X. Intern. Conf. on Phys. of Semicond., Cambridge, Massachusets, 504, 1970.
18. Brinkman,W.F., Rice,T.M., Anderson,P.W., Chui,S.T.: Phys. Rev. Letters 28, 961 (1972).

19. Murzin, V. N., Zayatz, V. A., Kononenko, V. L.: Proc. XI. Intern. Conf. on Phys. of Semicond., Warszawa, 679, 1972.
20. Halpern, J., Lax, B.: J. Phys. Chem. Solids **26**, 911 (1965).
21. Keldysh, L. V., Sylin, A. P.: FTT **15**, 5, 1532 (1973).
22. Keldysh, L. V.: UFN **100**, 3, 514 (1970).
23. Pokrovskii, Ja. E., Svistunova, K. J.: ZhETF, Piz. Red. **13**, 297 (1971).

Prof. Dr. V. Bagaev
Lebedev Physical Institute
Moscow B 3/2, USSR
Leninsky Prospekt 53

Theory of Electron-Hole Drops in Germanium and Silicon

T. M. RICE

Contents

1. Introduction

By shining light on a semiconductor, high densities of non-equilibrium carriers are created. These carriers quickly relax to the band extrema and in indirect band gap semiconductors, where luminescent recombination must be phonon assisted, the carriers can live for long times. Under these conditions an electron-hole plasma can be obtained at low temperatures and relatively high density. In recent years there has been numerous experiments which suggest that the electron-hole plasma undergoes a gas-liquid transition at low temperatures as a function of density. At low density the electrons and holes bind together as excitons. Two excitons can bind together to an excitonic molecule or biexciton [1, 2]. However, Keldysh [3] suggested that at high densities an electron-hole metallic liquid could be the stable. At that time Asnin and Rogachev [4] reported evidence of a conductivity transition with increasing optical excitation power. An unambiguous interpretation of conductivity experiments is not possible and subsequent work has concentrated on other experimental probes particularly luminescence experiments [5–7]. In this review the emphasis will be on theory and reviews of the experimental situation may be found elsewhere [8].

The most important question from a theoretical point of view is whether successful predictions can be made for the equation of state of the electron-hole system. This is a real challenge to modern many body theory because electrons and holes in a semiconductor such as Ge or Si can be accurately described by a Hamiltonian with effective masses

for electron and holes and a Coulomb interaction scaled by the static
dielectric constant. The ideal model of a single isotropic band extremum
for the electron and for the hole has been treated by several groups
[9–13]. In the case of equal masses for the electron and hole, a well
defined local minimum in the ground state energy versus density is found
but as discussed in Section 2, in none of the various methods employed,
is the metallic ground state bound relative to a dilute gas of excitons.
The best theoretical estimate to date is that in this case the ground
state is an infinitely dilute gas of excitonic molecules [11]. However in
semiconductors such as Ge and Si the band structure plays an important
role with degenerate bands and anisotropic effective masses. Both
effects are found to favor the metallic state greatly [10, 11, 14]. For Ge a
binding energy of approximately 50% of the binding of a single exciton
is obtained. The agreement between the theoretical estimates and the
latest experimental results for three parameters of the ground state,
viz., the binding energy, the equilibrium density and the compressibility,
will be discussed in detail in Section 2.

Recently there has been considerable interest in the surface structure
of the electron hole liquid [15–17]. Several calculations of the surface
energy have been reported and are reviewed in Section 3. Surface effects
on the bulk properties of an electron-hole drop whose radius is of the
order of a micron are negligible. The surface energy however determines
the low lying oscillations of the drop as a whole. The observation of such
oscillations has been suggested as a way of determining the surface
energy experimentally [15]. Another property which is sensitive to the
surface is the charge structure of a drop [16]. Theoretical calculations
suggest that the electrons are more tightly bound than the holes in Ge
[18]. This effect is counter balanced by the dipole layer and the charge on
the drop and conflicting estimates [16, 17] have been reported for the
overall sign of charge residing on a drop. Several experiments are
suggested to determine the charge structure of a drop.

2. Ground State Properties

The theory of the ground state of the electron-hole system has been
considered by a number of authors. A complete theory at arbitrary
density has not been developed to date and calculations have been done
in either the dense or dilute regimes. In the dilute regime an electron is
bound to a particular hole to form an exciton and these excitons can
bind further to form a larger complex, such as a biexciton or excitonic
molecule. By contrast in the dense regime it is assumed that the electrons
and holes separately form degenerate Fermi gases and while there is an

enhanced probability to find a hole at an electron and vice versa, they are separately itinerant and form a metallic liquid. Different methods have been applied in the two regimes and we will discuss them in turn.

a) Low Density — Molecular State

The Hamiltonian for the electrons and holes is

$$\mathcal{H} = -\sum_i V_i^2/2m_e - \sum_j V_j^2/2m_h - \sum_{i,j} e^2 [\varepsilon_0 |r_i^e - r_j^h|]^{-1}$$
$$+ \tfrac{1}{2} \sum_{i \neq j} e^2 [\varepsilon_0 |r_i^e - r_j^e|]^{-1} + \tfrac{1}{2} \sum_{i \neq j} e^2 [\varepsilon_0 |r_i^h - r_j^h|]^{-1} \tag{1}$$

where a simple band structure with isotropic masses m_e and m_h for the electrons and holes, has been assumed. It is straightforward to solve for the exciton or bound state of one electron and one hole. The problem is identical to the familiar Coulomb problem with a mass which is the reduced mass $\mu(\mu^{-1} = m_e^{-1} + m_h^{-1})$ and an electric charge scaled by $\sqrt{\varepsilon_0}$, where ε_0 is the static dielectric constant. The binding energy of the exciton (E^x) is $e^2/2\varepsilon_0 a_x$ and Bohr radius of the exciton is $a_x = \varepsilon_0/\mu e^2$. These are the characteristic energy and length in the electron-hole problem. The effective mass approximation used in Eq. (1) holds when $E^x \ll E_g$ and $a_x \gg a$ where E_g and a are the energy gap and the lattice parameter respectively. In Ge and Si these conditions are fulfilled.

Two excitons will have a long range attractive Van der Waals interaction and a short range repulsion. As with the interaction of two Hydrogen atoms the nature of the force is strongly spin dependent and the size of the attractive potential well is much larger when the two electrons are in a singlet state. In the equal mass limit there is symmetry between the electrons and holes and the attractive potential is maximized by having the holes also in a singlet state. Hylleraas and Ore [1], who studied the positronium molecule, $(m_h = m_e)$ introduced a wave function of the form

$$\Psi_{H-O} = 2\exp(-\tfrac{1}{2}k(r_{1a} + r_{1b} + r_{2a} + r_{2b}))$$
$$\cdot \cosh(\tfrac{1}{2}\beta k(r_{1a} - r_{1b} - r_{2a} + r_{2b})) \tag{2}$$

where 1, 2 and a, b refer to electrons and holes respectively. The parameters k and β were adjusted variationally. They found a binding energy for the molecule of 1.7% of E^x. The reason for the very small binding is the large cost in kinetic energy required to localize the excitons, when both electron and holes are light. Subsequently the Hylleraas-Ore calculation was criticized [19, 20] but a later reexamination [21] reaffirmed the original calculation. Recently variational calculations using a Hylleraas-Ore wavefunction modified by a function which depends on the hole-hole separation [20–22], have been carried out for the full range of electron-hole mass ratio, $0 \leq m_e/m_h \leq 1$. While the molecular binding energy for $m_e = m_h$ was improved to 2.9%, the overall magnitude of the molecular binding remains at a few percent of E^x for $m_e/m_h > 0.2$.

The interaction of two molecules has been considered by Brinkman and Rice [11] who assumed a pairwise interaction between excitons. This should be a good assumption in this case since the biexciton is weakly bound and therefore is spatially very extended. Because one must now average over all possible spin configurations between the excitons they find a large repulsive interaction between two molecules. They conclude that energy density curve in the low density regime the energy monotonically increases with density from the value of separated molecules.

In a semiconductor, such as Ge or Si the electrons have an additional degeneracy arising from the band structure. Wang and Kittel [23] have shown that this can greatly modify the energy of exciton complexes. However, their calculations are restricted to the infinite mass limit for the hole and in Ge and Si, $m_h \approx m_e$. It is far from clear [21] that larger complexes have a binding energy greater than the few percent of E^x one expects for a biexciton in Ge or Si.

b) Metallic Electron-Hole Liquid

At high densities, the excitons will overlap so strongly as to lose their individuality and one must treat the electrons and holes as two interpenetrating Fermi liquids. In the high density limit the dominant term in energy will be the kinetic energy E_K of the electrons and holes. For the nondegenerate band structure it is easily evaluated as

$$E_K = \tfrac{3}{5}(k_F^2/2m_e + k_F^2/2m_h) = \tfrac{3}{10}(k_F^2/\mu) \tag{3}$$

$$= 2.21/r_s^2 \tag{4}$$

expressed in units of E^x for electron. The parameter r_s is related to the density of electrons n by $n = 3/4\pi r_s^3 a_x^3$. At high densities the leading correction is the exchange energy, E_{ex}, between the electrons and the holes separately. This contribution arises from the reduction in Coulomb energy because of the spatial correlation imposed by the Pauli principle on electron (holes) in the same spin state. Its value is well known

$$E_{ex} = -3e^2 k_F/2\pi\varepsilon_0 = -1.832/r_s \tag{5}$$

The remaining terms in the energy expansion, E_{corr}, arise from the correlations between the electrons and holes and between electrons (holes) with antiparallel spins. These terms can only be approximately evaluated. Hanamura [9] used a high density expansion to estimate this energy. His result for E_{corr} is

$$E_{corr} = 0.497 \ln r_s - 0.364 . \tag{6}$$

He finds a minimum in the energy-density curve but with a minimum value at $r_s = 1.7$ of $E_G = -0.35$ which is substantially above -1, the energy of separated excitons. However it is known from experience with the single component electron gas that high density expansions are strictly limited to $r_s \lesssim 1$. In the electron gas the random phase approximation (RPA) as modified by Hubbard or Nozieres and Pines is believed to give a good description of the energy even in the intermediate density regime ($r_s \approx 1$–6). This approximation has been applied to the electron-hole liquid by Brinkman, Rice, Anderson and Chui [10, 11]. While these authors find a much lower minimum (-0.86) at a similar density, the minimum energy is still substantially above -1.

The biggest question about the RPA is whether the correlation between electrons and holes are properly treated since it is just such correlations which give rise to the exciton bound state at low densities. In terms of diagrams, the relevant terms are the repeated electron-hole ladder graphs. Various estimates indicate that such terms are not large. In particular for $m_e = m_h$ the Mott criterion for the occurrence of a real exciton bound state is $r_s \approx 10$ which is a much lower density than the predicted equilibrium value. Recently Vashishta, Bhattarchayya and Singwi (VBS) [12] have applied the Singwi-Tosi-Land-Sjolander (STLS) method to the electron hole liquid. In this method a set of self-consistent equations for the particle-particle correlation functions is derived by approximating the equations of motion. While it is not clear how to justify this procedure in terms of the usual diagramatic expansions, the STLS method has been successfully applied to the related problem of one position in an electron gas. The results of VBS for $m_e = m_h$ give a still lower value for $E_g = -0.99$ but do not give binding either. Inoue and Hanamara [13] have attempted a variational calculation with a correlated wave function. They are forced to make approximations in their evaluation so that their answer is not truly variational. Their final answer is very close to that of VBS. In the limit $m_h \gg m_e$ the calculations of Wigner and Huntington [24] on metallic hydrogen gave a value of -1.05, which is substantially above the energy of H_2 molecules but below that of H atoms. For finite values of the mass ratio m_e/m_h the energy of the metallic states rises as $(m_e/m_h)^{1/2}$ due to the zero point motion of the holes and this is sufficient to drive E_G above -1 by m_h/m_e

= 16. The conclusion from all of the theoretical calculations is that there is a well defined metallic liquid state but its energy probably lies above that of excitonic molecules.

It is not possible to obtain an example of the idealized band structure in nature but a large (111) strain on Ge leads to a band structure with single nondegenerate valleys for the electrons and holes. The bands are however highly ellipsoidal and far from spherical. This favors the metallic state since the mass average which enters the kinetic energy of the metal is the geometric average $(m_{de} = (m_{\parallel} m_{\perp}^2)^{1/3})$. This is substantially larger than the optical mass average $(m_{oe}^{-1} = \frac{1}{3}m_{\parallel}^{-1} + \frac{2}{3}m_{\perp}^{-1})$. In Ge for the electrons $m_{de} = 0.22$ electron masses while $m_{oe} = 0.12$. This anisotropy has the effect of lowering the energy of the metallic liquid in the modified RPA calculations [11] to $-1.06 E_B$ while VBS [12] obtained an even lower value (-1.17). Experimental studies by Bagaev et al. [25] and by Benoit a la Guilaume et al. [6] indicate a metallic state but the binding energy has not been measured.

In Ge and Si in the absence of strain the band structure is complex. In Ge there are four electron ellipsoids and a coupled set of hole bands. This leads to considerable complications in the calculations. BRAC [10, 11] calculated the kinetic and exchange energies exactly and obtained the values

$$E_a = 0.468/r_s^2 - 1.136/r_s + E_{corr} \tag{7}$$

expressed in Rydbergs defined by the optically averaged reduced mass. The large reduction in the kinetic energy E_K and smaller reduction in E_{ex} when compared to Eqs. (4) and (5) comes from the degeneracies and anisotropies in the band structure which both act to favor the metal. In the correlation energy BRAC included the degeneracies but used optically averaged masses. They obtained the curve shown in Fig. 1. The binding energy when compared to the exact exciton energy is 1.7 meV ($\approx 50\%$ of exciton energy). In a recent letter VBS reported substantially the same value using a modified STLS approximation for E_{corr}. Combescot and Nozieres have used the Nozieres-Pines approximation and were able to incorporate the anisotropy of the band structure in some details into the calculation of E_{corr}. This leads to larger binding energy of 2.0 meV for the metallic liquid. These represent fairly small differences in the correlation energy. At the equilibrium density in Ge, $E_K = +3.2$ meV, $E_{ex} = -4.8$ meV and $E_{corr} = -3.6$ meV. In Table 1 we list the theoretical results for Ge and for Si where they are available.

The three parameters of the ground state which have been measured are the binding energy, the equilibrium density and the compressibility. Two different methods have been used to obtain the binding energy. One method is by measuring the temperature dependence of either the

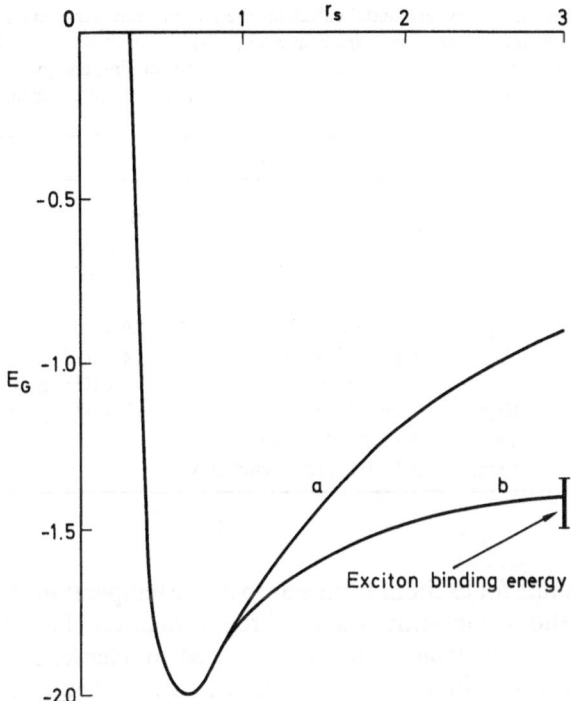

Fig. 1. The ground state energy E_G versus interparticle separation r_s, for the electron-hole liquid in Ge. The curve marked (a) is from Ref. 10 and 11. The curve marked (b) is that used in the surface calculation based on the functional form (9). The energy units are exciton Rydbergs and the parameter r_s is defined from the electron density n by $n = 3/4\pi r_s^3 a_x^3$ where a_x is the exciton Bohr radius. Exciton units are determined by the reduced optical mass average for electrons and holes. The observed exciton binding energy (3.6 meV) is shown

Table 1. The values obtained by different experiments on Ge

Measurements of binding energy in Ge	
1. Work Function by Evaporation	
Hensel et al. [26]	1.4 $\pm$ 0.2 meV
2. Luminescence	
Thomas et al. [7]	1.55 − 2.0 meV
Benoit a la Guillaume and Voos [29]	2.0 meV
3. Temperature Dependence of Threshold	
Pokrovskii [8]	1.5 meV
Hensel et al. [26]	1.4 meV
Lo et al. [27]	1.54 $\pm$ 0.25 meV
McGroddy et al. [28]	1.5 $\pm$ 0.1 meV

Table 2. Theoretical and Experimental values for the ground state parameters in Ge and Si. The theoretical estimates are from Brinkman et al. [10, 11], Combescot and Nozieres [14], and Vashista et al. [12]. The experimental values for the binding energy in Ge are from Table 1, in Si from Ref. [29]. The other experimental values in Ge are from Ref. [7]

Ground state; comparison of theory and experiment

		Ge	Si	
Energy	Theory	1.7 meV	5.7 meV	BRAC
		2.5 meV (2.0 meV)	6.3 meV	CN
		1.8 meV		VBS
	Expt.	1.5–2.0 meV	5.6 meV	
Density	Theory	1.8×10^{17} cm^{-3}	3.4×10^{18} cm^{-3}	BRAC
		2.0×10^{17} cm^{-3}	3.1×10^{18} cm^{-3}	CN, VBS
	Expt.	2.4×10^{17} cm^{-3}	3.7×10^{18} cm^{-3}	
Compressibility $= (n_0^3 E_G'')^{-1}$	Theory	5.5×10^{-18} cm^3/meV		BRAC
	Expt.	$3.7 \pm 0.8 \times 10^{-18}$ cm^3/meV		TPRH

evaporation rate for excitons from a drop or the temperature dependence of the threshold concentration for drop formation. For the former measurements, cyclotron resonance was used by Hensel et al. [26] to monitor the decay of drops with time at various temperatures. The latter experiments were performed by several groups using luminescence [8, 27], cyclotron resonance [26], and $p - n$ junction detectors [28]. All of these thermal measurements are in the range· 1.4–1.6 meV. The actual values obtained are listed in Table 1. A second set of measurements are spectroscopic measurements of the splitting between the upper edge of the luminescence from the electron-hole liquid and free exciton line. The values reported from these measurements have varied with time due to uncertainties in the interpretation of the exciton line shape but recent careful measurements [7, 29] give values in the range 1.55–2.0 meV. The upper value is obtained if the line broadening of the free exciton line is attributed to phonon lifetime effects which seems the most reasonable explanation at present. This value however leaves an unexplained discrepancy of 0.5 meV between the two sets of values. For Si only luminescence values are available. In Table 2 a comparison is given between the various theoretical and experimental values in Ge and Si.

The equilibrium value for the density can be obtained by fitting the luminescence line shape and the most recent value is quoted in Table 2. A third parameter which can be measured experimentally is the compressibility of the liquid. Thomas et al. [7] by a careful study of the temperature dependence of the luminescence line up to 4.2 K determined the thermal expansion, which, in a Fermi system, is related to the compressibility. The theoretical and experimental values for the latter are also quoted in Table 2.

It is clear from Table 2 that the overall agreement between the first principles theoretical estimates and the experimental values is satisfactory.

3. Surface Properties

In practice in Ge and Si the electron-hole liquid is in the form of drops whose diameter is typically of the order of $10\,\mu$. This has led to interest in the possible effects of the finite size and in suggestions to make use of the finite size to study surface properties. The study of the surface of the electron-hole liquid is of interest in its own right and a knowledge of surface effects is important for an understanding of the kinetics of drop formation. Surface studies in general have been much in vogue in the last few years and the electron-hole liquid offers a good test for the theories developed for the electronic surface. The technique most often used has been the energy density functional method. In this method the total energy is written as

$$\mathscr{E} = \int \left\{ \mathscr{E}_G(\varrho_e, \varrho_h) + \tfrac{1}{72} \sum_{i=e,h} (\nabla \varrho_i)^2 / m_i \varrho_i + E_{es} \right\} d\tau. \tag{8}$$

The first term is the ground state energy density of a uniform electron-hole liquid with electron and hole densities of $\varrho_e(r)$ and $\varrho_h(r)$. The last term is the electrostatic energy density. The remaining terms are the leading terms in a gradient expansion of the energy density.

For the case of local charge neutrality, i.e., $\varrho_e(r) \equiv \varrho_h(r) \equiv \varrho(r)$, it is possible to give an analytic solution to the minimization of the energy functional for a family of ground state energy functions.

$$\mathscr{E}(\varrho) = \varrho |E_0| \left\{ -2(\varrho/\varrho_0)^{1/n} + (\varrho/\varrho_0)^{2/n} \right\} - E_B. \tag{9}$$

Within this family the binding energy per particle E_0 and the equilibrium density ϱ_0 can be set at their corresponding values in the electron hole liquid. The third parameter n is a number which can be used to fit the local curvature around the minimum energy. The results as shown in Fig. 1 and demonstrate that a satisfactory fit can be made. The model energy function has been chosen to go to exact binding energy of the exciton in the low density limit. The various calculations discussed in Section 2 however go to zero in the low density limit since the treatment of the electron hole scattering is incomplete in all calculations.

Minimization of (8) with respect to $\varrho(r)$ leads to the result

$$-\tfrac{1}{18} M^{-1} \nabla^2 u + (\partial \mathscr{E}/\partial \varrho) u = E_0 u \tag{10}$$

where $u = \varrho^{1/2}$ and $M^{-1} = m_e^{-1} + m_h^{-1}$. E_0 is a Lagrange multiplier arising from the condition that number of particles is fixed. For the case of a planar surface the solutions of (10) must satisfy the boundary conditions

$$u \to u_0 = \varrho_0^{1/2}, \quad u' \to 0, \quad u'' \to 0 \quad \text{as} \quad x \to -\infty$$
$$u \to 0, \qquad\qquad u' \to 0, \quad u'' \to 0 \quad \text{as} \quad x \to +\infty \tag{11}$$

where ϱ_0 is equilibrium density obtained by minimizing the energy per particle $\mathscr{E}/\varrho$. Thus in the limit $x \to -\infty$

$$\partial(\mathscr{E}/\varrho)/\partial\varrho|_{\varrho_0} = 0 \tag{12}$$

and

$$E_0 = \partial\mathscr{E}/\partial\varrho|_{\varrho_0} = \mathscr{E}/\varrho|_{\varrho_0} . \tag{13}$$

These results are the condition that in the interior of the liquid $\varrho \to \varrho_0$, the equilibrium density, and that E_0 is the chemical potential to add an electron and a hole.

The surface energy S is found by taking the difference between the energy per unit area of the liquid and the energy the same number of particles would have if they were at the equilibrium density, i.e.

$$S = E - \int_{-\infty}^{\infty} E_0 \varrho(x)\, dx . \tag{14}$$

This can be shown to be equal to

$$S = \tfrac{1}{9} M^{-1} \int_{-\infty}^{+\infty} (u')^2\, dx$$
$$= \int_0^{\varrho_0} \{\tfrac{1}{18} M^{-1} (\mathscr{E}/\varrho - E_0)\}^{1/2}\, d\varrho . \tag{15}$$

One can integrate Eq. (10) once to give

$$\tfrac{1}{18} M^{-1} (u')^2 = \mathscr{E}(\varrho) - E_0 u^2 . \tag{16}$$

The solution of this equation is

$$\varrho = \varrho_0 (1 + \exp(x/na))^{-n} \tag{17}$$

where a, the length scale of the surface is given by $a^2 = 1/72 M |E_0|$. The surface energy is given by

$$S = (\varrho_0/(n+1)) |E_0/18 M|^{1/2} . \tag{18}$$

Analytic results of this form are useful in showing how the surface varies with the parameters of the bulk. Note however that surface energy is not related to the absolute value of the binding energy as would appear from (18) but rather to the compressibility which is $\sim E_0/n^2$.

In a recent publication the present author calculated the surface properties using the values calculated in BRAC, $\varrho_0 = 1.8 \times 10^{17}$ cm^{-3}; $E_0 = 0.64 E^x$ where E^x is the Rydberg of an exciton in Ge ($E^x = 2.65$ meV). Isotropic bands with the density of states masses were assumed leading to value of $M = 0.135$. The values obtained were for the surface energy

$$S = 0.061 \, E^x/a_x^2 = 8.3 \times 10^{-5} \text{ ergs/cm}^2 \tag{19}$$

and surface width D as measured by the separation between the points with $\varrho/\varrho_0 = 0.1$ and 0.9,

$$D = 0.76 \, a_x \tag{20}$$

and a_x is the exciton Bohr radius ($a_x = 177$ Å in Ge).

Similar values for surface energy were obtained by Sander et al. [15]. These authors used the energy density functional method also but used the Hartree-Fock energy, i.e. only the kinetic and exchange energies E_K and E_{ex} are included. They include the exchange correction to gradient term and solved the minimization problem variationally. Their results give a value close to that quoted above for the surface energy,

$$S = 0.08 \, E^x/a_x^2 . \tag{21}$$

Recently Reinecke and Ying [17] have examined more carefully the question of averaging over the anisotropic masses and the coupled valence band masses and within the RPA have obtained substantially smaller values for the average masses than the density of states masses. Thus they obtained rather larger values for the surface energy and thickness. Their values for the surface energy are in the range

$$S = 0.09 - 0.11 \, E^x/a_x^2 \tag{22}$$

depending on whether the results of Ref. [10] or [14] on the ground state energy were used.

This range of values will however have only a small effect on the binding energy of the particles in a drop. Consider a drop of radius 1 µ. It is straightforward to estimate the shift in the binding energy of such a drop. Using the standard formula for this shift of $8\pi r_s^3 a_x^3 S/3R$ where $R = 10^{-4}$ cm one estimates its magnitude as $2 - 4 \times 10^{-3} E^x$, corresponding to the range of values quoted above. Such a shift is only a 1% correction at most to the bulk binding energy. This is hardly a surprising result since a drop of this size contains of the order 10^6 electrons and holes, and the fraction on the surface is only 1%. The surface energy per particle on the surface has the same order of magnitude as the bulk

binding energy so a very crude estimate of the shift of the binding energy
in a finite drop gives the same result.

Sander et al. [15] pointed out that a measurement of the frequency
of the shape oscillation of a finite drop would determine the surface
energy. The lowest frequency mode which couples to an external electric
field corresponds to a quadrupole distortion of a spherical drop. There
is no dipole mode since that corresponds to the translation of the drop
as a whole and there clearly is no restoring force to such a displacement.
The frequency of the quadrupole mode ω_Q can be determined by hydro-
dynamic theory

$$\omega_Q^2 = 8 S/\varrho_0 R^3 . \tag{23}$$

This leads to a value of $\omega_Q \approx 10^9 \sec^{-1}$. Sander et al. [15] proposed to
observe this mode by applying an electric field and looking for a resonant
absorption at ω_Q. The symmetric breathing mode of a drop ω_S typically
has a value $\omega_S \sim 5 \times 10^{10} \sec^{-1}$, but such a mode does not couple to the
electric field in a neutral drop.

The theory presented so far has dealt with a locally neutral system.
However the kinetic energies of electrons and holes in Ge and Si differ
substantially and this will lead to departures from local charge neutrality.
The Fermi energy of electrons is substantially lower than that of the holes
and the self energies of both species are approximately equal [18]. Thus
in the absence of a dipole layer electrons will be more tightly bound than
holes. Thermal evaporation of holes will be more rapid than electrons
and lead to a net negative charge. However a dipole layer on the surface
of a drop will clearly change the relative binding of electrons and holes.
The question of the size and sign of the dipole layer has been investigated
by the present author [16] and by Reinecke and Ying [17]. The sign of
the dipole is such that the region of negative charge is on the inside and
that of positive charge is on the outside. The sign of dipole layer is
determined by the fact that the difference in binding energy between
electrons and holes is larger at higher densities. Thus energy is gained
by having the excess electrons in the higher density region on the inside
of the surface.

The magnitude of the dipole layer is clearly sensitive to the overall
width of the surface which in turn is sensitive to the value of the masses.
Thus there is a fairly large discrepancy between the values of the dipole
energy obtained by the author (0.13) and by Reinecke and Ying
(0.12–0.38). In the latter case the potential of dipole layer is sufficiently
large to make the electrons less tightly bound than holes. In the former
case estimates were obtained in range 20–500 for the charge on a drop of
diameter 10 μ. These values were estimated by requiring that in steady
state equal numbers of electrons and holes were evaporated.

However at low temperatures Auger processes can predominate over thermal evaporation process and it has been pointed out to the author by Herring that these may influence the charge on a drop. In an Auger process an electron and a hole recombine giving their energy and momentum to two other particles. (The three particle Auger process is not allowed in Ge because of the limitations imposed by energy and momentum conservation). The three Auger processes which involve the excitation of a high energy electron and hole, two high energy electrons and two high energy holes, may have different rates leading to net charging of a drop if the total emission rate for electrons and holes are different. It is difficult to make an estimate of the charge imbalance due to this process. However at temperatures of 3 K or above the thermal evaporation processes are much more important than the Auger processes and should determine the charge.

In principle it is possible to determine experimentally the dipole layer and the charge on a drop. If one measures the shift in the band-to-band transition inside the drop compared to outside one would determine the self-energy shift of a carrier. In particular an accurate measurement of the energy splitting of the Γ_8 and Γ_7 in Ge should show a reduction of ≈ 6 meV inside the drop. This reduction arises because the split-off hole level Γ_7 is not renormalized by the many body effects while the Γ_8 level of course has the full self-energy shift. The actual binding of the holes could be measured directly by using the transition between light and heavy hole bands to do the equivalent of a photoemission experiment. By turning an external source through the work function for holes one could determine the critical energy for photoemission. The photo emitted holes could be detected by several means such as cyclotron resonance. Finally the charge on a drop could be directly measured by observing the motion of an electron in an electric field [16].

4. Conclusions

The theory of the ground state properties of the metallic electron-hole liquid has now been studied by several groups using somewhat different methods. The overall agreement between the various methods and between the theoretical and experimental values for the parameters of the ground state is good. There is less unaminity however on the surface properties and on the charge structure in particular. Experimental evidence bearing on these questions would be most useful.

Acknowledgements. The author is grateful to his colleagues at Bell Laboratories for numerous conversations on the topic of the electron-hole liquid especially W. F. Brinkman, T. G. Phillips, J. C. Hensel, G. A. Thomas, and C. Herring.

References

1. Hylleraas,E.A., Ore,A.: Phys. Rev. **71**, 493 (1947). Also, Ore,A.: Phys. Rev. **71**, 913 (1947).
2. Lampert,M.A.: Phys. Rev. Letters **1**, 450 (1958).
3. Keldysh,L.V.: In: Ryvkin,S.M., Shmastev,V.V. (Eds.): Proceedings of the Ninth International Conference on the Physics of Semiconductors, Moscow 1968, p. 1303. Leningrad: Nauka 1968.
4. Asnin,V.M., Rogachev,A.A.: Zh. Eksper. i Teor. Fiz. Pisma Red. **9**, 415 (1969) [English transl. JETP Letters **9**, 248 (1969)].
5. Pokrovskii,Ya.E., Kaminsky,A., Svistunova,K.: In: Keller,S.P., Hensel,J.C., Stern,F. (Eds.): Proceedings of the Tenth International Conference on the Physics of Semiconductors, Cambridge, Mass. 1970, CONF-700501, p. 504. (U.S. AEC Division of Technical Information Springfield, Va. 1970)
6. Benoit a la Guillaume,C., Voos,M., Salvan,F.: Phys. Rev. B**5**, 3079 (1972); Phys. Rev. B**7**, 1723 (1973).
7. Thomas,G.A., Phillips,T.G., Rice,T.M., Hensel,J.C.: Phys. Rev. Letters **31**, 386 (1973).
8. Pokrovskii,Ya.E.: Phys. Stat. Solid (a) **11**, 385 (1972).
9. Hanamura,E.: In: Keller,S.P., Hensel,J.C., Stern,F. (Eds.): Proceedings of the Tenth International Conference on the Physics of Semiconductors, Cambridge, Mass. 1970, CONF-700501, p. 487. (U.S. AEC Division of Technical Information, Springfield, Va. 1970).
10. Brinkman,W.F., Rice,T.M., Anderson,P.W., Chui,S.T.: Phys. Rev. Letters **28**, 961 (1972).
11. Brinkman,W.F., Rice,T.M.: Phys. Rev. B**7**, 1508 (1973).
12. Vashishta,P., Bhattacharyya,P., Singwi,K.S.: Phys. Rev. Letters **30**, 1248 (1973).
13. Inoue,M., Hanamura,E.: J. Phys. Soc. Japan **34**, 652 (1973); **35**, 643 (1973).
14. Combescot,M., Nozieres,Ph.: J. Phys. C**5**, 2369 (1972) and Combescot,M. (Thesis, Univ. de Paris 1973).
15. Sander,L.M., Shore,H.B., Sham,L.J.: Phys. Rev. Letters **31**, 533 (1973).
16. Rice,T.M.: Phys. Rev. B **9**, 1540 (1974).
17. Reinecke,T.L., Ying,S.C.: Solid State Commun. **14**, 381 (1974).
18. Rice,T.M.: Nuovo Cimento (in press).
19. Sharma,R.R.: Phys. Rev. **170**, 770 (1968); **171**, 36 (1968).
20. Akimoto,O., Hanamura,E.: Solid State Commun. **10**, 253 (1972).
21. Brinkman,W.F., Rice,T.M., Bell,B.J.: Phys. Rev. B**8**, 1570 (1973).
22. Akimoto,O., Hanamura,E.: J. Phys. Soc. Japan **33**, 1357 (1972).
23. Wang,J.S-Y., Kittel,C.: Phys. Letters **42**A, 189 (1972).
24. Wigner,E., Huntington,H.B.: J. Chem. Phys. **3**, 764 (1935).
25. Bagaev,V.S., Galkina,T.I., Gogolin,O.V.: In: Keller,S.P., Hensel,J.C., Stern,F. (Eds.): Proceedings of the Tenth International Conference on the Physics of Semiconductors, Cambridge, Mass. 1970, CONF-700501, p. 500. (U.S. AEC Division of Technical Information, Springfield, Va. 1970).
26. Hensel,J.C., Phillips,T.G., Rice,T.M.: Phys. Rev. Letters **30**, 227 (1973).
27. Lo,T.K., Feldman,B.J., Jeffries,C.D.: Phys. Rev. Letters **31**, 224 (1973).
28. McGroddy,J.C., Voos,M., Christiansen,O.: Solid State Commun. **13**, 1801 (1974).
29. Benoit a la Guillaume,C., Voos,M.: Solid State Commun. **12**, 1257 (1973).

T. M. Rice   Present Address:
Bell Laboratories   Dept. of Physics
Murray Hill   Simon Fraser University
New Jersey 07974   Burnaby, B.C., Canada
USA

IV. Biexcitons and Droplets

Spectroscopic Study of Exciton-Exciton Interaction (Biexcitons, Drops) in Semiconducting Crystals

B. V. Novikov

The present paper is a survey of experimental work on exciton interaction carried out in the past three years in the "Optics of Solids Laboratory" at Leningrad University. Investigations of exciton interaction in this laboratory were initiated by the late Prof. E. F. Gross, who discovered the spectrum of exciton absorption in Cu_2O in 1951 [1]. Later, he paid great attention to the study of exciton interaction at high density. Gross and his collaborators [2] observed Bose condensation of excitons in CdSe crystals. These investigations are being continued in this laboratory. A short time ago Japanese investigators discovered Bose condensation of excitonic molecules also in CdSe crystals using pico-second laser light pulses for excitation [3].

The germanium crystal is of great interest for investigators with respect to the study of exciton interaction. These properties manifest themselves especially clearly here due to the long life-time of excitons. It was in germanium that the phenomena of metallization and con-densation of excitons [4] were first observed, and, later, also the emission from the condensed phase [5]. Interesting information, in the writer's opinion, may be obtained from the spectroscopic study of low-tempera-ture emission in germanium. It will be shown that this emission occurs at 2 K at a rather low excitation density. It has been possible to observe spectroscopically the dynamics of formation of a characteristic low-temperature emission and its development with increasing density.

At present, theory does not yet allow to predict with certainty the existence of one or other phenomenon related to excitons at a high density (excitonic molecules, condensation of excitons and molecules, etc.) in a particular semiconductor. Therefore, the study of new materials is most important. Thus, investigation of HgI_2 crystals has been carried out in our laboratory, which made it possible to discover excitonic molecule emission.

A peculiar recombinational emission in Ge, discovered in 1969 [5], which is observed at low temperature and, as a rule, at high excitation levels, is being studied intensively in a number of laboratories. At present, there are two principal views on the nature of this emission. Pokrovsky and collaborators [5], Bagaev and collaborators [6] and some others

[7] consider this emission as resulting from the carrier recombination in electron-hole drops. According to Rogachev and collaborators [8] and some other [9] investigators, the above emission is due to the radiative recombination of biexcitons and only at a high density there is an exciton condensation into a liquid phase. A similar situation is observed in silicon [10].

This paper deals with the spectrometrical features of low-temperature luminescence of pure germanium as well as germanium doped with various impurities.

The effect of group III and V element impurities on the electrical properties of germanium has been studied in great detail, but the luminescence of these impurities at low temperature is much less understood.

Luminescence of exciton complexes (bound excitons) was primarily studied in germanium single crystals doped with various donor (Sb, P, As)

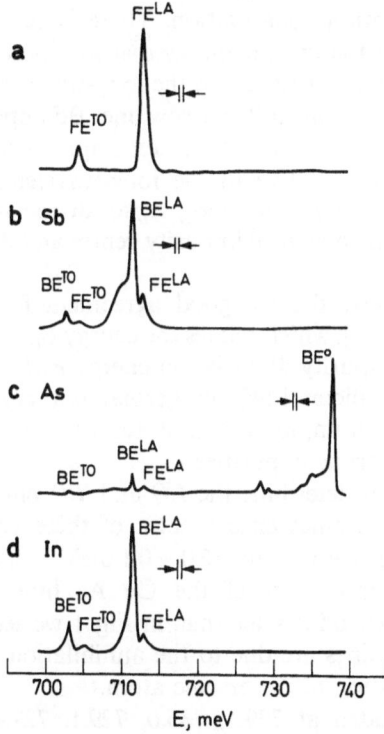

Fig. 1a–d. Luminescence spectra of pure and doped germanium crystals at 4.2 K. (a) Ge, $n_A + n_D < 10^{13}$ cm^{-3}; (b) Ge : Sb, $n_D = 5$ to 8.10^{15} cm^{-3}; (c) Ge : As, $n_D = 5$ to 8.10^{15} cm^{-3}; (d) Ge : In, $n_A = 3.10^{15}$ cm^{-3}

and acceptor (Ga, In) impurities [11]. The impurity concentration varied from 5.10^{13} to 5.10^{16} cm^{-3}. Luminescence was detected using a DFS-12 grating spectrometer with a dispersion of 10 Å/mm and a PbS photoresistor as detector. Measurements were done at 4.2 K. The luminescence was excited using an incandescent lamp or a helium-neon laser with a wave-length 0.63 μ and 1.52 μ. Excitation spectra were studied with a MDR-2 monochromator with a dispersion of 20 Å/mm.

Figure 1 shows the luminescence spectra of pure germanium and germanium doped with Sb, As and In at 4.2 K. The Ge : P luminescence spectrum is similar in its general features to the Ge : As spectrum and the Ge : Ga spectrum resembles the Ge : In spectrum.

The luminescence spectrum of pure germanium (Fig. 1) shows the emission FE from free indirect excitons with simultaneous excitation of LA and TO phonons. Transitions involving TA and LO phonons are forbidden [12] for excitons near the bottom of the exciton band and, therefore, at liquid-helium temperatures they do not appear in the luminescence spectrum. Zero-phonon transitions are forbidden by the law of quasimomentum conservation. There is substantial asymmetry in the bands due to the kinetic energy distribution of the free excitons.

In the doped samples (Fig. 1), on the long-wavelength side of the free exciton luminescence band, quite narrow lines BE appear due to radiative annihilation of excitons bound at neutral impurity atoms. The amount of shift of these lines relative to the long-wavelength tail of the free exciton emission band gives the energy E_D for dissociation of the complex into a free exciton and a neutral impurity centre and depends on the kind of impurity.

The data obtained show a good agreement for Ge crystals with the rule of Haynes [13], which relates the energy E_D of dissociation of the complex with the impurity dissociation energy $E_i (E_D \sim 0.1 E_i)$.

In contrast to silicon [14], in germanium crystals zero-phonon transitions at exciton-impurity complexes can only be observed for arsenic and phosphorous impurities.

It was possible to determine the LO and TO phonon energies from the bound exciton luminescence spectra of these crystals, which turn out to be equal to 27.6 ± 0.1 and 36.0 ± 0.1 meV, respectively.

In the zero-phonon part of the Ge : As luminescence spectrum traced on a more expanded scale than in Fig. 1 we see a number of new lines (Fig. 2). These lines are due to the annihilation of free and bound excitons with excitation of the arsenic atom.

These lines, located at 739.1, 735.0, 729.1, 727.4, and 726.4 meV, weaken in the same way with decreasing temperature, which probably indicates that they arise from the same recombination mechanism. The relative narrowness of these lines (half-width 0.2 meV) and their energy

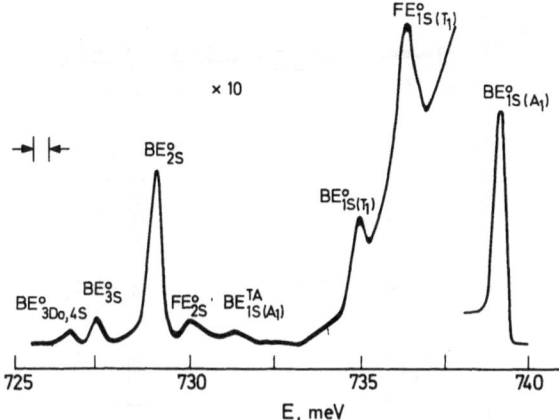

Fig. 2. Zero-phonon part of the luminescence spectrum from a germanium crystal doped with arsenic. $n_D = 5$ to 8.10^{15} cm^{-3}, $T = 4.2$ K

positions suggest that they arise from radiative annihilation of excitons bound at neutral arsenic atoms, in which the arsenic atom remains in one of its excited states. The shift of the observed lines relatively to the main zero-phonon line lying at 739.1 meV gives the energy difference between the ground and excited states of the impurity. So-called "two electron" transitions are observed in the luminescence of excitons bound to neutral centres in several other crystals [14, 15].

Figure 3 illustrates the scheme for the energy levels of the arsenic donor in germanium, which we obtained from the spectrum of two-electron transitions and data on infrared absorption [16]. The paper contains the results of a calculation of the energies of these levels done by Faulkner in the effective mass approximation [17] (Fig. 3).

Some of the discrepancies between our data and Faulkner's calculations arise from the limitations of the effective mass method and also apparently from certain errors stemming from the fitting of experimental and theoretical data in [17].

Figure 2, in addition to the above-mentioned lines, also shows the lines $FE^0_{1S(T_1)}$ (730.1 meV), whose appearance evidently arises from annihilation of free excitons near arsenic impurity centres in which quasimomentum and part of the energy are transferred to the impurity centre. This recombination mechanism for free excitons was first observed in silicon [14]. The energy interval between the BE^0_{2S} and FE^0_{2S} lines and between the $BE^0_{1S(T_1)}$ and $FE^0_{1S(T_1)}$ lines is close to the dissociation energy of an exciton-impurity complex formed with arsenic atoms. The substantial half-widths of these lines and the slower decrease of the

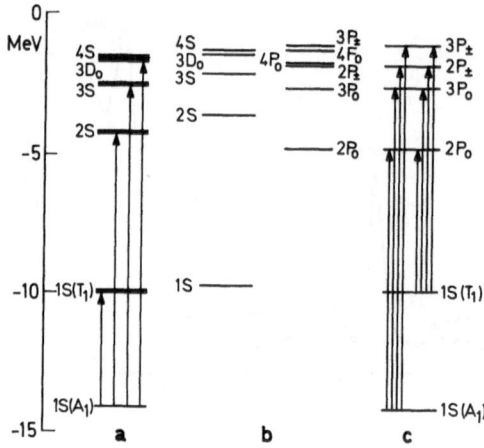

Fig. 3a–c. Energy level scheme for the As donor in germanium from data in the present work (a), IR absorption spectra [16] (c), and calculations in the effective mass approximation [17] (b)

intensity with increasing temperature in comparison to the bound exciton lines indicate that these lines are more closely involved with recombination of bound excitons.

Figure 4 gives luminescence spectra of a germanium crystal containing less that 10^{13} cm^{-3} residual impurities at the same rate of volume excitation and two different temperatures. With the temperature decreasing from 4.2 to 1.7 K, broad emission bands appear on the long-wavelength side of the FE free exciton emission bands first reported in [5] (we shall designate these bands with A and an index for the type of phonon the excitation of which involves an indirect transition, since there is no standard designation for these bands). The intensity of free exciton emission weakens, so that at 1.7 K the spectrum consists only of A-bands. The weak line observed between FE_{LA}- and A_{LA}-bands is due to the emission of excitons bound at the neutral residual impurities.

With excitation rate changing from 10^{18} to 10^{23} sec^{-1} cm^{-3}, the energy position, shape, half-width, as well as the relative intensity of A-bands corresponding to excitation of different types of phonons in the crystal lattice remains unchanged.

Figure 5 shows luminescence spectra of a germanium crystal containing 5.10^{15} cm^{-3} Sb donors at the same temperature and two different excitation levels.

Comparing these spectra, one can see that with a changing excitation level of this crystal the position, shape, half-width and the relative

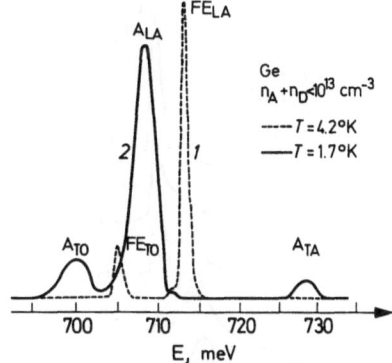

Fig. 4. Luminescence spectra of Ge (concentration of impurities $n_D + n_A < 10^{13}$ cm^{-3}) with one and the same volume excitation ($g = 10^{19}$ sec^{-1} cm^{-3}). 1 — at 4.2 K, 2 — at 1.7 K

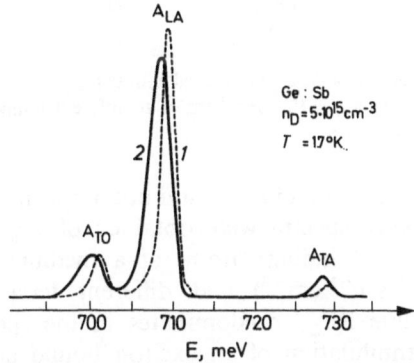

Fig. 5. Luminescence spectra of a Ge crystal at 1.7 K doped with Sb ($n_D = 5.10^{15}$ cm^{-3}) for two different excitation levels. 1–10^{18} sec^{-1} cm^{-3}, 2–10^{23} sec^{-1} cm^{-3}

intensity of the bands with excitation of different phonons undergo a marked change. A thorough comparison reveals that these spectral characteristics of the crystal A-emission, obtained at a high excitation level, do not practically differ from the corresponding characteristics of this emission in pure samples.

Because A-emission bands in the crystal containing impurity centres generally occupy a different energy position from that in pure crystals when the excitation is not too great, it was found interesting to observe in greater detail the process of the appearance of these bands in the

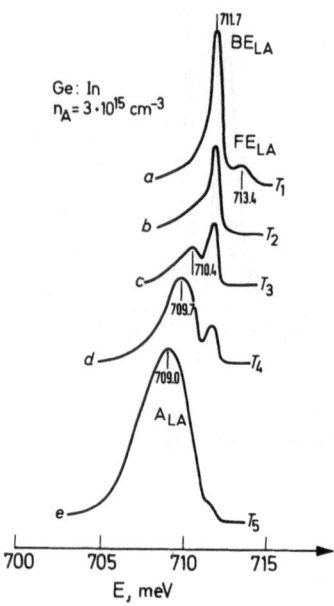

Fig. 6. Luminescence spectra of a Ge crystal, doped with In ($n_A = 3.10^{15}$ cm^{-3}) at one and the same volume excitation level $g = 10^{19}$ sec^{-1} cm^{-3} at different temperatures $T_1 > T_2 > T_3 > T_4 > T_5$; $T_1 = 4.2$ K; $T_5 = 1.7$ K

emission spectrum of germanium containing impurity centres. For this purpose, luminescence spectra were obtained of a germanium crystal containing 3.10^{15} cm^{-3} indium atoms at a medium level of volume excitation ($g = 10^{19}$ sec^{-1} cm^{-3}) and different temperatures (Fig. 6).

At 4.2 K, the line BE_{LA} predominates in the spectrum, which is due to radiative annihilation of an exciton bound at indium neutral atoms [11]. As the temperature decreases, this line acquires a long-wavelength "tail" from which a maximum emerges, its energy position shifting towards the long-wave length side. This energy shift cannot be explained by the overlapping of the "tails" of the two (BE_{LA} and A_{LA}) lines, for it continues even when the line BE_{LA} is much weaker than A_{LA} [1].

In crystals containing donor centres of arsenicum and phosphorus, an intensive A_0-band was observed which appeared in a zero-phonon transition with dispersion at the impurity atoms. A similar band was also observed in germanium crystals containing isoelectronic silicon atoms [18].

---

[1] The effect of doping on the A-band emission spectrum was studied by Bagaev et al. [37]. Their work, however, did not reveal any substantial change in the spectroscopic characteristics at the introduction of impurity centres into the crystals. In the light of the results presented here, it can be accounted for by a relatively high level of excitation used in [37].

It is believed that the results obtained can best be explained by assuming that emission in A-bands results from radiative recombination in electron-hole drops[2]. Indeed, the presence of a large number of condensation centres, which can be impurity centres must cause the appearance of numerous small electron-hole drops. Distortion of the potential in the surface layer of sufficiently small drops must result in an energy dependence of the particles bound at an electron-hole drop on its size, with a decrease in the size of drops accompanied by an increase in the average energy of the particles in the drop. This, in turn, will bring about a shift of the drop emission band towards greater energies with decreasing excitation intensity, which is observed experimentally in all the samples containing 10^{14}–10^{16} impurities per cm^{-3} (Fig. 5). The maximum energy shift of the A-band in our experiments was about 2 meV. It should be noted that a very similar phenomenon was observed in silicon crystals containing boron impurities [19].

In pure germanium crystals, where the impurity concentration does not exceed 10^{13} cm^{-3}, at the same temperatures and excitation densities at which emission was studied in doped crystals, no shift of the energy position or change of the A-band emission half-width was observed. It is probably due to the fact that with a small number of condensation centres and at the same excitation levels, the smallest average size of the drop becomes so large that the near-surface layer does not significantly affect the average energy of particles in the drop.

It is of interest to note that changing the excitation density of crystals containing impurity centres results in changes of relative intensities of the components with excitation of different phonons and their half-width (Fig. 5). Thus, with excitation density increasing from 10^{18} to 10^{23} sec^{-1} cm^{-3}, the intensity ratio of TA- and LA-components increased almost twofold, while the intensity ratio of TO- and LA components increased by only 10%. The half-width of the lines changed (from 1.6 to 2.6 meV). These facts are also difficult to explain in terms of the biexciton concept, but they can be interpreted as a shift of the Fermi level in drops of small size and the resulting decrease in the intensity of the extremum-forbidden zones of indirect transition with excitation of a TA-phonon.

To study the exciton capture by impurity centres, investigations were carried out of the photoconductivity spectra of germanium crystals

---

[2] It should be noted that at low excitation densities the drops may have dielectric character. Excitons and biexcitons forming such drops may preserve their individuality. At higher densities the drops become metallized and the interaction between electrons and holes changes substantially. It is thought that a number of experiments can be explained consistently by assuming the formation of dielectric drops from excitons and their subsequent metallization.

containing 10^{13}–10^{16} donor centres per $1\,\mathrm{cm}^{-3}$ at 4.2 K. They were concerned with indirect exciton transitions, where the absorption coefficient is small (under $10\,\mathrm{cm}^{-1}$) and the surface effects can be neglected [20].

Taking into account the possibility of ionization of excitons and impurity centres [21] by collisions already in electric fields of 3–6 V/cm, special attention was paid to the value of the electric field strength at which photoconductivity was studied. The field strength in the samples was determined by means of point contacts.

In all the samples studied, the photoconductivity spectrum at 4.2 K and at fields exceeding 4 V/cm had the usual character (Fig. 7c), correlating completely with the well-known absorption coefficient of the germanium indirect edge. In this spectrum one can clearly see the "steps" created at the formation of indirect excitons in the crystal with excitation of TA and LA phonons.

In germanium samples containing about 10^{14} per cm^{-3} small donors at a field strength of about 1 V/cm, a photoconductivity spectrum was observed, which differed substantially from the photocurrent spectrum in Fig. 7b. In this spectrum, at the beginning of the thresholds corresponding to the formation of excitons with excitation of TA, LA, and TO phonons, photocurrent maxima are observed with a half-width 5–10 meV whose long-wavelength edges correspond exactly to the beginning of the absorption threshold. In still weaker fields these maxima are practically absent too (Fig. 7a). In view of the fact that this structure of the photoconductivity spectra is absent in the most pure crystals, it was supposed that its appearance is due to ionization of impurity centres by exciton capture[3]. This phenomenon is considered theoretically in the work of Trlifaj [22], who shows that the most effective interaction between excitons and impurity centres occurs at low exciton velocities.

At indirect transitions, formation of free excitons with different kinetic energies is known to be possible, with excitation of the crystal by light having an energy corresponding to the indirect edge, accompanied by the formation of excitons having small kinetic energies. These "cold" excitons can ionize effectively the neutral donor impurity centres, creating free electrons which increase the crystal conductivity. With increasing energy of the excitons created by light, the photocurrent changes a little, for during their lifetime (10^{-5}–10^{-6} sec) the excitons can thermalize, their velocities decreasing to values which, according to Trlifaj, are effective for the interaction with impurity centres. This

---

[3] In Ge:P and Ge:As, photoconductivity was detected in the form of narrow maxima also in the region of bound exciton lines, which was probably due to Auger recombination process. It should also be noted that the photocurrent dependence on the intensity of exciting light in the region of indirect exciton transitions has a superlinear character.

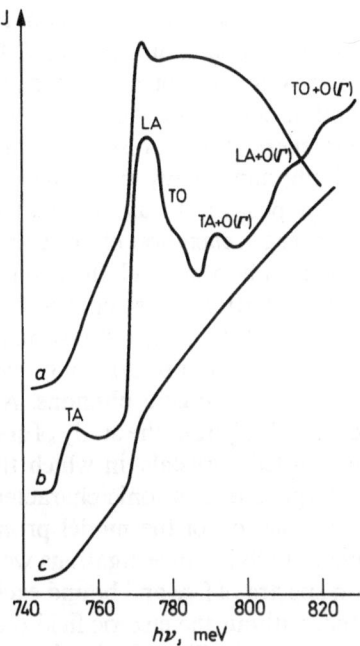

Fig. 7a–c. Photoconductivity spectrum of Ge $(n_A + n_D < 10^{13})$ at 4.2 K. (a) $E = 0.05$ V/cm; (b) $E = 1.5$ V/cm; (c) $E = 4.0$ V/cm

consideration provides a qualitative explanation of the character of the spectrum at weak fields.

At stronger fields there is a greater probability of exciton ionization by collision with free current carriers. There is not enough time for excitons to thermalize during their lifetime and, at the interaction with impurity centres, they still possess the kinetic energy which they had at their formation. In this case the photoconductivity spectrum will reveal a dependence of the cross-section of exciton capture by impurity centres on exciton velocity $\sigma(v)$. At small kinetic energies the photocurrent is at its maximum and drops sharply with increasing kinetic energy and, consequently, exciton velocity. This results in the appearance of maxima at the beginning of the exciton absorption threshold. .

Further increase in the electric field intensity is responsible for the free and bound exciton ionization by collisions becoming a more probable process than exciton capture at impurity centres, and the spectral distribution of photoconductivity has the usual character determined by the absorption coefficient and by the photoactive decay of excitons because of ionization by collision with free current carriers.

Particularly interesting, in the writer's opinion, are the three weaker short wavelength maxima designated in Fig. 7 as $TA + O(\Gamma)$, $LA + O(\Gamma)$ and $TO + O(\Gamma)$. Their position is not connected with any changes of the germanium absorption coefficient. These maxima turned out to be shifted towards greater energies with respect to the TA, LA and TO maxima by an optical phonon energy from the centre of the B-zone (37.3 meV). This makes it possible to assume that their appearance is connected with a rapid energy relaxation of hot excitons with excitation of long-wavelength optical phonons and their subsequent interaction with impurity centres with creation of free current carriers. The presence of these maxima also means that energy relaxation of hot excitons in covalent germanium crystals by optical long wavelength phonons proceeds much faster than by acoustic phonons. A similar result was obtained earlier in the work [23] from the study of spectra of free exciton luminescence excitation in CdS crystals, in which the binding between separate atoms is to a large extent of ionic character.

In order to find confirmation of the model proposed as explaining the structure in photoconductivity, investigations were also made of the luminescence excitation spectra of free and bound excitons in germanium (Fig. 8). As was expected, without the electric field the excitation spectra had no clearly defined structure. They had a broad maximum in the region of indirect exciton transitions characteristic of the quasi-equilibrium case, when the lifetime of excitons is much longer than their energy relaxation time. A sharp decrease at short-wavelengths is probably due to an increasing surface effect with increasing absorption coefficient.

The excitation spectrum of drop luminescence in germanium crystals was also investigated. This is probably the only opportunity to observe the excitation spectrum ascribed to the exciton interaction effects at high intensities. The excitation spectrum of drop luminescence differs little from those of the free and bound excitons. On the one hand, it shows that the excitons contribute substantially to the formation of drops and, on the other hand, it confirms the existence of a quasi-equilibrium state between excitons and drops. Thermalized excitons participate in the process of drop formation.

The exciton spectrum of HgI_2 crystals has been known for a long time [24]. However, the exciton properties of this compound have not been studied in so much detail as in CdS, GaP and Ge crystals. Meanwhile, detailed information on the exciton properties is necessary in order to interpret the results on the spectra obtained at high excitation levels.

First, we shall briefly consider the data on the exciton spectra at ordinary excitation levels [25, 26]. In the absorption spectrum of sufficiently thin single crystals of HgI_2, a group of lines is observed at 4.2 K,

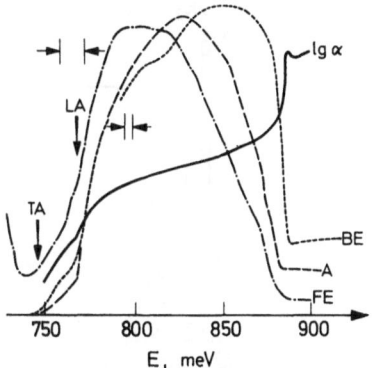

Fig. 8. Luminescence excitation spectra (ES) of Ge at 4.2 K. FE — ES of free exciton, BE — ES of bound exciton, A — ES of "drops"

the most intensive one being at $\lambda = 5297\,\text{Å}$ $(n = 1)$. On the short-wave side of it there are two much weaker lines, $\lambda = 5246\,\text{Å}$ and $5239\,\text{Å}$ $(n = 2, 3)$. All the lines are polarized with $E \perp C$ (where C is the optical axis of the crystal) and are described well by the formula:

$$E_n = E_g - E_i/n^2 \qquad n = 1, 2, 3$$

where $E_g = 2.369$ eV and $E_i = 0.029$ eV. Estimates based on these data give an effective mass of the exciton of $m_{\text{eff}} = 0.13\,m_0$ and a radius of the ground state of $r_{\text{exc}} \sim 25\,\text{Å}$. At $T = 4.2$ K there are also two narrow absorption lines in the long-wave part of the spectrum, $\lambda_1 (5310\,\text{Å})$ and $\lambda_2 (5321\,\text{Å})$. The line $5321\,\text{Å}$ is ascribed by all investigators to the formation of an exciton bound at its own defect [25–27].

At 4.2 K and the usual excitation level, the HgI_2 crystal is characterized by a rich linear luminescence spectrum (Fig. 9). Some of the emission lines are in resonant coincidence with those of absorption.

The shortest wave length in the luminescence spectrum is the line $5297\,\text{Å}$ $(n = 1)$. The most intensive in the luminescence spectrum at 4.2 K and usual excitation levels are the lines $\lambda_2 (5221\,\text{Å})$ and $\lambda_x (5217\,\text{Å})$. At higher temperature the intensity of the lines λ_2 and λ_x drops sharply (they disappear from the spectrum at $T \sim 40$ K), while the intensity of the line λ_1 at first increases slightly and then slowly decreases. The line λ_1 is observed in the spectrum up to 77 K. The heating of the crystal is accompanied by a general shift of the lines towards the long wave side of the spectrum. The temperature coefficient in the interval 20–77 K, according to our data, is $-2.6\ 10^{-4}$ eV/deg [25]. On the long wave length side of the exciton emission resonance line $(n = 1)$ there are a number of equidistant lines which are due to the exciton radiative

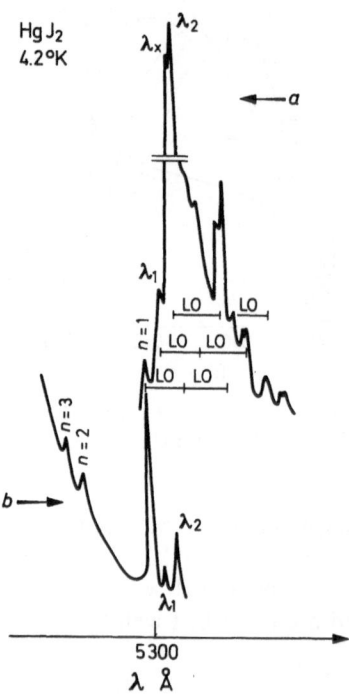

Fig. 9a and b. Densiometric luminescence (a) and absorption (b) spectra of HgI_2 single crystal at 4.2 K

annihilation with a simultaneous generation of one or several longitudinal optical phonons. The value of the LO phonon is $115 \pm 5 \, \text{cm}^{-1}$. Similar vibrational series are characteristic of the lines λ_1, λ_2, and λ_x.

We have also studied the excitation spectra of the luminescence lines λ_1 and λ_2. The maximum emission value in these lines is shown to be reached when the exciting light corresponds to the free exciton line ($n = 1$, 5297 Å). It confirms the important role of the excitons in the formation of the emission processes in the lines λ_1 and λ_2.

In the low-temperature photocurrent spectra, minima of the photocurrent curves corresponded to the exciton absorption lines $n = 1, 2, 3$. As is shown in our papers [28], the state of the surface during the near-surface generation of excitons determined the nature of the processes of the photoelectrically active exciton decay.

HgI_2 is an exceedingly interesting object for the investigation of high-density excitons, for at 4.2 K the position of the exciton level $n = 1$ ($\lambda = 5297$ Å) in this crystal almost coincides with the wave-length of the second harmonic of the neodymium laser, so that it becomes possible to excite strongly the crystal at the frequency of the exciton transition. We have investigated the luminescence of HgI_2 crystals excited mainly by such a laser. The exciting light in the two neodymium rods used had wavelengths of 5288 and 5298 Å. For a comparative investigation, use was made of a nitrogen laser with $\lambda_{exc} = 3371$ Å, a pulse duration of 10 nsec and a pulse peak power of 100 kW.

The pulse power of the neodymium laser ranged from 0.01 to 1.2 MW/cm^2; its duration was about 50 nsec. According to approximate estimates, the maximum exciton concentration (in this case) reached 10^{18} cm^{-3}. The luminescence spectra were photographed with an instrument having a dispersion 31 Å/mm. When excited by a nitrogen laser the spectra were continuously recorded photoelectrically by a recorder. The dispersion of the instrument in this case was 5 Å/mm. The samples, which were single-crystal plates with the optical axis in the plane of the investigated surface, were cooled by direct immersion in liquid helium. More than 30 samples were studied.

At an excitation density less than 0.07 MW/cm^2, the luminescence spectrum of the HgI_2 crystal is in general similar to the spectrum obtained by excitation with a mercury lamp [25]. As a rule we have observed the two strongest emission lines of the bound excitons, $\lambda_x = 5317$ Å and $\lambda_2 = 5321$ Å. At an excitation density $W = 0.07$ MW/cm^2 (Fig. 10), a new emission band, $M(\lambda = 5333$ Å), of 1 meV width appears in the spectrum [29]. With further increase of the excitation intensity, the M band becomes much stronger and predominates in the spectrum (Fig. 10). The band is broadened and its maximum shifts towards the long wave length side of the spectrum. At a density of $W = 0.3$ MW/cm^2, the maximum of the band is located at $\lambda = 5337$ Å, and the width amounts to 3–4 meV. At the same excitation there appear in the spectrum, in addition to the M band, also a new band $P(\lambda = 5368$ Å) 2 meV wide and two weaker bands at 5383 Å and 5412 Å, of approximately the same width (Fig. 10). A subsequent increase of the excitation to $W = 1.2$ MW/cm^2 causes further growth in the intensity of all the new bands. Starting with $W = 0.1$, all spectra contain the narrow line of the bound exciton ($\lambda = 5354$ Å). By its wavelength, this line corresponds to the radiative annihilation of the bound exciton with a simultaneous generation of the LO phonon ($\lambda_2 - LO$).

Let us examine in greater detail the properties of the M band. The high intensity of the M band has made it possible to investigate the dependence of its intensity on that of the exciting light. We have found that up to $W = 0.3$ MW/cm^2 the M band is amplified superlinearly with increasing excitation, and then increases more slowly in the interval $W = 0.3-1.2$ MW/cm^2. A superlinear dependence of the luminescence intensity on the excitation is observed for interactions between excitons

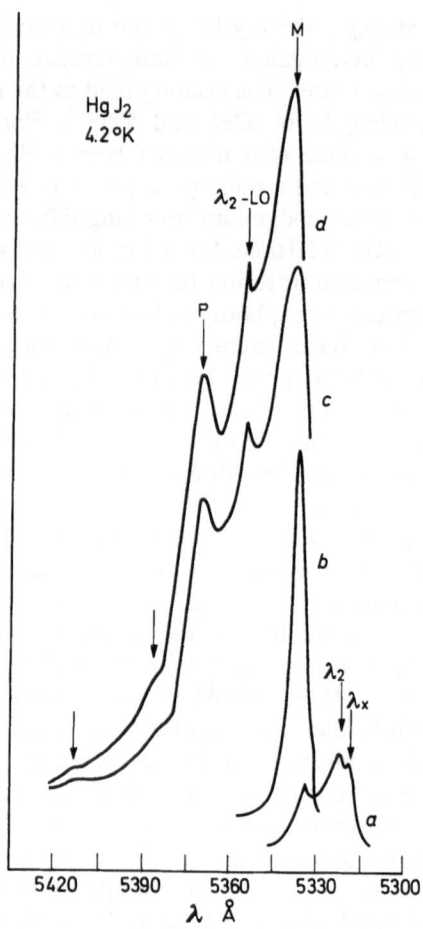

Fig. 10a–d. Luminescence spectra of HgI_2 crystal at $T = 4.2\,K$ excited with a laser of wave length 5288 Å with a power of 0.07 (a), 0.2 (b), 0.3 (c), and 1.2 MW/cm² (d)

[30]. This fact is illustrated by Fig. 11 showing the dependence of the luminescence intensity of the M band and the λ_2 line on the intensity of the exciting light. The M band is shifted 15 meV towards the long wave length side of the exciton line $n = 1$ ($\lambda = 5297$ Å). It can therefore be assumed that the M band is due not to Auger recombination of the free excitons, but more likely to radiative decay of the exciton molecule, as a result of which one exciton becomes annihilated and the second remains in the non-dissociated state. A similar process is observed, e.g., in CuCl crystals [31]. In this model, the short-wave edge of the M band

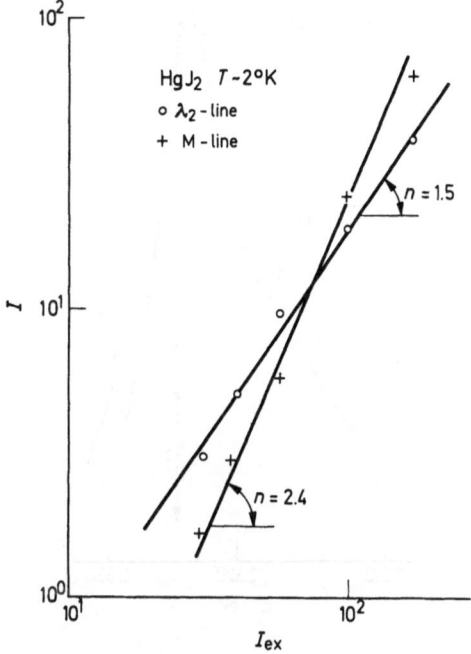

Fig. 11. Dependence of the M and λ_2 emission lines intensity I on the intensity of exciting light I_{ex} at $T = 4.2$ K (excitation with nitrogen laser)

is separated from the free-exciton line by a distance equal to the binding energy E_i^M of the excitons in the molecule.

The M band has an asymmetrical Maxwellian-like contour with a sharp drop on the short wave length side of the spectrum. This is precisely the shape expected for the considered radiative biexciton decay, when the biexciton momentum is carried away by a non-dissociated exciton. The shift of the maximum of the M band towards the long wave length side may indicate an increase in the effective temperature of biexcitons. This temperature estimated by the shape of the M band contour amounts to $T \sim 25$ K at medium excitation intensities (Fig. 12). At the same time, the lattice temperature does not exceed $T \sim 10$ K, as can be deduced from the spectral position of the $\lambda_2 - LO$ line [26].

In the model proposed the biexciton binding energy is about 0.5 of the exciton binding energy (E_i). This value is considerably greater than the one predicted theoretically [32]. It can be supposed that the binding of excitons into a molecule in HgI_2 occurs from a lower energy

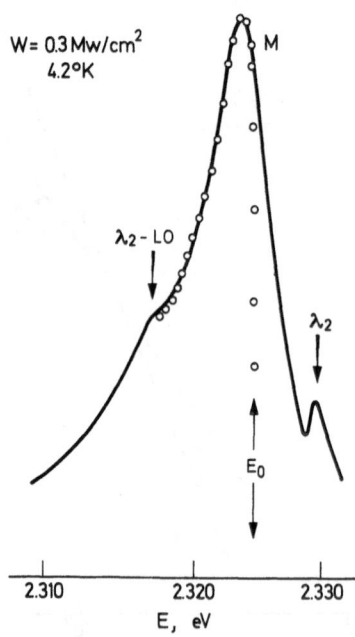

Fig. 12. The M band shape in HgI$_2$ single crystal excited by the nitrogen laser at 0.3 MW/cm^2 power, measured at 4.2 K, showing the Maxwell distribution at 25 K

level than $n = 1$. Certain considerations allow to interpret the line λ_1 (5310 Å) as belonging to the free exciton spectrum. In this case the molecule binding energy will go down to 11 meV, while the biexciton binding energy will be 0.32 of the exciton binding energy. Both the values (11 and 15 meV) exceed those that can be expected from theoretical considerations [32]. However, in a recently published theoretical paper of Gutlyansky and Khartsiev [33] it is shown that the anisotropy of the carrier effective masses can substantially increase the binding energy of the biexciton, making it comparable with that of the exciton. These considerations are probably realized in germanium, where $m_\parallel / m_\perp = 20$. It can be assumed that the anisotropy of masses can also be responsible for an increase in the binding energy in other crystals, in particular, in the investigated HgI$_2$ crystals.

Recent investigations have made it possible to find out important details characterizing the shift of the M band towards the long wave length side at higher excitation levels. It turned out that at medium excitation levels the M band can be a doublet, M_1 and M_2, with a distance between the components of 2 meV (5333–53339 Å; Fig. 13). Its short

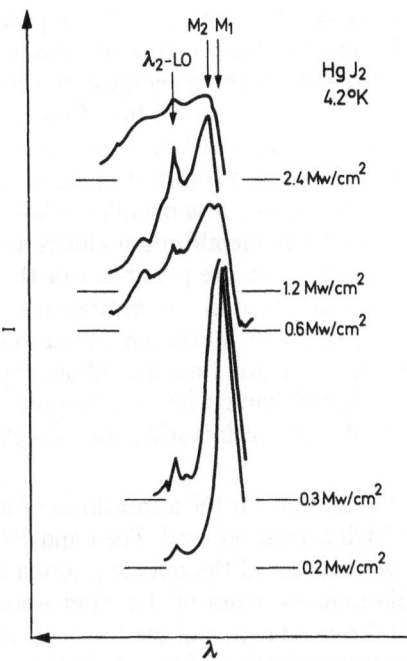

Fig. 13. Densiometric luminescence spectra of HgI_2 in the spectral region of M_1, M_2 doublet at different excitation levels (0.2 MW/cm² to 2.4 MW/cm²)

wave length component is observed at low intensities, while the long wave length component predominates in the spectrum at high intensities. The shift of the band is due to the changing intensity of the doublet components. The bands have similar spectral forms: a sharp drop on the short wave length side and a more gradual fall on the long wave length side. It should also be noted that the M_2 band is several times broader than the M_1 band. With the intensities of the laser light approaching the threshold of crystal destruction, the bands become very much diffused, yet their short-wave length edge remains clear enough.

The structure of the biexciton emission lines was observed in CuBr [34] and CuCl [35]. In CuBr it was ascribed to the biexciton formation from the singlet and triplet exciton states. However, in the absorption and emission spectrum of the free exciton of HgI_2, no splitting was observed corresponding to the splitting in the biexciton line (2 meV). It should also be noted that with increasing excitation intensity the long wave length component of the doublet (M_2) predominates in the spectrum. When considering the nature of the M_2 line one should not rule out the possibility of the formation of complexes of greater size than biexcitons at big enough exciton concentrations. At the formation of polyexcitons and drops the binding energy of the complex grows with the increasing number of particles in it [36]. If the processes similar to those considered above for the biexciton (M_1) still continue, the result must be a shift of the emission line towards the long wave length side by an amount of the binding energy of a new complex. With increasing exciton density in the complex (drop), metallization effects and the appearance of electron-hole plasma emission are possible. For the excitons in HgI_2 these can be expected at exciton concentrations of $n \sim 10^{18}$ cm^{-3} ($nr_{ex}^3 \sim 1$), which corresponds to maximum excitation powers. Indeed, in these cases all the new emission lines are diffuse and in the spectrum there remains only a broad band with a maximum at about 5340 Å, the short wave length edge of which is still clear enough and is located at about 5330 Å.

Let us consider the reasons for the appearance of other bands in the HgI_2 spectrum at a high excitation level. The band P (5368 Å) is shifted from the M line by an amount of the optical phonon and is its phonon replica. It is possible that processes of this kind were not observed in biexcitons earlier. In the same region of the spectrum one can expect the occurrence of Auger recombination of two excitons, as a result of which one exciton becomes annihilated, giving away a part of its energy for the dissociation of the second.

The 5412 Å band is shifted by an amount E_i towards the long-wave length side away from the M band and can correspond to Auger recombination of the biexciton, in which one exciton in annihilated radiatively and the other dissociates into a free electron and a hole. It is interesting to note that the weak band with $\lambda = 5383$ Å is located in that part of the spectrum where the next biexciton emission process can occur, namely, annihilation of one exciton with simultaneous excitation of the other one into the state $n = 2$.

The results presented in our paper show a number of new spectrometrical properties of drop luminescence in germanium detected by investigation mainly at low excitation levels.

In HgI_2 crystals a line was observed which is connected with the radiative biexciton decay, as a result of which one exciton becomes annihilated, giving away energy for the decomposition of the molecule and the acceleration of the other exciton. Phenomena were also observed which resulted from other interaction processes between excitons and molecules. The writer supposes that in order to interpret some phenomena both in germanium and in HgI_2, it is possible to consider drop states in which excitons are not yet dissociated (insulating drops).

The writer thanks his colleagues I. Akopjan, M. Pimonenko, B. Rasbirin and N. Sokolov, with whom he was able to discuss again some of the results comprising the contents of the present paper.

References

1. Gross, E. F., Karryev, N. A.: Dokl. Akad. Nauk SSSR **84**, 471 1952.
2. Akopyan, I. Kh., Gross, E. F., Razbirin, B. S.: Zh. Eksperim. i Teor. Fiz. Pis. Red. **12**, 366 (1970).
3. Kuroda, H., Shionoya, Sh., Saito, H., Hanamura, E.: Technical Report of ISSP Ser. A N 562, 1972.
4. Asnin, V. M., Rogachev, A.: Zh. Eksperim. i Teor. Fiz. Pis. Red. **7**, 464 (1968).
   Rogachev, A. A.: Proc. Int. Conf. Semicond. Moscou 1968, p. 431.
5. Pokrovsky, Ya. E., Svistunova, K. I.: Zh. Eksperim. i Teor. Fiz. Pis. Red. **9**, 435 (1969).
6. Bagaev, V. S., Galkina, T. I., Gogolin, O. V., Kedysh, L. V.: Zh. Eksperim. i Teor. Fiz. Pis. Red. **10**, 309 (1969).
   Bagaev, V. S., Galkina, T. I., Penin, N. A., Stopachinsky, V. B., Churaeva, M. N.: Zh. Eksperim. i Teor. Fiz. Pis. Red. **16**, N 3, 120 (1972).
7. Vavilov, V. S., Sajac, V. A., Mursin, V. N.: Zh. Eksperim. i Teor. Fiz. Pis. Red. **10**, 304 (1969).
   Benoit à la Guillaume, Voos, M.: Solid State Commun. **11**, 1585 (1972).
8. Rogachev, A. A.: Izv. Akad. Nauk SSSR Ser. Fiz. **37**, N 2, 229 (1973).
9. Benoit à la Guillaume, Salvan, F., Voos, M.: Proc. Internat. Conf. Luminescence Amsterdam 1970, p. 315.
10. Pokrovsky, Ya. E.: Phys. State Solids (a) **11**, 385 (1972).
    Asnin, V. M., Lomasov, Ya. N., Rogachev, A. A.: Fiz. Tverd. Tela **14**, 3457 (1972).
11. Gross, E. F., Novikov, B. V., Sokolov, N. S.: Fiz. Tverd. Tela **14**, 443 (1972).
12. Benoit à la Guillaume, Parodi, O.: Proc. Int. Conf. Semicond. Prague 1961, p. 426.
    Elliot, R. I.: Phys. Rev. **108**, 1384 (1957).
13. Hayns, I. R.: Phys. Rev. Letters **4**, 361 (1960).
14. Dean, P. I., Hayns, I. R., Flood, W. F.: Phys. Rev. **161**, 711 (1967).
15. Dean, P. J., Cuthbert, I. D., Thomas, D. G., Lynch, R. T.: Phys. Rev. Letters **18**, 122 (1967).
    Henry, C. H., Nassau, K.: Phys. Rev. **2** B, 997 (1970).
    Schainer, W., Yep, T. O.: Solid State Commun. **9**, 421 (1971).
16. Reuszer, J. H., Fisher, P.: Phys. Rev. **135**, A 1125 (1964).
17. Faulkner, R. A.: Phys. Rev. **184**, 713 (1969).
18. Gross, E. F., Sokolov, N. S., Titkov, A. N.: Fiz. Tverd. Tela **14**, 2004 (1972).
19. Kaminsky, A. S., Pokrovsky, Ya. E., Alkeev, N. V.: Zh. Eksperim. i Teor. Fiz. **59**, 1937 (1970).
20. Gross, E. F., Ilinsky, A. V., Novikov, B. V., Sokolov, N. S.: Zh. Eksperim. i Teor. Fiz. Pis. Red. **12**, 259 (1970).
21. Asnin, V. M., Rogachev, A. A., and Ryvkin, S. M.: Fiz. Tekh. Poluprov. **1**, 1740 (1967).
    Zavarickaja, E. I.: Tr. Fiz. Inst. Akad. Nauk USSR **37**, 41 (1966).
22. Trlifaj, M.: Czech. J. Phys. **9**, 446 (1959).
23. Gross, E. F., Permogorov, S. A., Travnikov, V., Selkin, A. B.: J. Phys. Chem. Solids **31**, 2595 (1970).
24. Nikitine, S., Sieskind, M.: Compt. Rend. **240**, 1324 (1955).
    Sieskind, M.: J. Phys. **17**, 821 (1956).
    Gross, E. F., Kapliansky, A. A.: Zh. Tekh. Fiz. **245**, 659 (1957).
    Gross, E. F., Kapliansky, A. A., Novikov, B. V.: Dokl. Akad. Nauk SSSR **110**, 761 (1956).

25. Novikov,B.V., Pimonenko,M.M.: Fiz. Tekh. Poluprov. **4**, 2077 (1970).
28. Novikov,B.V., Pimonenko,M.M.: Fiz. Tekh. Poluprov. **6**, 771 (1972).
27. Kleim,R., Raga,F., Nikitine,S.: Proc. Intern. Conf. Lumin. Budapest 1966, p. 1496.
28. Griegoriev,R.V., Novikov,B.V., Cherednichenko,A.E.: Phys. Stat. Solids **28**, K85 (1968).
    Novikov,B.V., Ilinsky,A.V., Lieder,K.F., Sokolov,N.S.: Phys. Stat. Solids **48**, 473 (1971).
29. Akopyan,I.Kh. Novikov,B.V., Pimonenko,M.M., Razbirin,B.S.: Zh. Eksperim. i Teor. Fiz. Pis. Red. **17**, N7, 419 (1973).
30. Hayns,J.R.: Phys. Rev. Letters **17**, 860 (1966).
    Mysyrowicz,A., Grun,J.B., Levy,R., Bivas,A., Nikitine,S.: Phys. Rev. Letters **26** A, 615 (1968).
    Nikitine,S., Haken,H.: Izv. Akad. Nauk SSSR Ser. Fiz. **37**, N2, 220 (1973).
31. Souma,H., Goto,T., Ohta,T., Ueta,M.: J. Phys. Soc. Japan **29**, N3, 697 (1970).
32. Akimoto,O., Hanamura,E.: J. Phys. Soc. Japan **33**, N6, 1537 (1972).
33. Gutlyansky,E.D., Khartsiev,V.E.: Solid. State Commun. **12**, N1, 1087 (1973) and in press.
34. Souma,H., Koike,H., Sunzuki,Kaoru,K.,Ueta,M.: J. Phys. Soc. Japan **31**, 1285 (1971).
35. Grun,J.B., Nikitine,S., Bivas,A., Levy,R.: J. Lunin **1,2**, 241 (1970).
36. Gogolin,O.V., Rashba,E.I.: Zh. Eksperim. i Teor. Fiz. Pis. Red. **17**, N12, 690 (1973).
37. Alekseev,A.S., Bagaev,V.S., Galkina,T.I., Gagolin,O.V., Penin,N.A.: Fiz. Tverd. Tela 12, 3516 (1970).

Dr. B. V. Novikov
Institute of Physics
Molecular Physics Department
Leningrad State University
Leningrad, USSR

Exciton Condensation in Germanium

A. A. ROGACHEV

Contents

Electrons and holes in semiconductors may be considered as particles possessing an effective mass and interacting by a potential approaching the Coulomb law $e^2/\varepsilon r$ at distances exceeding a few lattice constants (ε being the dielectric constant of the crystal). Such an interaction involves inevitably the possibility of formation of electron-hole "atoms" (excitons), exciton molecules (biexcitons) and a condensed phase of electrons and holes [1–26]. Because of the high values of dielectric constants and small electron and hole effective masses, the characteristic atomic unit of length of the system, the effective Bohr radius $r_h = \varepsilon \hbar^2/e^2 m^*$ (where $m^{*-1} = m_e^{-1} + m_h^{-1}$, m_e and m_h being the effective masses of the electron and hole), is several tens of lattice constants. Therefore the characteristic energies in the exciton system turn out to be three to four, and the concentrations five to six orders of magnitude smaller than in real atomic systems. Indeed, the region of "metallic" concentrations ($r_h n^{1/3} \approx 0.3$) in the exciton system is reached at very low concentrations (10^{12}–10^{17} cm^{-3} for typical semiconductors). The effective mass anisotropy, the possibility of varying the electron to hole mass ratio and the many valley structure make semiconductors a very convenient vehicle for studying the properties of many-electron systems.

The most suitable subjects for investigation of many-electron phenomena are indirect semiconductors characterized by large carrier lifetimes. In germanium the lifetime of current carriers (10^{-5}–10^{-4} sec) is larger than the times of carrier thermalization and of exciton and biexciton formation, resulting in a possibility of attaining thermodynamic equilibrium between the current carriers, excitons and biexcitons. For direct semiconductors these times are at best comparable. Theoretical estimates show the exciton binding energy in both the exciton molecule and condensed phase to be at least a few tenths of the exciton ionization energy [31, 32, 54, 55] making such states observable at helium temperature.

Many authors have reported on the experimental observation of biexcitons and the exciton condensate in germanium and silicon [3–26]. Experimental data obtained by different authors exhibit considerable discrepancies. One can distinguish two major viewpoints on the nature and experimental manifestation of many-electron phenomena in these semiconductors. In the first, the broad long wavelength lines in the luminescence spectra of germanium ($h\nu_{max} = 0.709$ eV) and silicon ($h\nu_{max} = 1.08$ eV) are due to radiation recombination of exciton molecules [3–9]. The second hypothesis ascribes these lines to the exciton condensate.

Assuming the first viewpoint, the binding energy of the exciton molecules in germanium turns out to be $3.5 \cdot 10^{-3}$ eV, the density of electron-hole pairs in the condensate being $(2-4) \cdot 10^{16}$ cm^{-3}. The appearance of a condensed phase it noted only at average concentrations of electron-hole pairs exceeding 10^{15} cm^{-3} whereas the lines belonging to the exciton molecules are observed at sufficiently low temperatures already at a concentration of 10^{12} cm^{-3}.

According to the second viewpoint, one has to assume the binding energy of exciton molecule to be less than $0.5 \cdot 10^{-3}$ eV, while the density of the condensed phase turns out to be $2.6 \cdot 10^{17}$ cm^{-3}. The condensation process is assumed to be so fast as to permit practically all excitons to condense below 2 K even at average concentrations less than 10^{12} cm^{-3}. A comprehensive review of experimental data supporting this viewpoint is given in the paper by V. S. Bagaev in the present book.

In this review, we will assume the line peaked at 0.709 eV and its TO and TA phonon replicas to belong to excitonic molecules, although this interpretation meets more or less serious difficulties. The first two sections of the review deal with the experimental data concerning the metallization and condensation of excitons in germanium and with some properties of the exciton condensate. The third section reviews experimental data leading us to the conclusion that condensation proceeds from a dielectric phase representing a gas of exciton molecules. The last section contains a brief comparison of existing viewpoints on the nature of the condensed phase and a review of theoretical papers on the subject. A meaningful comparison of results obtained by different authors or even by the same author in different experiments is made difficult by the absence of reliable data on the average concentration of electron-hole (bound or free) pairs at which the experiment was carried out. In the majority of cases where the electron-hole pair concentrations are specified, they were obtained through a preliminary measurement of photoconductivity in the samples under study at room temperature under the same conditions of illumination as used in the subsequent measurements performed at liquid helium.

1. Conductivity and Microwave Absorption by the Exciton Condensate in Germanium

A number of papers [27–30] discuss the transition to the metallic type of conduction at high exciton densities in germanium. Studies were carried out on samples with thicknesses much less than the carrier diffusion length, the area between the contacts being illuminated uniformly by a pulsed light source. The only conductivity found at average exciton concentration less than 10^{16} cm^{-3} is due to electrons and holes which are not bound into excitons. The magnitude of this conductivity is weakly dependent on concentration. At a concentration of $2 \cdot 10^{16}$ cm^{-3} the conductivity increases sharply (Fig. 1). The magni-

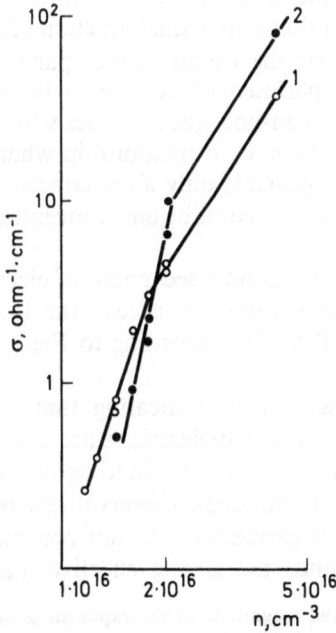

Fig. 1. Conductivity of germanium as function of concentration of electron-hole pairs. 1 $T = 4.2$ K, 2 $T = 2$ K [29]

tude of the conductivity in the region of steep growth fluctuates strongly [29]. The data of Fig. 1 are averaged over a larger number of light pulses. After the steep growth, the conductivity rises smoothly with concentration while fluctuations typical of the first region disappear. The experimental dependence of conductivity on concentration follows here the law

$$\sigma \sim n^3/T^2 . \tag{1}$$

At a concentration of $2 \cdot 10^{17} \, \text{cm}^{-3}$ and $T = 2 \, \text{K}$ the carrier mobility reaches $10^6 \, \text{cm}^2/\text{V sec}$.

The region of steep conductivity growth has been interpreted [27–30] as due to Mott transition in the exciton system. Further studies to be discussed later showed the metal-dielectric transition to be accompanied here by a vapor-liquid type phase transition.

Relation (1) is very close[1] to one that could be expected for the conductivity of a strongly degenerate electron-hole gas [33]. As is well-known, in this case

$$\sigma = e\mu n \sim [E_{Fn} E_{Fp} n/(kT)^2] \sim n^{7/3}/T^2 \tag{2}$$

where E_{Fn} and E_{Fp} are Fermi energies of the electrons and holes, respectively.

The reason for the observed carrier mobility increase with concentration consists in that only a small fraction of carriers contained in a layer of width kT near the Fermi surface participate in the scattering process. Thus the experiments lead one to believe that the degree of degeneracy of the electron-hole gas increases with average concentration within the region of the σ vs. n relationship where formula (1) is valid. Naturally this can be possible only if the concentration in the electron-hole plasma exceeds the equilibrium concentration in the assumed condensed phase.

This implies that if a condensed phase of electrons and holes really does exist, its density should not exceed the lowest concentration at which formula (1) still holds. According to Fig. 1, this concentration is close to $2 \cdot 10^{16} \, \text{cm}^{-3}$.

There are arguments [34] indicating that a degenerate electron-hole gas can still become a dielectric, since even a strongly screened Coulomb interaction can result in a binding of the electrons and holes, located near the Fermi surfaces. Observations of clearly pronounced metallic conduction in germanium do not contradict the results of the paper in question, since the strong effective mass anisotropy results

---

[1] Some disagreement in the magnitude of the exponent is here apparently due to the fact that formula (2) was derived for a strongly degenerate electron-hole gas when $n^{1/3} r_h \gg 1$, while in fact $n^{1/3} r_h \approx 1$ even for the highest concentrations attained in the experiment in question.

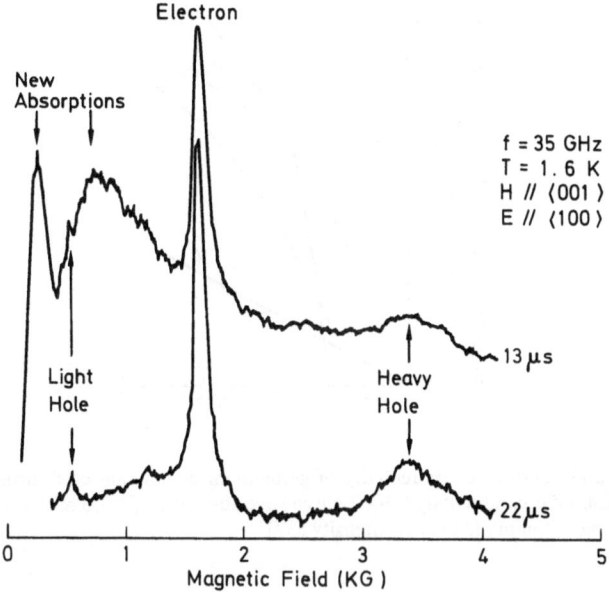

Fig. 2. Time-resolved cyclotron resonance spectra at 1.6 K [38]

in a practically zero gap in the electron and hole spectrum [35]. It is not clear, however, to what extent the results of the calculations can be used in the "metallic" region of concentrations ($r_h n^{1/3} \approx 1$).

Microwave conductivity measurements [36–39] show only a very weak absorption due to unbound carriers to exist at electron-hole pair concentrations below 10^{15} cm^{-3}. In this concentration region one observes undisplaced cyclotron resonance lines [38, 39], which indicates that the free carrier concentration is much less than the average concentration of electron-hole pairs. At a higher average concentration a strong microwave absorption appears the intensity of which sharply rises with the concentration. Simultaneously in a magnetic field a broad magnetoplasma resonance line appears. This new line coexists with the slightly broadened cyclotron resonance lines within a relatively wide concentration range. Such a shape of the spectrum is to be expected for a two-phase system consisting of regions with low and high free carrier concentration. The carrier concentration in the strongly conducting phase is shown to be 10^{17} cm^{-3} within an order of magnitude [38], which does not disagree with the earlier estimates. Figure 2 presents cyclotron resonance spectra before and after the threshold of metallic

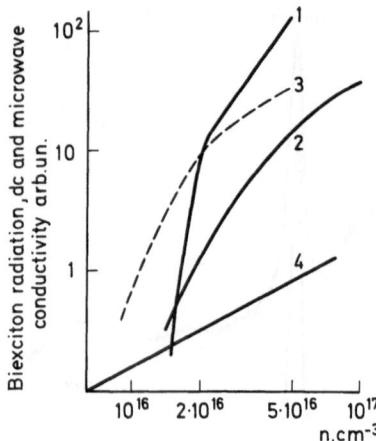

Fig. 3. D.c. and microwave conductivity of germanium as function of electron-hole pair concentration. 1 d.c. conductivity [29], 2 microwave absorption [36], 3 microwave absorption [37], 4 biexciton luminescence intensity [37]

conduction. The result of d.c. and microwave conductivity measurements are summarized in Fig. 3.

The simplest experiment revealing the two-phase character of the exciton system at high excitation levels is the observation of giant photocurrent fluctuations in n-p junctions [11–13]. The idea of the experiment is as follows. The appearance of an electron-hole drop near a reverse-biased n-p junction should result in a current fluctuation with a total charge equal to the number of electrons and holes in the drop. The change in the shape of current pulses through an n-p junction exhibited with increasing intensity of 1 μsec-long intense light pulses used to produce nonequilibrium current carriers is shown in Fig. 4. At sufficiently low excitation levels there are no photocurrent fluctuations, the current pulse shape being determined by the diffusion and recombination of current carriers in the sample. After the onset of current fluctuations, their intensity rises sharply (Fig. 4). At sufficiently high excitation levels, when the condensed phase fills up the whole sample, the fluctuations disappear again.

The luminescence spectra recorded simultaneously with the observation of fluctuations exhibit only the emission lines belonging to the biexcitons. The experimental data presented in this section indicate that the metallic exciton phase appears at concentrations above 10^{15} cm^{-3} while the biexciton emission line is seen at much lower concentrations. Taking into account probable errors in the determination

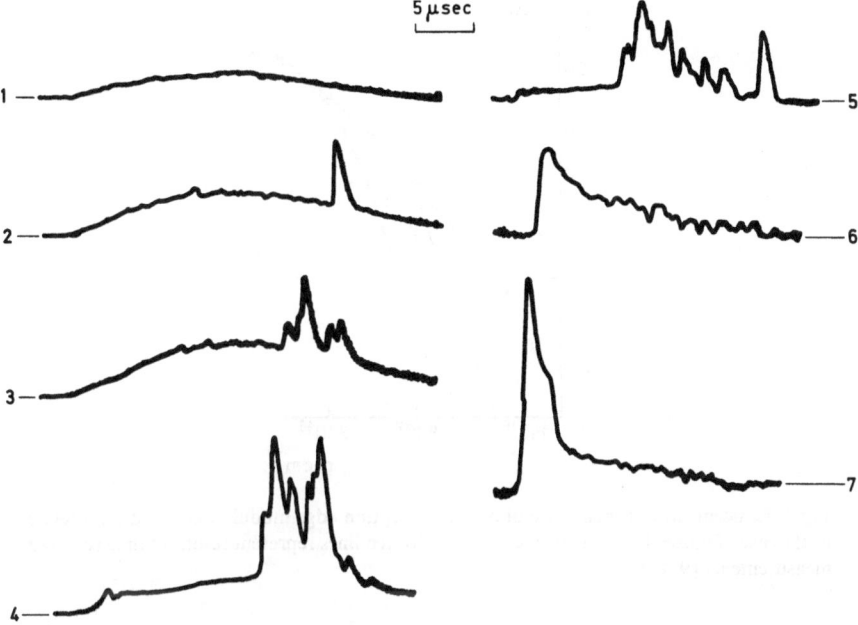

Fig. 4. Photocurrent oscillograms on n-p junction made with successive increases of excitation level. Electron-hole pair concentration, cm^{-3}: 1) $1 \cdot 10^{15}$; 2) $2 \cdot 10^{15}$; 3) $2.2 \cdot 10^{15}$; 4) $2.5 \cdot 10^{15}$; 5) $3.3 \cdot 10^{15}$; 6) $5 \cdot 10^{15}$; 7) $1 \cdot 10^{16}$ [12]

of the average electron-hole pair concentration, the density of the condensed phase derived from these experiments lies apparently within the limits $2 \cdot 10^{16}$–10^{17} cm^{-3}. The next section deals with the results of optical measurements permitting a substantial improvement of the accuracy of the determination of the condensed phase density.

2. Optical Studies

The existence of a phase transition and the density of the condensed phase were investigated [8–10] using the dependence of the shape of the direct absorption edge on current carrier concentration. At sufficiently low carrier concentrations the predominant contribution to the change of the edge shape comes from the screening of the Coulomb interaction by free current carriers [30, 40, 41]. At high concentrations, when the Coulomb interaction becomes practically "switched off", the shape of the spectra is affected considerably by the filling of the valence band

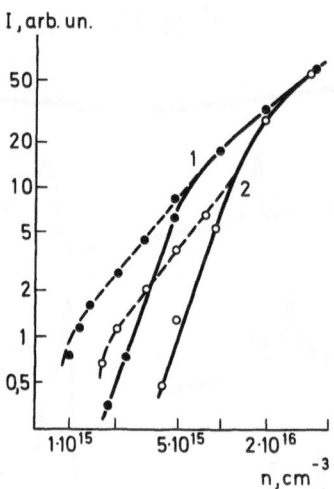

Fig. 5. Concentration dependence of direct absorption edge modulation. Solid lines relate to the case of pulsed light intensity changing. Broken lines represent results of time resolved measurements [9, 49]

edge. Since the shape of the absorption edge is in this case related fairly simply to the position of the Fermi level of holes, their concentration can be readily determined [10].

The modulation of the absorption coefficient resulting from pulsed light-induced electron-hole generation exhibits a sharp rise at pair concentration exceeding 10^{15} cm^{-3} (Fig. 5). Throughout the region of sharp rise the shape of the modulation spectra remains constant while the modulation value increases by more than 2 orders of magnitude. At concentrations above $2 \cdot 10^{16}$ cm^{-3} the rate of modulation growth slows down, the shape of the spectra becoming substantially dependent on concentration. The constancy of the modulation spectrum shape indicates a division of the sample into regions with a high and constant concentration of free current carriers and those with a low free carrier concentration. Another proof of the existence of two phases was obtained in experiments in which the spectra corresponding to different average electron-hole concentration were obtained by making measurements a certain interval of time t after the end of the exciting light pulse[2]. Naturally in the case of a homogeneous system the result of these experiments and those where the concentration was varied by changing the intensity of exciting light should coincide. However, because of the

---

[2] The concentration of electron-hole pairs in the sample decays with time by a law close to $n \sim n_0\, e^{-t/\tau}$. The time constant τ was derived from the decay of luminescence in the sample.

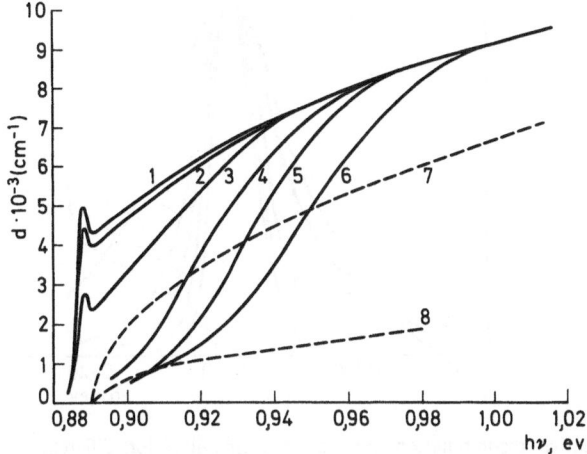

Fig. 6. Direct absorption edge spectra recorded at different concentrations of electron hole pairs (cm⁻³), $T = 1.8$ K. 1) $3 \cdot 10^{15}$; 2) $1 \cdot 10^{16}$; 3) $3 \cdot 10^{16}$; 4) $6 \cdot 10^{16}$; 5) $1.5 \cdot 10^{17}$; 6) $3 \cdot 10^{17}$; 7) The shape of direct absorption edge of germanium calculated without taking into account Coulomb interaction of electrons and holes. 8) The same for light hole band only [10]

existence of electron drops such coincidence is not observed, since while the drop formation is a sufficiently slow process with a rate depending strongly on supersaturation, the drops which have already appeared will relax with time constant close to the lifetime of electron-hole pairs in the drops. Figure 5 shows that such a phenomenon does indeed take place. In Fig. 6 we present the absorption spectra obtained at high electron-hole concentration, and in Fig. 7 the emission spectra recorded under the same conditions [10]. Curves 2 and 3 in Fig. 6 refer to the concentration range where dielectric and metallic regions exist in the sample simultaneously. This is indicated by the simultaneous existence in the spectra of the exciton line and by the considerable decrease of absorption probability in the continuum. This conclusion follows from the works [40, 41] showing that the decrease of absorption probability in the continuum at energies separated from the direct transition edge by a few exciton binding energies occurs at much higher concentrations of free carriers than the disappearance of the exciton line does. Curves 4 to 6 correspond to concentrations at which the whole sample is filled up with the degenerate electron-hole gas, where an increase in the intensity of exciting light produces increased concentration at each point in the sample. The magnitude of the spectral shift toward shorter wavelengths characterizes the Fermi energy of the holes. The

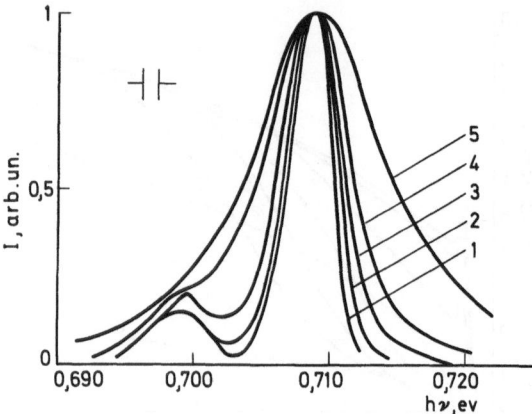

Fig. 7. Radiative recombination spectra of germanium for different concentrations, $T = 1.8$ K. 1) $2 \cdot 10^{16}$; 2) $5 \cdot 10^{16}$; 3) $1 \cdot 10^{17}$; 4) $2 \cdot 10^{17}$; 5) $4 \cdot 10^{17}$ [10]

absorption spectrum considered represents a superposition of transitions between the light and heavy hole bands and C-band at the Γ point. The concentration corresponding to curve 6 was estimated to be $3 \cdot 10^{17}$ cm^{-3} [10]. Since the total filling up of the sample by condensate occurs at concentration about 6–7 times lower than $3 \cdot 10^{17}$ cm^{-3}, the condensate density turns out to be $(4–6) \cdot 10^{16}$ cm^{-3} within a factor of about 2.

An attempt was made [10] to relate the average concentration at which metallic conduction appears to the degree of filling up the sample by the electron-hole drops. The jump in the conductivity is found to occur when 1/5 of the volume of the samples is filled up. This result is in a fair agreement with percolation theory [40]. However, to what extent this theory is applicable in this case is not clear, since in thin samples (of 10 µm) used in Ref. [10] the most probable "center" of condensation should apparently be the sample surface, while the percolation theory describes rather the case of uniform distributions of the condensation centres.

The radiative recombination spectra presented in Fig. 7 also permit us to estimate the maximum condensate density. If the line peaked at 0.709 eV were to belong to electron-hole drops with a density $2.6 \cdot 10^{17}$ cm^{-3}, its shape would be expected to remain unchanged until the average concentration of electron-hole pairs exceeds this value. This results from the fact that up to this concentrations the addition of particles to the system will produce only an increase in the volume of the condensate which cannot affect the shape of the emission spectra substantially.

On the contrary, if the line is associated with biexcitons, line broadening should be observed at much lower concentrations. At any rate, it should become noticeable at $2 \cdot 10^{16}\,\mathrm{cm}^{-3}$ where the metal-dielectric transition occurs [29]. The data of Fig. 7 show that the broadening of the biexciton line is already observable at $10^{16}\,\mathrm{cm}^{-3}$, its width doubling at a concentration of $2 \cdot 10^{17}\,\mathrm{cm}^{-3}$.

Studies were made [43, 44] of the polarization of recombination radiation from uniaxially stressed germanium. These studies showed that the 0.709 eV-line cannot belong to the exciton condensate. At a sufficiently low temperature the major part of excitons and biexcitons in a uniaxially stressed crystal will occupy the lowest level arising in the splitting of the ground state. It is essential that the magnitude of splitting sufficient for such a redistribution of particles should be close to kT. Since the split levels correspond to different degrees of radiation polarization, the preferential occupation of the lowest level by excitons or biexcitons can be established by studying the polarization of the corresponding recombination radiation as function of the temperature and of the direction and magnitude of stress. Radiative recombination in germanium occurs mainly by transitions through the Γ-minimum of the C-band involving emission of a longitudinal acoustic phonon. Therefore if the exciton ground state were fourfold degenerate in the hole momentum, then the shift of the C-band extrema due to uniaxial stress would not produce polarization of exciton luminescence, which would occur only due to the valence band splitting. However, because of the anisotropy in the exciton effective mass, the exciton ground state is split by $\Delta_c \approx 1$ meV, so that substantial polarization appears already at stresses where the splitting in the C-band is of order kT, while that of the valence band is negligible. To obtain such a splitting of the C-band in the $\langle 111 \rangle$ direction at helium temperatures, stresses of a few tens of kg/cm^2 are sufficient. Similar considerations are valid also for bi-excitons, at least when the exchange energy of holes is close to or less than Δ_c [44]. Far more simpler is the interpretation of the polarization of radiation, connected with the TO phonon. In this case optical transitions take place through the L point and initial splitting Δ_c is not needed to obtain a polarization.

In contrast to biexcitons, the polarization of luminescence from the condensed phase should be observed at much higher stresses, when the valence band splitting is comparable to the Fermi energy of electrons and holes. Assuming the 0.709 eV line to be associated with the con-densate, its Fermi energy should be $5 \cdot 10^{-3}$ eV, which exceeds by tens of times the magnitude of kT at helium temperature. Noticeable polariza-tion of radiation in this case can be expected to appear only at stresses of a few hundreds of kg/cm^2. Figure 8 shows the degree of polarization

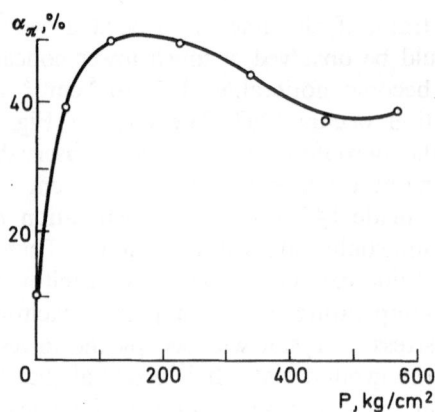

Fig. 8. Polarization of LA biexciton line as function of stress applied in the [1$\bar{1}$1] direction, $T = 2$ K [43]

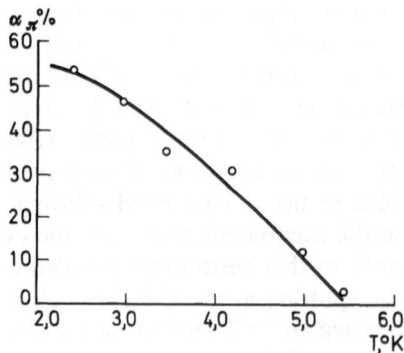

Fig. 9. Temperature dependence of biexciton recombination radiation polarization $P = 100$ kg/cm^2 [43]

of the 0.709 eV line in germanium vs. stress applied in the ⟨111⟩ direction at 1.8 K. The radiation is seen to become strongly polarized already at very low stresses. The polarization reaches the 50 % level at 100 kg/cm^2 corresponding to a C-band splitting by less than $1 \cdot 10^{-3}$ eV. At the same time one observes a strong temperature dependence of the degree of polarization (Fig. 9). An increase in the excitation level produced a decrease in the degree of polarization. At stresses lower than 200 kg/cm^2 and concentrations greater than $2 \cdot 10^{16}$ cm^{-3} the luminescence turns out to be practically completely depolarized.

Thus the results of studying luminescence polarization in uniaxially stressed germanium represents one more confirmation of the 0.709 eV line not being associated with the exciton condensate.

The next section deals with experiments showing that the 0.709 eV line results from radiative recombination of the exciton molecules.

3. Biexcitons in Germanium

The main proof of the existence of biexcitons in germanium and silicon has been obtained by studying the temperature and concentration dependences of the corresponding recombination lines [3–6]. At present there are no direct experimental data on the cross section for exciton formation in germanium. Measurement of the conductivity due to the electrons and holes not bound in exciton states yields a lower limit for this quantity of 10^{-11} cm^2. Studies of recombination radiation with continuously operating light sources are usually carried out at average electron-hole pair concentrations of the order of 10^{14} cm^{-3} corresponding to an exciton formation time not greater than 10^{-8} sec. Therefore when studying the kinetics of formation of biexcitons and of the exciton condensate, the exciting radiation can be considered to generate excitons. Since the rate of biexciton formation is proportional to n_{ex}^2 while that of their thermal dissociation or annihilation is proportional to n_b, the concentration of biexcitons should be proportional to the square of that of excitons.

Under steady-state conditions, the concentrations of biexcitons and excitons are related in a simple way:

$$n_b = \sigma v n_{ex}^2 [1/\tau_b + \sigma v (N_{ex}^2/N_b) \exp(-E_M/kT)]^{-1} \tag{3}$$

where E_M is the biexciton dissociation energy, N_{ex} and N_b are the densities of states in the exciton and biexciton bands, τ_b and σ being the biexciton lifetime and cross section of formation, respectively.

At a sufficiently high temperature, $1/\tau_b$ can be neglected, and (3) reduces to the well-known thermodynamic formula for bimolecular reactions:

$$n_b = n_{ex}^2 (N_b/N_{ex}^2) \exp E_M/kT \tag{4}$$

whose comparison with experiment permits us to derive E_M.

At low temperatures formula (3) can be used to estimate the biexciton formation cross section σ.

If the excitation is produced by short light pulses with a duration much shorter than the exciton and biexciton lifetime τ_b, one can derive simple expressions describing the relaxation process in the limiting

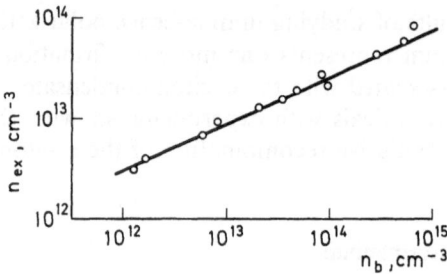

Fig. 10. Free exciton concentration as function of biexciton concentration in germanium at $T = 4.2$ K [49]

cases of $n_b \gg n_{ex}$ or $n_{ex} \gg n_b$. If, for instance, $n_b \gg n_{ex}$ and the temperature is sufficiently high so that the thermal dissociation time of biexcitons $(N_b/N_{ex}^2 \sigma v) \exp E_M/kT$ is much smaller than the time of their annihilation τ_b, then the curves of the biexciton and exciton concentration relaxation take on the form:

$$n_b = n_{b0} e^{-t/\tau_b}$$

$$n_{ex} = (n_{b0} N_{ex}^2/N_b)^{1/2} \exp(-E_M/2kT) \exp(-t/2\tau_b)$$

$$(5)$$

the relationship (4) being valid at each moment of time.

The total number of excitons and biexcitons in a sample is proportional to the intensity of the corresponding emission lines. Experiments where current carriers are produced by weakly absorbed light lend themselves most easily to an interpretation. In this case the distribution of excitons and biexcitons in the sample is not distorted by the diffusion processes. In the case of surface excitation of samples by strongly absorbed light, quantitative data can be obtained only for samples with thicknesses not exceeding the carrier diffusion length. In reference [6] nonequilibrium carriers were excited by two-photon absorption of radiation with photons of an energy of 0.53 eV generated by a pulsed dysprosium laser. Figure 10 presents the dependence of n_{ex} on n_b with n_b varying by more than three orders of magnitude, indicating it to be quadratic. The temperature dependence of $n_{ex}^2 n_b^{-1}$ shown in Fig. 11 corresponds to a binding energy E_M of the exciton molecule of $(3.3-3.6) \cdot 10^{-3}$ eV. According to (4), the quantity $n_{ex}^2 n_b^{-1}$ should not depend on the excitation level, remaining the same for all germanium samples. However, experiments show this quantity to be dependent on experimental conditions, although only to a small extent (Fig. 11). A possible explanation of this fact consists in inaccurate determination of the electron-hole pair concentration. In germanium, the quantity

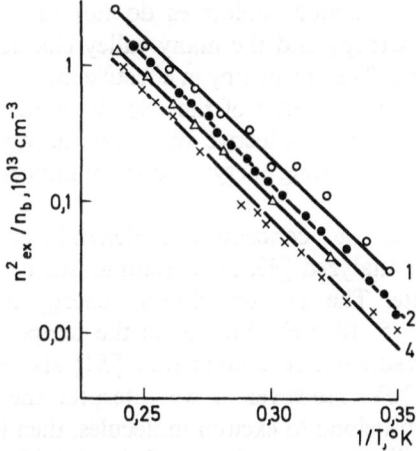

Fig. 11. Temperature dependence of $n_{ex}^2 n_b^{-1}$ in germanium for different concentrations of electron-hole pairs (cm^{-3}). 1) $1.5 \cdot 10^{14}$; 2) $1 \cdot 10^{14}$; 3) $5 \cdot 10^{13}$; 4) $3 \cdot 10^{13}$ [6]

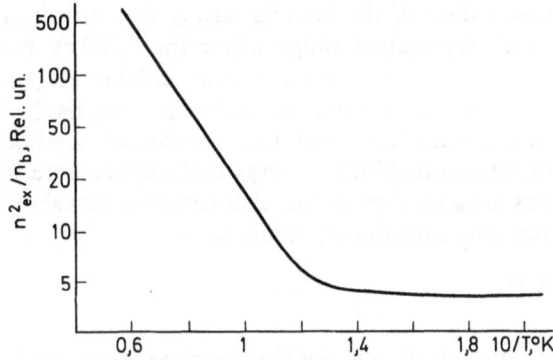

Fig. 12. Temperature dependence of $n_{ex}^2 n_b^{-1}$ for silicon [26]

$n_{ex}^2 n_b^{-1}$ ceases to be temperature-dependent at $T = 2.5$ K, hence by formula (3), the cross section for biexciton formation is 10^{-12} cm^2. Figure 12 shows the corresponding data for silicon. The magnitude of E_M in silicon turned out to be $5.5 \cdot 10^{-3}$ eV [26]. The experimental values for the binding energy of the exciton molecules in germanium and silicon exceed by about an order of magnitude the theoretical predictions obtained for an isotropic effective mass [45, 46]. Comparing the theory with experiment here one should bear in mind, however, that the existing calculations for the exciton molecules do not take into account the effective mass anisotropy and the many-valley character of the germanium band structure. The anisotropy in effective mass was shown [31, 32] to increase the binding energy of the exciton molecule, however, the anisotropy in the electron effective mass can hardly account for the observed values of the binding energy of exciton molecules in germanium and silicon.

The temperature and concentration dependences of the 0.709 eV line intensity were analyzed [23, 24] assuming this line to be due to the exciton condensate. The exciton binding energy in the condensate turned out to be $1.5 \cdot 10^{-3}$ eV. Studies of the temperature dependence of the cyclotron resonance relaxation time [54] also yield a close lying value of $1.4 \cdot 10^{-3}$ eV. However, if we interpret these data assuming the 0.709 eV line to belong to exciton molecules, then in order to obtain the biexciton binding energy, these values should be doubled. The resulting value $(2.8 - 3) \cdot 10^{-3}$ eV does not differ much from the value $(3.5-3.6) \cdot 10^{-3}$ eV discussed earlier and obtained from essentially similar measurements. A small discrepancy is apparently due to differences in the treatment of experimental data.

From these values of the binding energy one can draw some conclusions on the temperature range where the 0.709 eV line should be observed both if it belongs to the exciton condensate and to the biexcitons. When the exciton condensate appears, the level of the exciton chemical potential should lie lower than the exciton band bottom by the magnitude of the exciton binding energy in the condensate, E_c. Assuming the exciton gas to be ideal we obtain that condensation should begin at a critical exciton concentration n_{cr} equal to

$$n_{cr} = N_{ex} e^{-E_c/kT} \tag{6}$$

The "critical" concentration n'_{cr} for the biexciton formation is determined from the condition of equal numbers of excitons and biexcitons in the system. In accordance with (4), we get

$$n'_{cr} = (N^2_{ex}/N_b) e^{-E_b/kT} \approx N_{ex} e^{-2E_c/kT} \tag{7}$$

At $T = 4.2$ K we have $n_{cr} = 3 \cdot 10^{15}$ cm^{-3}, and $n'_{cr} = 10^{14}$ cm^{-3}. All the available experimental data show the 0.709 eV line at 4.2 K to appear at a concentration of 10^{13}–10^{14} cm^{-3}. Thus the probability that this line belongs to the exciton condensate with a binding energy of $1.5 \cdot 10^{-3}$ eV should be considered to be very low. The time dependence of the appearance and relaxation of free excitons and exciton molecules has also been studied [47, 48]. The decay time of free exciton emission at $T = 4.2$ K and low excitation levels was found to exceed by a factor of two that of the exciton molecule emission while the rise of both these emissions with the sample excited by short light pulses was much smaller than the lifetime. The results of these experiments agree with the above value of the cross section for biexciton formation. The shape of the 0.709 eV line was discussed [17, 24, 49] within the framework of both the biexciton and condensate model. An assumption is made that in radiative annihilation of one of the excitons making up a molecule, part of the energy can be transferred to another exciton [58], resulting in a wide emission band rather than in a narrow line[3]. For the long wavelength part of this band, there appears more likely to be the mechanism proposed originally by Haynes [3], according to which, in annihilation of one of two excitons another becomes ionized. This process should produce a long wavelength "tail" of the emission. This "tail" has been observed [6]. The binding energy of a biexciton in germanium being close to that of an exciton, the effective radius of the biexciton can be taken equal to the Bohr radius of the exciton, r_h. Then the average momentum of electron and hole in the molecule will be close to r_h^{-1}, and the average recoil energy determining the halfwidth of the emission line will be close to the exciton binding energy. The experimental data available agree with this conclusion.

Within the framework of the condensate hypothesis the shape of this line reflects the Fermi distribution of electrons and holes in the drop. One succeeds in obtaining agreement between the calculated and experimental spectra assuming the electron-hole pair concentration in the condensate to be $2.6 \cdot 10^{17}$ cm^{-3} [23, 24, 50].

The spectral shape of the line associated with emission of a TA phonon ($h\nu_{max} = 0.728$ eV) was also studied [50]. In contrast to the 0.709 eV line, this line is forbidden in the dipole approximation. If we attribute this line to the emission of the electron-hole drops, then the width of the line yields for the electron-hole pair concentration in the condensate a value of $1.8 \cdot 10^{17}$ cm^{-3}.

An analysis was made [5, 6] of the effect of nonequilibrium long wavelength phonons on the kinetics of biexciton formation. Long wavelength phonons produced in the thermalization of light-generated current carriers have lifetimes exceeding by far the lifetime of the current carriers. Such phonons accumulating at high concentrations in the sample shift the equilibrium in the direction of free exciton formation. In this case the biexciton model predicts a linear dependence of n_b on n_{ex} which was indeed observed experimentally [5, 6].

---

[3] This is possible since radiative recombination in germanium occurs with participation of a phonon carrying away the major part of quasimomentum.

4. Other Experiments and Discussion

The most convincing evidence for the exciton condensate origin of the 0.709 eV line in germanium comes from the observation of light scattering appearing simultaneously with this line [23–25]. An analysis of the scattering data yields the value $7.6 \cdot 10^{-4}$ cm for the drop radius at an average electron-hole concentration of $5 \cdot 10^{13}$ cm^{-3} [23], and (8–13) $\cdot 10^{-4}$ cm at about the same average concentration. At sufficiently low temperatures (2–2.5 K) the emission spectrum exhibits only the 0.709 eV line, which fact was noted in Ref. 25 and other papers dealing with the kinetics of generation of this radiation and discussed in the preceding section. The authors [23–25] did not measure the free exciton concentration, however the absence of the corresponding emission line in the spectra obtained probably implies that their concentration is at least 2–3 orders smaller than the average concentration of the electron-hole pairs. Thus the free exciton concentration required to maintain the existence of electron-hole drops of this size is close to 10^{12} cm^{-3}. Equating the rate of the arrival of excitons into a drop to that of their recombination in it, we obtain for the exciton concentration required for the existence of a drop of radius r_d:

$$n = \tfrac{1}{3}(r_d^2 N_d/D\tau_d) \tag{8}$$

where N_d is the concentration of electron-hole pairs in the drop, D the exciton diffusion coefficient, and τ_d the lifetime of current carriers in the drop.

At $N_d = 2.6 \cdot 10^{17}$ cm^{-3}, $D = 100$ cm^2/sec and $\tau_d = 10^{-5}$ sec, $r_d = 10^{-3}$ cm we get $n = 10^{14}$ cm^{-3}, which exceeds the expected value by 2 orders of magnitude. This difficulty can be overcome by assuming the drops to form near potential wells caused by the crystal lattice tension. The depth of such wells should be of the order of $1 \cdot 10^{-3}$ eV. The position of the "drop" line in the emission spectrum should be accordingly shifted to longer wavelengths by this value. This shift should be different in different samples while depending on excitation level in a given sample. Experimental observation of such shifts should apparently not present considerable difficulties.

Some publications [20, 21, 51] reported on the observation of far infrared absorption near 150 μm appearing simultaneously with the 0.709 eV line. The results of these experiments can at present be interpreted equally well within both the condensate and biexciton hypothesis for these lines. Performing these measurements at somewhat higher excitation intensities would be of considerable interest, since at a drop radius in excess of $\lambda/2\pi n$ (where λ is the wavelength in air, n the refractive index of germanium) the plasma absorption spectrum shifts markedly toward longer wavelengths.

Ref. [21] reported the shape of the far infrared absorption spectra to be independent of the intensity of exciting light although it was here apparently somewhat higher than in Refs. [23–25]. If this is indeed so, then the data of Refs. [19, 20] should be considered as a confirmation of the biexciton rather than of the condensate hypothesis of the origin of the 0.709 eV line.

It is possible to reach an agreement between far infrared and light scattering measurements by assuming that at low temperatures formation of smaller drops takes place and that increasing of excitation intensity leads to raise the drop number rather than the drop radius. Such an assumption is in contradiction with results of [25] in which the increase of drop radius was observed when the temperature decreased. But in more recent investigation the same authors came to the opposite conclusion.

At present the most promising experiment to determine the correct hypothesis on the nature of the condensed state would apparently be a development of the cyclotron resonance and plasma absorption studies. These experiments reveal in the most direct way possible the two-phase character of the system, providing a possibility of obtaining very reliable estimates of current carrier concentration in both phases.

The energy and density of the condensed phase have been calculated theoretically [52–55]. Some authors [54, 55] took into account the peculiarities in the band structure of germanium and silicon. The results of these calculations show the condensed state in many-valley semiconductors to possess a larger binding energy and a higher density than in the case of semiconductors with simple bands. The reasons for this are as follows. The kinetic energy of electrons in germanium is smaller than that in the corresponding one-valley semiconductor by a factor of $4^{2/3}$ while the exchange energy decreases insignificantly. The exchange energy decreases because the exchange effect occurs only between electrons in the same valley. In germanium this results in the factor $4^{-1/3}$, another factor of 0.84 arising because of a strong anisotropy of the electron effective mass. The structure of the valence band also favors an increase in the binding energy mainly because the effective spin of the hole is 3/2 and not 1/2 as in the case of the simplest band structure. A decrease in the kinetic energy favors a strong growth of the correlation energy. The ground state energy expressed in the units $E_{ex} = e^2/2\varepsilon r_h$ has the form [54]:

$$E_k = 0.468 \; r_s^{-2} - 1.136 r_s^{-1} - E_{corr}$$

where $r_s = r_h^{-1}(\frac{4}{3}\pi n)^{-1/3}$, $r_h = \varepsilon\hbar^2/e^2 m^*$, m^* is the reduced exciton mass which is $0.046 m_0$ for germanium.

The correlation energy was calculated by various modifications of the random phase approximation [56, 57]. The RPA was modified to take the spin of holes and electrons more accurately into account. The conventional procedure results in large errors at large k because of particles with parallel and antiparallel spins being considered in the same way. At r_s corresponding to the minimum binding energy of the condensate the correlation energy is approximately 1.5, which exceeds by far both the kinetic and exchange energy.

The binding energy of the exciton condensed state in germanium was found [54] to be $1.7 \cdot 10^{-3}$ eV, and the density, $1.8 \cdot 10^{17}$ cm^{-3}, Other authors [55] using an improved method of calculating the correlation energy and somewhat different data on the effective mass of heavy holes obtained $2.5 \cdot 10^{-3}$ eV and $2 \cdot 10^{17}$ cm^{-3}, respectively. These figures agree well with experiment if the 0.709 eV line is ascribed to the exciton condensate.

Experience gained in calculations of the binding energy of metals shows that the large value of correlation energy obtained for systems of this kind does not necessarily mean that the validity of calculation of Refs. [54, 55] should be questioned. However, in order for it to be valid, it is of crucial importance that there is no possibility of exciton molecule formation with a sufficiently high binding energy. Experimental data presented in Section 3 of this review appear to bear out the existence of such molecules.

The recognition of the existence of exciton molecules in germanium and silicon will require revision of the present theory of the exciton condensate.

The author is grateful to V. M. Asnin, J. C. McGroddy, G. E. Pikus and N. I. Sablina for fruitful discussions.

References

1. Gross, E. P.: Nuovo Cimento, Suppl. **4**, 672 (1956).
2. Lampert, M. A.: Phys. Rev. Letters **1**, 451 (1958).
3. Haynes, J. R.: Phys. Rev. Letters **17**, 860 (1966).
4. C. Benoît à la Guillaume, Salvan, F., Voos, M.: Proc. Int. Conf. Luminescence Amsterdam 1970, p. 315.
5. Asnin, V. M., Rogachev, A. A., Sablina, N. I.: Fiz. Tech. Poluprov. **4**, 808 (1970).
6. Asnin, V. M., Zubov, B. V., Murina, T. M., Prokhorov, A. M., Rogachev, A. A.: JETP **62**, 737 (1972).

7. Keldysh, L. V.: Proc. of the Ninth Int. Conf. on the Physics of Semiconductors, Moscow, 1968. Leningrad: Nauka 1969.
8. Asnin, V. M., Rogachev, A. A.: JETP Letters 11, 162 (1970).
9. Asnin, V. M., Rogachev, A. A.: Proc. III. Internat. Conf. Photoconductivity, Stanford University 1969, p. 13.
10. Asnin, V. M.: Fiz. Tverd. Tela 15, 3298 (1973).
11. Asnin, V. M., Rogachev, A. A., Sablina, N. I.: JETP Letters 11, 162 (1970).
12. Asnin, V. M., Rogachev, A. A., Sablina, N. I.: Fiz. Tverd. Tela 14, 399 (1972).
13. C. Benoît à la Guillaume, Voos, M., Salvan, F., Laurant, I. M., Bonnet, A.: Compt. Rend. 272, 236 (1972).
14. Pokrovsky, Ya. E., Svistunova, K. I.: JETP Letters 9, 435 (1969).
15. Kaminsky, A. S., Pokrovsky, Ya. E.: JETP Letters 11, 381 (1970).
16. Pokrovsky, Ya. E., Svistunova, K. I.: Fiz. Tekh. Poluprov. 4, 491 (1970).
17. Pokrovsky, Ya. E., Kaminsky, A. S., Svistunova, K. I.: Proc. X. Internat. Conf. Phys. Semicond, Cambridge, Massachusetts, 1970, p. 504.
18. Bagaev, V. S., Galkina, T. I., Gogolin, O. V., Keldysh, L. V.: JETP Letters 10, 309 (1969).
19. Bagaev, V. S., Galkina, T. I., Gogolin, O. V.: Proc. X. Internat. Conf. Phys. Semicond, Cambridge, Massachusetts, 1970, p. 500.
20. Vavilov, B. S., Zayats, V. A., Murzin, N. I.: JETP Letters 10, 304 (1969).
21. Vasilov, B. S., Zayats, V. A., Murzin, N. I.: Proc. X. Internat. Conf. Phys. Semicond, Cambridge, Massachusetts, 1970, p. 509.
22. Ashkinadze, B. M., Crecu, I. P., Ryvkin, S. M., Yaroshetsky, I. D.: JETP 58, 507 (1970).
23. Pokrovsky, Ya. E., Svistunova, K. I.: JETP Letters 13, 297 (1971).
24. Pokrovsky, Ya. E.: Proc. XI. Intern. Conf. Phys. Semicond. Warsaw 1972, p. 69.
25. Sibeldin, H. I., Bagaev, B. S., Tsvetkov, V. A., Penin, N. A.: Fiz. Tverd. Tela 15, 177 (1973).
26. Asnin, V. M., Lomasol, Yu. N., Rogachev, A. A.: Fiz. Tverd. Tela. 14, 3457 (1972).
27. Asnin, V. M., Rogachev, A. A., Ryvkin, S. M.: Fiz. Tekh. Polupr. 1, 742 (1967).
28. Asnin, V. M., Rogachev, A. A.: JETP Letters 7, 467 (1968).
29. Asnin, V. M., Rogachev, A. A.: JETP Letters 14, 494 (1972).
30. Rogachev, A. A.: Proc. IX. Internat. Conf. Phys. Semicond. Moscow 1968, p. 407. Leningrad: Nauka 1969.
31. Brinkman, W. F., Rice, T. M., Bell, B.: In press.
32. Akimoto, O.: In press.
33. Barber, W. G.: Proc. Roy. Soc. 158, 383 (1937).
34. Keldysh, L. V., Kopaev, Yu. V.: Fiz. Tverd. Tela. 6, 2791 (1964).
35. Kozlov, A. N., Maksimov, L. A.: JETP 48, 1184 (1965).
36. Ashkinadze, B. M., Sultanov, F.: JETP Letters 16, 271 (1972).
37. Gladkov, P. S., Zhurkin, B. G., Penin, N. A.: Fiz. Tekh. Polupr. 6, 1919 (1972).
38. Sanada, T., Okyama, T., Otsuka, E.: Solid State Commun. 12, 1201 (1973).
39. Gladkov, P. S., Ginodman, V. B., Zhurkin, B. G., Mikchalev, V. G., Penin, N. A., Shipulo, G. P.: Short. Commun. Phys. 12, 17 (1971).
40. Asnin, V. M., Rogachev, A. A.: Phys. Stat. Sol. 20, 755 (1967).
41. Asnin, V. M., Rogachev, A. A., Erictavy, G. L.: Phys. Stat. Sol. 29, 443 (1968).
42. Skal, A. S., Shklovsky, B. Y., Efros, A. L.: JETP Letters 28, 242 (1973).
43. Asnin, V. M., Lomasov, Yu. N., Rogachev, A. A.: JETP Letters 18, 242 (1973).
44. Bear, G. L., Pikus, G. E.: JETP Letters 10, 745 (1973).
45. Hylleras, E. A., Ore, A.: Phys. Rev. 71, 493 (1947).
46. Hanamura, E., Akimoto, O.: Sol et Com. 10, 253 (1972).
47. Asnin, V. M., Rogachev, A. A., Sablina, N. I.: Fiz. Tekh. Polupr. 5, 1846 (1971).
48. Zubov, V. B., Kalinushkin, V. P., Murina, T. M., Prokhorov, A. M., Rogachev, A. A.: Fiz. Tekh. Polupr. 7, 1614 (1973).

49. Rogachev, A. A.: Proc. Intern. Conf. Luminescence, Leningrad 1972. Izv. Akad. Nauk. SSR Ser. Fiz. **37**, 229 (1973).
50. C. Benoît à la Guillaume, Voos, M.: Phys. Rev. **13.7**, 1728 (1973).
51. Murzin, V. N., Zayatz, V. A., Konepenko, N. L.: Proc. XI. Intern. Conference Phys. Semicond. Warsaw, 1972.
52. Hanamura, E.: Proc. X. Internat. Conf. Semicond. Cambridge, Massachusetts, 1970. p. 487.
53. Inoue, M., Hanamura, E.: J. Phys. Soc. Japan **34**, 652 (1973).
54. Brinkman, W. F., Rice, T. M.: Phys. Rev. **37**, 1508 (1973).
55. Combescot, M., Nozieres, P.: J. Phys. C**5**, 2369 (1972).
56. Hubbard, J.: Proc. Roy. Soc. London A**243**, 336 (1957).
57. Nozières, P., Pines, D.: Phys. Rev. **111**, 442 (1958).
58. Grun, J. B., Nikitine, S., Bivas, A., Levy, R.: J. Luminescence **1**, 244 (1970).

Prof. Dr. A. A. Rogachev
A. F. Ioffe Physical-Technical Institute
Academy of Sciences of USSR
Leningrad/USSR

V. Special Optical Properties of Excitons at High Density

Gigantic Oscillator Strengths Inherent in Exciton Complexes

E. I. RASHBA

Contents

1. Introduction

In spectra of semiconductors with direct transitions the intrinsic absorption, i.e. exciton bands, is preceded by a number of narrow and weaker lines [1, 2]. For brevity, in what follows we shall designate these lines as a forespectrum. The presence of forespectra is a general rule. They have been observed in many semiconductors, and in some cases (for instance, in CdS and CdSe) the forespectra consist of about a dozen of lines. The analysis of the experimental data on CdS [3] showed that some forespectrum lines arise due to various defects of the crystal lattice (impurities, structure defects) to which the excitons may be bound. Such entities are called "impurity excitons", or exciton-impurity complexes. Later [4] the assignment of some forespectrum lines was proposed as corresponding to definite types of defects. Similar bands have been earlier discovered and interpreted in the ionic crystal spectra [5] (there they are known as α- and β-bands).

Very high intensity is an important peculiarity of the forespectrum lines. Very often their intensity is about $\sim 10^{-3} \div 10^{-2}$ of that of the exciton lines even in "pure" crystals, i.e. in the undoped crystals, where the impurity concentration could not have been extremely high. This anomalously high forespectrum intensity was sometimes even the cause of certain confusions; for instance, some CdS forespectrum lines were interpreted first [6] as intrinsic exciton bands.

Soon, however, the high oscillator strength has been shown [7] to be a natural property of impurity excitons in semiconductors. In paper [7] the results obtained earlier for Frenkel excitons were extended to Wannier-Mott excitons: gigantic changes in the oscillator strength of the impurity molecular excitons have been theoretically predicted [8] and experimentally discovered [9] under the conditions when the impurity exciton level approaches the edge of the exciton band.

The effect under consideration may be explained as follows: the wave function of the exciton weakly bound to the defect involves a large region around the defect, and, speaking the classical language, the oscillations of the electronic polarization are coherent over the whole region. Thus, in case the transition to the bottom of the exciton band is allowed, the impurity exciton oscillator strength is proportional to the volume covered by the envelope function of the impurity exciton, or, which is just the same, it is proportional to $E_i^{-3/2}$, where E_i is the binding energy of the impurity exciton with a defect. A model calculation given below results in the following approximate formula for the impurity exciton oscillator strength per single defect:

$$f_i = (E_0/E_i)^{3/2} f_{ex}, \tag{1}$$

where f_{ex} is the oscillator strength of the intrinsic exciton absorption per primitive cell, and E_0 is the value characterizing the exciton band width, which is of atomic order of magnitude ($\sim 1 \div 10 \, \text{eV}$) in semiconductors.

In a number of semiconductors $E_i \sim 10^{-2} \, \text{eV}$, and, therefore, $(E_0/E_i)^{3/2} \sim 10^4$. Consequently, f_i should be large and exceed f_{ex} by several orders of magnitude. Such gigantic oscillator strengths and short times of the radiative recombination $\tau_i \sim 10^{-9}$ sec connected with them have been already discovered in some semiconductors; the experimental data are discussed below.

It follows from the foregoing that the arising of the gigantic oscillator strengths in itself is not connected with any specific properties of the defect potential – it is only of importance that E_i should be small enough. It becomes then obvious that they must appear not only when the exciton is bound to a statical defect, but also when it is bound to some other quasiparticles.

At present two types of such complexes involving excitons may be mentioned. The first type includes purely electronic complexes, for instance, biexcitons [10]. The second one includes the exciton-phonon complexes, i.e., the exciton and optical phonon bound states: it has been proved recently that such entities may exist (see review papers [11, 12]).

When real optical phonons are present in the crystal, there must be an exciton-phonon complex photoproduction channel, the exciton created by light being bound to one of these phonons already in the photo-transition process. When excitons are present in the crystal, a similar channel of the biexciton production should exist. The frequency of the corresponding transitions must be somewhat less than the exciton absorption frequency, and it is essential to note that the oscillator strength in this channel must be gigantic. Since at present it is possible to create considerable concentrations of excitons, and, according to [13], of non-equilibrium optical phonons, the processes under consideration may be thought to be accessible for the experimental observation. We hope that the discussion of these problems given below would stimulate the corresponding experiments.

2. Exciton-Impurity Centers

a) Theory

In principle, the theory of the optical transitions in the exciton-impurity centers is very simple. It may be found in the papers [7, 14, 15]; their results being essentially identical.

The semiconductors with direct allowed transitions are of greatest interest for us, since the oscillator strength for the impurity excitons is maximum for them. Let us first confine ourselves to the defects of the isoelectronic impurity, or the charged donor (or acceptor) type – the influence of such a defect on the exciton may be reduced to a certain potential acting upon the electron and the hole.

Let $\Psi(k_e, k_h)$ stand for the envelope function (i.e., the wave function of the effective-mass method) of the exciton in the momentum representation; the bands, for simplicity, are taken to be nondegenerate. Since for the allowed transitions the interband matrix element practically does not depend on the momentum, and due to momentum conservation at the optical transition the total momentum of the arising quasiparticles is equal to zero, the transition matrix element describing the creation of a band or impurity exciton will be the following:

$$M \propto \sum_{k_e} \Psi(k_e, -k_e). \tag{2}$$

In the coordinate representation we have

$$M = C \int dr_e \, \Psi(r_e, r_e). \tag{3}$$

Here $\Psi(r_e, r_h)$ is the wave function of the exciton in a coordinate representation, and C is a certain constant depending on the probability of the interband optical transition.

It is easy to express the oscillator strength of the impurity exciton in terms of the oscillator strength of the band exciton. For the band excitons with the momentum K we have:

$$\Psi_K(r_e, r_h) = \Phi_{ex}(r) \cdot V^{-1/2} \exp(iKR), \tag{4}$$

$r = r_e - r_h$, R is the coordinate of the center-of-mass, V is the volume of the crystal. The matrix element does not vanish at $K = 0$ and equals [16]:

$$M_{ex} = C V^{1/2} \Phi_{ex}(0). \tag{5}$$

Using (3) and (5) it is easy to relate the oscillator strength of the impurity exciton f_i (per one center) to that of the intrinsic exciton absorption f_{ex} (per primitive cell):

$$f_i = v^{-1} \Phi_{ex}^{-2}(0) \left[\int dr_e \, \Psi_i(r_e, r_e) \right]^2 f_{ex}, \tag{6}$$

here v is the volume of the primitive cell. In (6) we neglected the difference in frequencies of the impurity and band exciton lines (ω_i and ω_{ex}).

It goes without saying that $\Psi_i(r_e, r_h)$ cannot be found in a general form, since it is a wave function of a three-particle problem. Therefore, it is reasonable first to make a rough estimation of the right hand side of (6) to get an impression about the order of its value. Suppose the regions of the electron and hole localization in the complex are of the same order of magnitude as the Bohr radius of the exciton a_{ex}. Then due to the normalization conditions $\Psi_i(0, 0) \sim a_{ex}^{-3}$, the integral in the numerator is of the order of 1, whereas $\Phi_{ex}^2(0) \sim a_{ex}^{-3}$; therefore,

$$f_i \sim (a_{ex}^3/v) f_{ex}. \tag{7}$$

Equation (7) represents the basic qualitative result: since for the large radius centers $a_{ex}^3/v \gg 1$, then $f_i \gg f_{ex}$. As a characteristic value of this factor we may give the following estimation $a_{ex}^3/v \sim 10^4$; therefore, the difference between f_i and f_{ex} is rather large.

The factor a_{ex}^3/v indicates the number of primitive cells inside the ψ-cloud of the complex. This makes it possible to understand the physical sense of Eq. (7): coherent oscillations of the electronic polarization arise in the whole region involved, and, as a result, an essential part of the total oscillator strength of the exciton absorption confined in the volume $\sim a_{ex}^3$, displays itself in the absorption on the "impurity" frequency. As was mentioned in [17], the exciton-impurity center behaves here like an antenna.

Note that $f_{ex} \propto \Phi_{ex}^2(0)$ contains the reverse factor $v/a_{ex}^3 \ll 1$, and that is why $f_{ex} \ll 1$ for the large radius band excitons. In the formula for f_i these factors are cancelled, and usually $f_i \sim 1 \div 10$.

There is one model, quite reasonable from the physical point of view, which allows to calculate very easily the right side of Eq. (6). It corresponds to an exciton weakly bound to the impurity center by a short-range potential. Apparently, it best of all describes the excitons bound to isoelectron centers. When the exciton is weakly bound to the center, i.e., $E_i \ll E_{ex}$, where E_{ex} is the exciton ionization energy, it may be regarded as a quasi-particle which is moving in the field of the center as a single entity, the radius of this state exceeding that of the internal motion in the exciton. Using the deuteron approximation in which the wave function is completely determined by the coupling energy, we obtain ($\hbar = 1$):

$$\Psi_i(r_e, r_h) = \Phi_{ex}(r)\, \varphi_i(R), \quad \varphi_i(R) = \sqrt{\varkappa/2\pi}\,(e^{-\varkappa R}/R), \varkappa = |2mE_i|^{1/2}, \qquad (8)$$

where m is the exciton mass. Taking account of (8) Eq. (6) may be transformed into

$$f_i = v^{-1}\left(\int \varphi_i(R)\, d R\right)^2 f_{ex}. \qquad (6a)$$

Equation (1) follows immediately from (6a) and (8) with

$$E_0 = (2/m)\,(\pi/v)^{2/3}. \qquad (9)$$

The mass m should be replaced in (9) by $(m_1 m_2 m_3)^{1/3}$ for ellipsoidal bands. It is convenient to rewrite (1) and (9) as follows

$$f_i = 8(\mu E_{ex}/mE_i)^{3/2}\,(\pi a_{ex}^3/v)\, f_{ex}, \qquad (10)$$

where μ is the exciton reduced mass. The estimation following from (10) agrees quite well with (7). Equation (10) is often used for handling the experimental data owing to its simplicity.

The spectra of excitons bound to ionized donors are very interesting from the experimental point of view: in this case the complex contains a positive Coulomb center, an electron, and a hole. The approximation (8) is rather rough for such centers, and, therefore, variational calculations have been made (with 5-parameter [18], and 19-parameter [19] functions). According to these calculations the bound state of the complex exists at $m_e/m_h < (m_e/m_h)_{cr} = 0.426$; at larger m_e/m_h the hole is transferred into a continuous spectrum. At $m_e/m_h \gtrsim 0.2$

$$f_i \approx 50\,(\pi a_{ex}^3/v)\, f_{ex}. \qquad (11)$$

On the other hand, if we make use of Eq. (10) and the fact that at $m_e/m_h \ll 1$ the binding energy will be $E_i \approx 0.2\,E_{ex}$, we shall obtain a formula of type (11), but with almost a doubled numerical coefficient. Thus, at small m_e/m_h the application of function (8) results in values of

f_i which are nearly twice as much as the true ones; so, it is natural, that the description of the qualitative picture is correct.

The difference, however, increases when m_e/m_h approaches the critical value; in this range the complex converts into a neutral donor, a weakly bound hole moving around it. Since in approaching the ionization threshold the overlapping of the electron and hole ψ-functions decreases progressively, the oscillator strength decreases, too. The same tendency follows from the numerical calculations also; for instance, at $m_e/m_h \approx 0.4$ the ratio f_i/f_{ex} is less by an order, than it should be according to Eq. (10). An analytical behaviour of f_i in the immediate vicinity of the dissociation threshold is not yet investigated. It can be summarized from the foregoing that the total dependence of f_i on E_i for the ionized donor appears to be opposite to that which follows from (10): the oscillator strength decreases monotonously (but does not increase!) with the decreasing of E_i. Therefore, care should be taken in using the numerical results obtained from (10) in applying them to the ionized donors.

As for the numerical values f_i/f_{ex} for the ionized donors resulting from the quantitative calculations [19], they are large for the most interesting materials. For CdS and ZnO this ratio ranges between $4 \cdot 10^4 \div 10^5$ and $10^4 \div 4 \cdot 10^4$ depending on the used numerical values of the parameters of these materials (with this the values of f_i are in the limits $10^2 \div 3 \cdot 10^2$ and $30 \div 110$, respectively).

Excitons bound to neutral donors are more complex entities; in this case the complex is a four-particle one and involves a positive Coulomb center, a hole, and two electrons. The quantitative theory of the optical spectra of such centers is not likely to be developed, though some calculations of the binding energy have been performed [20]. It is of importance that the lowest ionization potential of such a center corresponds to the removing of the exciton as a whole. One may therefore hope that a model solution based on function (8) when applied to the four-particle centers will happen to be more precise, than for the three-particle centers, and formula (10) will be fairly efficient for estimating the oscillator strengths.

The transition matrix element (3) for semiconductors with direct forbidden transitions is determined by the integral $\int dr_e [V_{r_h} \Psi(r_e, r_h)]_{r_e=r_h}$. The calculation details are, thus, changed, though all the basic results are retained. On the other hand, in the crystals with indirect transitions the oscillator strengths for the band and impurity excitons are comparable.

Gigantic oscillator strengths manifest themselves naturally not only in absorption, but also in other optical effects. First of all it concerns the radiative lifetime of the excited state, since $\tau_i \propto f_i^{-1}$. The general formula connecting τ_i and f_i is complicated for anisotropic crystals. If we assume, however, that ε_1, the real part of the dielectric constant, is isotropic, then

$$\tau_i = \tfrac{3}{4}(m_0/e^2 f_i) \frac{\varepsilon_1 + d(\varepsilon_1 \omega)/d\omega}{\kappa^2 d\kappa/d\omega}, \qquad \kappa = n\omega/c, \qquad n^2 = \varepsilon_1 ; \tag{12}$$

m_0 is the mass of the free electron. Equation (12) can be obtained if the well-known Einstein arguments are applied to dispersive media, and the proper expression [21] for the electromagnetic energy density is used. In this case the impurity absorption may be anisotropic (as is the case in the compounds $A_{II}B_{VI}$); only the impurity contribution into ε_1 should not be large.

The oscillator strength f_i is determined by the imaginary part of ε, connected with the impurity absorption, by the relation

$$f_i = (m_0 \omega/2\pi^2 e^2 N_i) \int \varepsilon_2(\omega) \, d\omega ; \tag{13}$$

the integration, as usual, is carried out over the whole impurity band. N_i is the impurity concentration.

Equation (12) takes into account the dispersion of $\varepsilon_1(\omega)$ which may be fairly essential, since the region under consideration is positioned in the vicinity of the intrinsic absorption edge. If, however, the dispersion is neglected, Eq. (12) may be simplified as follows

$$\tau_i = 3m_0 c^3/2e^2 n\omega^2 f_i . \tag{14}$$

Equations (12) and (14) lead to the estimation $\tau_i \sim 1$ nsec for typical values of the parameters. Such small radiative lifetimes provide a high yield of the low-temperature fluorescence, since the rate of the radiative recombination appears to be higher than that of the principle non-radiative processes (including Auger-processes [22], which predominate very often in the indirect band crystals). They also highly affect the criterion of the generation regime.

In case the exciting light frequency is close to the energy of the impurity excitons, the light scattering by the impurities becomes the resonant one and its cross-section highly increases. As usual, the mag-nitude of the cross section is proportional to the square of the oscillator strength for the transition into the intermediate state. Since here is involved as intermediate the state of the impurity exciton for which the oscillator strength is large, the scattering cross-sections are gigantic. This statement can be equally referred to the Rayleigh and Raman scattering (including spin-flip scattering, phonon scattering and so on).

For an example we give here the cross-section of the resonance spin-flip Raman scattering on neutral donors in crystals of the hexagonal CdS type. It is supposed that 1. the light frequency ω is so close to that of the impurity exciton ω_i, that all other intermediate states may be neglected, 2. the spin-orbit splitting of the valence band is large, 3. the incident light propagates along the hexagonal axis c, the backward scattering being investigated and, 4. the magnetic field H is tilted under the angle θ to the axis c. Then the differential cross-section is [17]:

$$d\sigma/d\Omega = (d\sigma/d\Omega)_0 \sin^2 \theta, \quad (d\sigma/d\Omega)_0 = (r_0^2/4) f_i^2 \left(\frac{\omega}{\omega - \omega_i}\right)^2,$$

$$r_0 = e^2/m_0 c^2 ;$$

(15)

here $(d\sigma/d\Omega)_0$ is the Rayleigh scattering cross-section in the same geometry. From (15) it is seen that $d\sigma/d\Omega$ may exceed considerably the Thomson cross-section not only due to the resonance factor, but also to the factor f_i^2.

b) Experimental Data

It should be noted that in a brief discussion of the experimental data on the gigantic oscillator strengths given below, we do not see our aim in making a detailed quantitative comparison of the theory with the experiment. As could be seen in the foregoing, the present theory may claim only a description of the basic qualitative regularities and the estimation of the absolute values of the oscillator strengths. The accuracy of the experiments is still relatively low also, and their handling in some cases is of a tentative character. Our purpose is to illustrate to what extent the experiment confirms the theoretical statements formulated in Section 2a, and just how the effect under consideration is pronounced in various materials. Namely this aspect of the matter is important for us below.

Most detailed data are available now on the CdS spectra. High intensity of lines corresponding to various complexes in this crystal has been interpreted in [7] as the first experimental manifestation of the presence of the gigantic oscillator strengths. Then the cross-section of the resonance spin-flip Raman scattering of the 4880 Å argon laser line has been measured on neutral donors [17]. It happened to be about $4 \cdot 10^{-19}$ cm^2/ster, that according to (15) with $\omega - \omega_i \approx 5$ meV results in the value $f_{I_2} \approx 9$ for the I_2 line corresponding to the neutral donors. The next achievement is connected with the direct experimental determination of τ_i for the two four-particle complexes; it has been performed with high accuracy by means of a special technique which made it possible to separate the times of the formation of complexes and their radiative decay [15]. It has been found that $\tau_{I_1} = 1.03$ nsec, and

Table 1. Summary of the experimental and theoretical values of f_i and τ_i for four types of the exciton-impurity complexes in CdS. Theoretical values of f_i are calculated by formulas (1) and (9), whereas τ_i is determined by formula (14) where f_i (theor.) is used. In the calculations the following values of parameters were used: $m = 1.6\,m_0$, $v = 49.4$ Å, $f_{ex} = 2.56 \cdot 10^{-3}$, $n = 3.05$

Complex	ω_i eV	E_i meV .	f_i exper.	f_i theor.	τ_i nsec exper.	τ_i nsec theor.
I_1	2.5359	17.8	0.7 [23]	1.9	1.03 [15]	1.85
I_2	2.5457	8.3	$\begin{cases} 9 & [17] \\ 5.4 & [23] \end{cases}$	6.2	0.5 [15]	0.56
I_3'	2.5472	6.5	6 [23]	8.5	—	0.41
I_3''	2.5483	5.4	18 [23]	11.1	—	0.32

$\tau_{I_2} = 0.5$ nsec (I_1 is a neutral acceptor). Both values are in quite satisfactory agreement with $\tau_{I_1} = 1.9$ nsec and $\tau_{I_2} = 0.56$ nsec, calculated by (10) and (14); the second value also well agrees with $\tau_{I_2} \approx 0.4$ nsec, found by means of the experimentally determined $f_{I_2} \approx 9$ [17]. And at last, direct measurements of f_i were made using the absorption spectra [23], and here both neutral centers and ionized donors (I_3' and I_3'') were investigated; on the whole, an agreement has been achieved with other experimental data and theory. In Table 1 the experimental data obtained by different authors are compared with theory. As can be seen from Table 1 the agreement of theory and experiment is quite satisfactory, it is even better than could have been expected, if we take account of the fact that the derivation of formula (10) is based on a very simplified model.

It should be also noted, that the available data on the resonance Raman scattering with the emission of the long-wave LO-phonons also indicate a great contribution of the exciton – impurity complexes I_1 and I_2 as intermediate states [24]. These data confirm the existence of the gigantic oscillator strengths, but are of a qualitative character only. In [25] using function (8) a formula has been obtained for the cross-section of the LO-Raman-scattering, and the conclusion has been made that the Raman scattering observed by the authors is connected with the impurity excitons: since, however, the experiments [25] were performed at room temperature, the identification of complexes responsible for this scattering could not have been made.

The available data on ZnO are much less complete, and the assignment of lines is not so much reliable as is the case with CdS. Nevertheless, already at an early stage of investigation the assignment of the intense line (3669.7 Å) as a line of the impurity exciton bound to the ionized donor [26] was proposed based on the large oscillator strength $(f_i/f_{ex} \sim 10^4)$ corresponding to it. Earlier this line due to its high intensity was attributed to the ground state of the A-exciton [27]. The direct measurements of τ_i for a number of bands lead to the values of the order of a nanosecond which agrees with high f_i [28]. Low $\tau_i < 2.8$ nsec were, for instance, obtained for the lines I_b and I_c (3672.0 Å) which correspond to ionized donors; so far as for such centers the fast non-radiative Auger-processes are not possible, the corresponding τ_i may be considered to be defined by a radiative decay.

Very interesting results were found for InP, the line of the impurity exciton $\omega_i = 1.4165$ eV being characterized by $\tau_i \approx 0.55$ nsec [29]. The comparison of theory with experiment for InP presents some difficulties owing to a complicated valence band structure and low accuracy available. Nevertheless it turned out to be possible under reasonable assumptions to achieve an agreement between theoretical and experimental values of τ_i with an accuracy up to the factor 2. According to [29] here $f_{ex} \approx 8 \cdot 10^{-5}$, $f_i \approx 13$, and, consequently, $f_i/f_{ex} \approx 1.6 \cdot 10^5$.

An intensive emission caused by impurity excitons has been also observed in GaN [30].

In GaAs at 2 K a distinctly pronounced structure corresponding to the impurity exciton has been found somewhat below the intrinsic exciton band in the spectra of reflection (and photoreflection) [31]. Even though the used samples had an impurity concentration $N_i \sim 10^{15}$ cm^{-3}, this structure dominated in the spectrum. Since in GaAs the binding energy of complexes is small ($E_i \approx 1$ meV at $E_{ex} = 4.7$ meV) this fact may be explained by assuming the ratio f_i/f_{ex} to be especially large. The latest experiments [32] completely substantiate this assumption. For the excitons bound to neutral donors the measured value $\tau_i = 1.07$ nsec ($f_i \approx 7$) agrees with the theoretical $\tau_i \approx 1.8$ nsec. For the excitons bound to ionized donors and neutral acceptors close lying values have been obtained (0.8 nsec and 1.6 nsec, respectively). For these three centers the values $f_i/f_{ex} \approx 3 \div 6 \cdot 10^5$ were found. The maximum values of the ratio f_i/f_{ex} determined experimentally by the present refers, thus, to GaAs and InP.

3. Electron-Hole Complexes

The previous Section dealt with the spectra which up to now were repeatedly observed and whose nature is quite clear. On the other hand, in this and the next Sections we shall mainly investigate those spectra which are not yet well studied or even have not been observed yet at all, but which should also possess a high oscillator strength.

Let us consider the entities consisting exclusively of electrons and holes, and at the first stage we restrict ourselves to biexcitons only. There are a qualitative analysis and numerical calculations [33], which show that at simple isotropic bands the biexciton binding energy E_b measured in units E_{ex} is a monotonous function of the mass ratio $\xi = m_e/m_h$. At $\xi = 0$ it is maximum $E_b \approx 0.35\,E_{ex}$ (an analog of the hydrogen molecule), and then it fast decreases when ξ increases, and at $\xi \to 1$ it tends to $E_b \approx 0.018\,E_{ex}$ (an analog of the positronium). Since E_b/E_{ex} does not change with the substitution of $\xi \to 1/\xi$, the biexciton must exist in the whole region of values of ξ. Apparently, this fact indicates that also at a complicated band structure there are very favourable conditions for the biexciton formation, but their binding energy (at least in a macroscopic approximation) is large at a very small mass ratio only. It may happen, surely, that in some cases the binding energy will increase due to the core interaction.

Thus, we shall proceed from the assumption that there is a bound state of two excitons, and that there is an exciton with the momentum K in the initial state of the crystal. Then an exciton light absorption is possible at which the created exciton appears to be bound to the original exciton at once, i.e., during the light absorption the exciton with the momentum K transforms into a biexciton with the same momentum (the photon momentum is neglected). If $\Psi_K(r_e r'_e, r_h r'_h)$ is the biexciton wave function, the matrix element of this transition will be equal to

$$M_K = 2C \int \Psi_K^*(r_e r'_e, r_e r'_h)\, \Psi_K(r'_e, r'_h)\, dr_e\, dr'_e\, dr'_h. \tag{16}$$

This formula corresponds to formula (3) for the impurity excitons. Making use of it we may express the oscillator strength f_b of the biexciton creation in this process through f_{ex} similarly to (6).

However, since the function $\Psi_K(r_e r'_e, r_h r'_h)$ is complicated, it is reasonable to investigate the limiting case $E_b \ll E_{ex}$, in which the biexciton may be considered to consist of two weakly interacting excitons. Since the lowest potential of the biexciton ionization corresponds to its dissociation into two excitons, this approximation should be quite satisfactory. Then the biexciton wave function [similar to (8)] may be approximately represented as

$$\Psi_K(r_e r'_e, r_h r'_h)$$

$$\approx 2^{-1/2} \left[\Phi_{ex}(r_e - r_h)\, \Phi_{ex}(r'_e - r'_h) - \Phi_{ex}(r_e - r'_h)\, \Phi_{ex}(r'_e - r_h) \right] \tag{17}$$

$$\cdot \varphi_b(r_h - r'_h)\, V^{-1/2} \exp\{iK(r_h + r'_h)/2\},$$

where φ_b is a function describing the relative motion of excitons in the biexciton. It is still described by formula (8) with the only difference that now $\varkappa = (mE_b)^{1/2} = a_b^{-1}$ (since the reduced biexciton mass $m/2$ enters into the equation for φ_b). Calculating the matrix element and taking into account that $E_b/E_{ex} \ll 1$ we come to the formula [34]

$$f_b(K) = 2v^{-1} |\int \varphi_b(\varrho) \exp(-iK\varrho/2) d\varrho|^2 f_{ex}. \tag{16a}$$

If we use the explicit form φ_b, then

$$f_b(K) = f_b(0) [1 + (a_b K/2)^2]^{-2}, \tag{18}$$

$$f_b(0) = 16(2\mu E_{ex}/mE_b)^{3/2} \pi a_{ex}^3 v^{-1} f_{ex}.$$

These formulas are similar to (6a) and (10); the extra factor 2 in (16a) is connected with the Bose behaviour of the excitons. Therefore, there should arise gigantic oscillator strengths for the biexcitons as is the case with the impurity excitons. It should be noted that $f_b(K)$ depends on the momentum K.

A most important peculiarity of the biexciton excitation spectra consists in that the transition frequency depends on K due to the difference between the exciton and biexciton masses (with increasing K the transition frequency decreases by $K^2/4m$). Thus, unlike the narrow impurity exciton lines, the biexciton band should have a noticeable width and should be disposed on the long-wave side of the threshold frequency $\omega_b = \omega_{ex} - E_b$. In case the energy distribution of excitons is of the Boltzmann type, the form of this band is defined by the relation (Boltzmann constant $k = 1$):

$$\sigma(\Omega) \propto \sqrt{\Omega}(1 + \Omega/E_b)^{-2} \exp(-2\Omega/T), \quad \Omega = \omega_b - \omega. \tag{19}$$

The spectrum is shown schematically in Fig. 1. It is seen from (19) that the band width is determined by the least of the values E_b and T.

The radiative decay lifetime of the biexciton corresponding to the inverse process is determined by formula (12) with the substitution $f_i \to f_b$; the form of the emission spectrum is illustrated in Fig. 1.

Let us now discuss the experimental results. In all the cases we know, the biexciton bands were looked for in the fluorescence spectra. The biexciton concentration is considerable (that is, it may be comparable with the exciton concentration N_{ex}) at $T \gtrsim T_{cr} = E_b/\ln(Q/N_{ex})$ only, where Q is the statistical factor of the exciton band. Usually, at moderate exciton concentrations $Q \gg N_{ex}$, that diminishes T_{cr}. It is then natural that the observation of biexcitons with small E_b is difficult. Apparently, CuCl ($E_{ex} \approx 0.187\,eV, E_b \approx 0.043\,eV$) and CuBr ($E_{ex} \approx 0.107\,eV, E_b \approx 0.029\,eV$) are now the only crystals with direct transitions, in which the existence of biexcitons is reliably shown (for the discussion of the results and

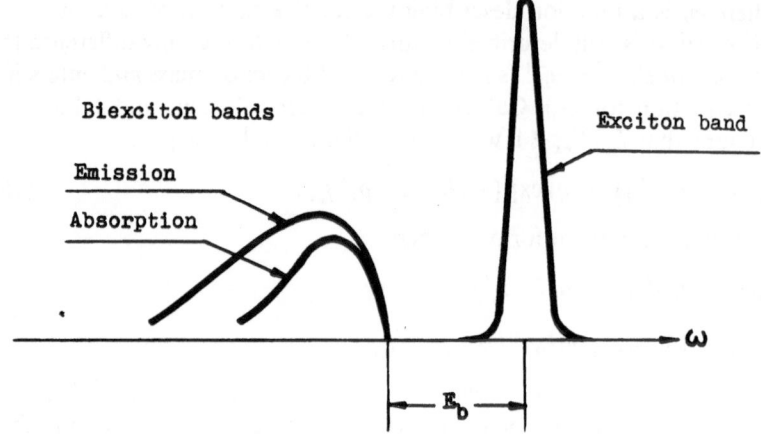

Fig. 1. Absorption spectrum corresponding to the transformation of excitons into bi-excitons, and the emission spectrum corresponding to opposite transitions

references see [35]); from the given data it is clear that E_b is fairly large in them. According to (18) we get $f_b/f_{ex} \sim 10^2$ for these crystals, i.e., the oscillator strength for the biexciton considerably exceeds f_{ex} (it is likely to be of importance for the biexciton stimulated emission in CuCl [36]).

Formula (19) indicates that the absorption spectra may happen to be more convenient for the observation of biexcitons with small E_b, than the fluorescence spectra. Indeed, the band width is limited by the value $\sim E_b$ due to the denominator even at high T, and, therefore, much weaker limitations are imposed upon T (the absence of thermal dissociation of excitons). As for the band intensity, it should be high at the expense of the gigantic oscillator strength.

In addition to the transitions to the biexciton discrete level the transitions to the continuous biexciton spectrum are also possible. If the frequency is counted from the exciton band $\Omega = \omega - \omega_{ex}$, the oscillator strength of absorption, per an interval of frequency Ω, induced by the exciton with the momentum K, is equal to:

$$f_b(\Omega) = (2a_b)^3 v^{-1} (E_b/\Omega)^{3/2} \frac{\{1 + (E_b/\Omega)(a_b K/2)\}^{1/2}}{\Omega + E_b[1 + (a_b K/2)^2]} f_{ex}, \qquad (20)$$

this expression should be averaged over the initial exciton distribution. The comparison of (20) and (18) shows that $f_b(\Omega)$ contains the same large factor a_{ex}^3/v, that is, it is also gigantic. The absorption starts at

$$\Omega = - E_b(a_b K/2)^2$$

and is extended into the region of higher frequencies. At the frequency $\Omega = 0$ it has a nonintegrable singularity. This singularity is cut off by the exciton damping and the retardation effects [37].

The absorption under consideration is a broadening of the exciton band caused by the exciton interaction, and attention should be paid largely to the fact that the intensity of "wings" is extremely high. It is obvious that the probability of the opposite transitions: the radiative recombination of one exciton accompanied by the transfer of its momentum and part of the energy to the second exciton will be also high; the form of the spectrum was studied in [37]. Formulas similar to (20) can be obtained for a number of other allied systems, and the order of magnitude of the oscillator strength depends little on the fact whether the bound state of the corresponding complex exists. Thus, the order of magnitude of the transition probabilities will not change, provided this "second" exciton will be substituted by an electron, or some other quasiparticle. Note that the laser effect due to the exciton-electron interaction has been already observed in CdS [38]. More complicated processes are also possible, they involve dissociated states of the exciton (i.e., electron-hole pair) [38, 39]; the theory of such transitions was developed in the lowest approximation only.

In crystals with indirect transitions comparable oscillator strengths correspond to both biexcitons and excitons. However, in the crystals with indirect transitions the additional extrema may exist corresponding to direct transitions (for instance, the minimum at $k = 0$ in Ge). Then to the photoproduction of the direct exciton near the indirect one with the formation of a biexciton would correspond a gigantic oscillator strength. At the same time such biexcitons, of course, cannot be observed in the fluorescence spectrum, in case it is sufficient time for the quasi-equilibrium in an excited state to be established.

There may arise an interesting situation, if the polyexciton concept will be corroborated [40], according to which in many-valley crystals the formation of entities consisting of many excitons corresponds to the minimum energy of the system. The number of excitons in the poly-excitons must be large, and, consequently, the reduction of kinetic energy when the direct exciton joins should be small. Therefore, the width of the absorption band corresponding to such a transition should be also small — of the order of the line width of the impurity excitons.

Recently a wide investigation has been devoted to the electron-hole metallic phase in Ge and Si [41], which exists in a form of droplets [42] in a wide interval of concentrations. However, the determination of regions in which droplets and aggregates (for instance, biexcitons) are present is still open to discussion [43]. The detection of the absorption bands corresponding to the binding of the direct exciton to various

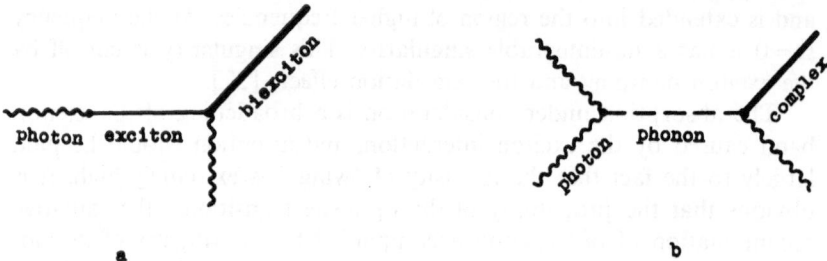

Fig. 2a and b. Various non-linear processes. (a) Two-quantum creation of biexcitons. (b) Two-photon creation of the exciton-phonon complex assisted by Raman scattering

aggregates could be of use for solving this problem, since it is natural that for the metallic phase such a band should be absent.

The question arises whether the gigantic oscillator strengths can be displayed in non-linear optical phenomena. A most simple process of such a kind is the two-quantum creation of biexcitons (Fig. 2a). It mainly proceeds through the one exciton intermediate state, and thus, the second stage of the process consists in the binding of the second exciton to the virtual exciton with the formation of a biexciton; therefore, a statement has been recently made [44] that in this process the gigantic oscillator strengths of the kind considered in the foregoing arise. In an approximation adopted above we get the following expression for the matrix element of the transition:

$$\sum_{\text{int}} \frac{\langle \text{biex}| p | \text{int}\rangle \langle \text{int}| p |0\rangle}{\omega_{\text{int}} - \omega_b} = 2\sqrt{2V} \, C^2 \Phi_{\text{ex}}^2(0) \int \varphi_b(r) \, dr / E_b \,. \tag{21}$$

It is convenient to compare the two-quantum absorption with the one-quantum one in the same area of spectrum, that is, with a usual exciton absorption. Its matrix element according to (5) is proportional to $\Phi_{\text{ex}}(0)$, and, therefore, the value of factor $\Phi_{\text{ex}}(0) \int \varphi_b(r) \, dr / E_b$ is decisive. The numerator of this expression is

$$\Phi_{\text{ex}}(0) \int \varphi_b(r) \, dr \sim \Phi_{\text{ex}}(0)/\varphi_b(0) \sim (a_b/a_{\text{ex}})^{3/2} \sim (E_{\text{ex}}/E_b)^{3/4} \,.$$

Though, under our assumptions this ratio is large, it, nevertheless, is much less than the factor $(a_b^3/v)^{1/2} \sim (E_0/E_b)^{3/4}$, determining the matrix element of the gigantic one-photon absorption; so, the direct analogy with the gigantic oscillator strengths considered above does practically not hold. The principal algebraic factor determining the large magnitude of the matrix element of the two-quantum transition arises due to the resonance denominator, the non-dimensional factor determining this matrix element being equal to:

$$\Phi_{ex}(0) \int \varphi_b(r) \, dr (E_0/E_b) \sim (\Phi_{ex}(0)/\varphi_b(0)) (E_0/E_b) \sim (E_{ex}/E_b)^{3/4} (E_0/E_b) \,.$$

Numerical estimations [44] indicate that the two-photon absorption should be comparable with a usual exciton absorption in a feasible range of intensities.

4. Exciton-Phonon Complexes

Beginning from 1966, in the spectra of magnetopolarons, impurity centers and excitons there have been found some features caused by interaction with optical phonons and possessing some characteristic peculiarities. From the analysis of these features some authors came to the conclusion that under certain conditions some excited states of these systems may be interpreted as complex involving a corresponding quasiparticle and a phonon. These results are discussed in [11, 12].

But here we are interested in the consequences which follow from the possibility of the exciton-phonon and magnetoexciton-phonon complex formation. If such weakly bound entities may exist then in the presence of real optical phonons in the lattice (equilibrium or nonequilibrium [13]) the light absorption, followed by an entity formation of a phonon present in the initial state and of an optically created exciton, should possess a gigantic oscillator strength. Correspondingly, the opposite process to the radiative decay of the entity should have a small characteristic time. The resonant three-photon process including the Raman scattering of one photon with the emission of a virtual polarization phonon and the subsequent absorption of the second photon with the formation of complex (Fig. 2b) should possess the high intensity too. All formulas are similar to those given in Sections 2 and 3.

It has been shown theoretically that a large exciton mass and the closeness to the resonance between the exciton excitation energy (e.g., $E_{2p} - E_{1s}$) and the phonon freqency ω_0 [45] favour the exciton-phonon complexes formation. The last condition is not a necessary one, and at a considerable coupling with phonons the complexes may arise in the absence of the resonance also [46]. Unfortunately, convincing experimental evidence of the formation of such entities is not yet available.

In strong magnetic fields the conditions under which the entities are formed become more favourable due to an increase of the exciton transverse mass. At the resonance between ω_0 and the cyclotron frequency of the electron or the hole they should arise even under weak coupling conditions [47]. Such entities were really observed in InSb [48].

The complexes were apparently observed also on TlBr and TlCl [49] (which is a more complicated case, since the magnetic field is not "strong"). Under certain conditions the entities may arise off resonance also [50].

5. Concluding Remarks

The general scheme of the processes considered above is the following. There is a certain entity: an impurity center, exciton, phonon, and so on. In the process of optical absorption an exciton joins this entity; thus it transforms into a more complicated entity. It has been shown that in all the cases the gigantic oscillator strengths correspond to such "overgrowing" processes.

Elsewhere we employed the elementary theory which allowed us to calculate the absorption of light with the formation of various types of entities. It can be easily refined by introducing a radiative line width $\Gamma_i = 1/\tau_i$. But at $\tau_i \sim 1$ nsec the width is $\Gamma_i \sim 10^{-3}$ meV, that is much less than the width due to other mechanisms (for instance, inhomogeneous broadening). The contribution of the impurity excitons in ε_1 will cause the light scattering at impurities.

In principle, it is in this region of the spectrum in the vicinity of exciton bands, which is very essential for us, that the approach based on polaritons is a most rigorous one. With such an approach the formulation of the problem changes considerably, as can be seen by the example of impurity excitons. This case is illustrated in Fig. 3. Polaritons, normal modes of perfect crystals, cannot be absorbed by local centers, when the inner dissipation mechanisms in them are absent. Therefore, the problem of the light absorption with the formation of an impurity exciton is replaced by that of the resonance scattering of polaritons by an impurity level, which is now quasidiscrete (since it is positioned in the region of the continuous spectrum of polaritons). In other words, the crystal with an impurity is considered as an ideal scatterer rather than an absorber, and the attenuation of light at the impurity exciton frequency is interpreted as a peak of the Rayleigh scattering rather than an absorption peak. This concept has been applied in [51] to the Wannier-Mott excitons, and in [52] to the Frenkel excitons. The width of the scattering peak coincides with the radiative width Γ_i, whereas f_{ex} and f_i are still connected by the relation (10). It is essential that this relation remains valid not only when the exciton-phonon coupling is weak ($E_{lt} \ll E_i$, E_{lt} is a longitudinal-transverse splitting), that is, the elementary theory deliberately holds true, but for the inverse limiting case as well [51]. Thus, the account of the polariton effects does not lead to a somewhat considerable change in the results in the region of a discrete level. It appears

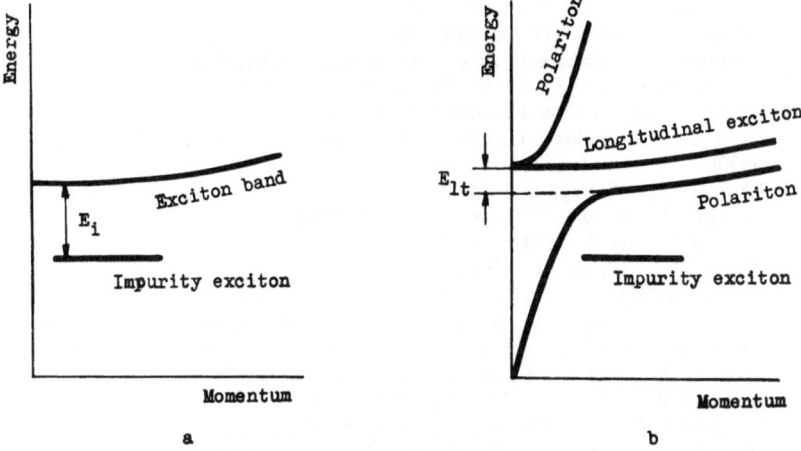

Fig. 3a and b. Energy spectrum of a crystal including the impurity exciton level. (a) Elementary scheme corresponding to the limit $f_{ex} \to 0$. (b) Scheme of the spectrum based on the polariton concept

to be much more important for a description of a continuous spectrum, since it causes a cut-off of the divergence near the bottom of the exciton band [37] (compare with Section 2).

A similar approach has been developed for biexcitons [37]; it may be also generalized to exciton-phonon complexes and to more complicated processes. Successful experiments in this field may stimulate further development of the theory.

References

1. Gross, E.F.: Nuovo Cimento Suppl. **3**, 672 (1956).
2. Nikitine, S.: Phil. Mag. **4**, N 37, 1 (1959).
3. Broude, V.L., Eremenko, V.V., Rashba, E.I.: Dokl. Akad. Nauk SSSR **114**, 520 (1957).
4. Thomas, D.G., Hopfield, J.J.: Phys. Rev. **128**, 2135 (1962).
   Reynolds, D.C., Litton, C.W., Collins, T.C.: Phys. Status Solidi **12**, 3 (1965).
5. Seitz, F.: Rev. Mod. Phys. **26**, 7 (1954).
6. Gross, E.F., Yakobson, M.A.: Dokl. Akad. Nauk SSSR **102**, 485 (1955).
7. Rashba, E.I., Gurgenishvili, G.E.: Fiz. Tverd. Tela **4**, 1029 (1962) [Sov. Phys. – Solid State **4**, 759 (1962)].
8. Rashba, E.I.: Opt. Spektrosk. **2**, 568 (1957).
9. Broude, V.L., Rashba, E.I., Sheka, E.F.: Dokl. Akad. Nauk SSSR **139**, 1084 (1961) [Sov. Phys. – Dokl. **6**, 718 (1962)].
   Sheka, E.F.: Fiz. Tverd. Tela **5**, 2361 (1963) [Sov. Phys. – Solid State **5**, 1718 (1964)].

10. Moskalenko, S. A.: Opt. Spektrosk. **5**, 147 (1958).
    Lampert, M. A.: Phys. Rev. Letters **1**, 450 (1958).
11. Levinson, Y. B., Rashba, E. I.: Rep. Progr. Phys. **36**, 1499 (1973).
12. Levinson, Y. B., Rashba, E. I.: Usp. Fiz. Nauk **111**, 683 (1973).
13. Vella-Coleiro, G. P.: Phys. Rev. Letters **23**, 697 (1969).
    Litton, C. W., Reynolds, D. C., Collings, T. C., Park, Y. S.: Phys. Rev. Letters **25**, 1619 (1970).
14. Khas, Z.: Czech. J. Phys. B **15**, 346 (1965).
15. Henry, C. H., Nassau, K.: Phys. Rev. **1**, B 1628 (1970).
16. Elliott, R. J.: Phys. Rev. **108**, 1384 (1957).
17. Thomas, D. G., Hopfield, J. J.: Phys. Rev. **175**, 1021 (1968).
18. Zinets, O. S., Sugakov, V. I.: Fiz. Tekh. Polupr. **1**, 880 (1967).
19. Skettrup, T., Suffczynski, M., Gorzkowski, W.: Phys. Rev. **4**, B 512 (1971); — Suffczyn-ski, M., Gorzkowski, W., Skettrup, T.: In: Physics of Impurity Centers in Crystals, Tallin, 1972 (Academy of Sciences of the Estonian SSR), p. 91.
20. Sharma, R. R., Rodriguez, S.: Phys. Rev. **159**, 649 (1967).
    Bednarek, S., Adamowski, J.: Phys. Letters A **41**, 347 (1972).
    Munschy, G., Carabatos, C.: Phys. Status Solidi (b) **57**, 523 (1973).
21. Landau, L. D., Lifshitz, E. M.: Elektrodinamika sploshnykh sred (Electrodynamics of Continuous Media), Gostekhizdat, 1957.
22. Sheynkman, M. K.: Fiz. Tverd. Tela **5**, 2780 (1963).
    Nelson, D. F., Cuthbert, J. D., Dean, P. J., Thomas, D. G.: Phys. Rev. Letters **17**, 1261 (1966).
23. Timofeev, V. B., Yalovets, T. N.: Fiz. Tverd. Tela **14**, 481 (1972).
24. Scott, J. F., Leite, R. C. C., Damen, T. C.: Phys. Rev. **188**, 1285 (1969).
    Damen, T. C., Shah, J.: Phys. Rev. Letters **27**, 1506 (1971).
25. Colwell, P. J., Klein, M.: Solid State Commun. **8**, 2095 (1970).
26. Park, Y. S., Litton, C. W., Collins, T. C., Reynolds, D. C.: Phys. Rev. **143**, 512 (1966).
27. Thomas, D. G.: J. Phys. Chem. Solids **15**, 86 (1960).
28. Skettrup, T., Lidholt, L. R.: Solid State Commun. **6**, 589 (1968).
29. Heim, U.: Phys. Status Solidi (b) **48**, 629 (1971).
30. Dingle, R., Sell, D. D., Stokowski, S. E., Ilegems, M.: Phys. Rev. **4**, B 1211 (1971).
31. Shay, J. L., Nahory, R. E.: Solid State Commun. **7**, 945 (1969).
32. Hwang, C. J., Dowson, L. R.: Solid State Commun. **10**, 443 (1972).
    Hwang, C. J.: Phys. Rev. **8**, B 646 (1973).
33. Wehner, R. K.: Solid State Commun. **7**, 457 (1969).
    Adamowski, J., Bednarek, S., Suffczynski, M.: Solid State Commun. **9**, 2037 (1971).
    Akimoto, O., Hanamura, E.: Solid State Commun. **10**, 253 (1972).
34. Gogolin, A. A., Rashba, E. I.: Zh. Eksperim. i Teor. Fiz. Pis. Red. **17**, 690 (1973).
35. Nikitine, S., Haken, H.: Izv. Akad. Nauk SSSR, Ser. Fiz. **37**, 220 (1973).
36. Shaklee, K. L., Leheny, R. F., Nahory, R. E.: Phys. Rev. Letters **26**, 888 (1971).
37. Gogolin, A. A.: Fiz. Tverd. Tela **15**, 2745 (1973).
38. Benoit à la Guillaume, C., Debever, J. M., Salvan, F.: Phys. Rev. **177**, 567 (1969).
39. Ryvkin, S. M., Grinberg, A. A., Kramer, N. I.: Fiz. Tverd. Tela **7**, 2195 (1965).
40. Wang, J. S.-Y., Kittel, C.: Phys. Letters A **42**, 189 (1972).
41. Asnin, V. M., Rogachev, A. A., Ryvkin, S. M.: Fiz. Tekh. Polupr. **1**, 1742 (1967).
42. Keldysh, L. V.: Proc. of the 9-th Intern. Conf. on the Physics of Semiconductors, Moscow, 1968, p. 1303. Leningrad: Nauka 1968.
43. Pokrovski, Ya.: Proc. 11 th Intern. Conf. on Physics of Semiconductors, Warsaw 1972, Vol. 1, p. 69.
    Benoit à la Guillaume, C.: Ibid., p. 659.
    Rogachev, A. A.: Izv. Akad. Nauk SSSR, Ser. Fiz. **37**, 227 (1973).

44. Hanamura, E.: Solid State Commun. **12**, 951 (1973).
45. Hermanson, J. C.: Phys. Rev. **2**, B 5043 (1970).
46. Rashba, E. I.: Zh. Eksperim. i Teor. Fiz. Pis'ma Red. **15**, 577 (1972) [Sov. Phys. – JETP Letters **15**, 411 (1972)]
47. Rashba, E. I., Edel'shtein, V. M.: Zh. Eksperim. i Teor. Fiz. **61**, 2592 (1971) [Sov. Phys. – JETP **34**, 1379 (1972)].
48. Johnson, J. E., Larsen, D. M.: Phys. Rev. Letters **16**, 655 (1966).
49. Kurita, S., Kobayashi, K.: J. Phys. Soc. Japan **30**, 1645 (1971).
50. Levinson, Y. B.: Zh. Eksperim. i Teor. Fiz. Pis'ma Red. **15**, 574 (1972) [Sov. Phys. – JETP Letters **15**, 408 (1972)].
51. Hopfield, J. J.: Phys. Rev. **182**, 945 (1969).
52. Sugakov, V. I.: Opt. Spektrosk. **24**, 477 (1968).

Prof. Dr. E. I. Rashba
L.D. Landau Institute for Theoretical Physics
Academy of Sciences of the USSR
Moscow/USSR

Interaction between Excitons at High Concentration[*]

R. LÉVY, A. BIVAS, J. B. GRUN, and S. NIKITINE

Contents

1. Introduction

In direct band gap semiconductors high concentrations of free carriers and excitons are created by intense laser light excitations. These excitons may be bound in pairs to form biexcitons, as observed in CuCl, CuBr, CdS, and CdSe. When such high densities of quasiparticles are achieved, the collisions between them become numerous. The broadening of the emission line due to the radiative decay of biexcitons is due to elastic collisions between excitons and biexcitons. Inelastic collisions between excitons and biexcitons, and between electrons and excitons also become very important as radiative recombination processes of these quasiparticles. At these concentrations, multiple-particle interaction effects have also to be taken into account. Such effects have been observed in the radiative recombination process due to inelastic collisions of excitons. They seem also to explain the drastic changes in the properties of excitons, such as the decrease in their binding energy, when high excitation intensities are used.

[*] This work was carried out within the frame of the agreement on French-German scientific cooperation between C.N.R.S., Paris, Université Louis Pasteur, Strasbourg, and Universität Stuttgart.

2. Inelastic Collisions

Exciton–exciton and electron–exciton interactions were first suggested
and studied theoretically by Benoît à la Guillaume et al. [1] in 1969.
Biexciton interactions have been observed recently by Saito et al. [2].

a) Exciton–Exciton Collisions

A new emission band called the P line (Fig. 1) was first observed in CdS
by Benoît à la Guillaume [1] after electron beam excitation and was
identified as an exciton–exciton collision recombination radiation. Later
on, Magde and Mahr [3] also observed it in CdS, CdSe, and ZnO after
laser light excitation.

The recombination process first proposed to explain the appearance
of the P emission line in CdS is an exciton–exciton inelastic collision

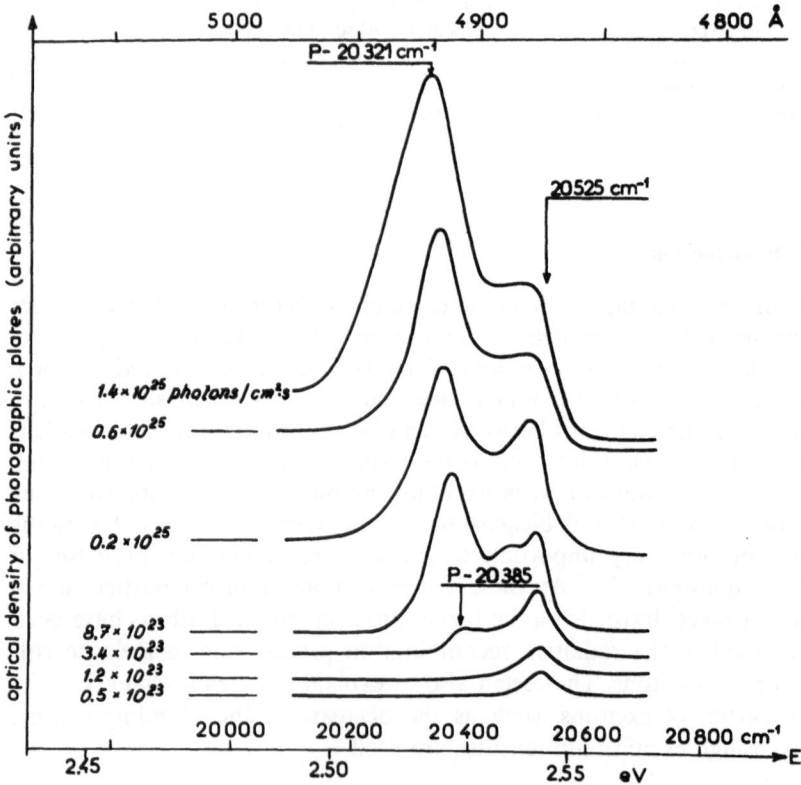

Fig. 1. Emission spectra of a thick sample of CdS for increasing ultraviolet light excitations at 4.2 K

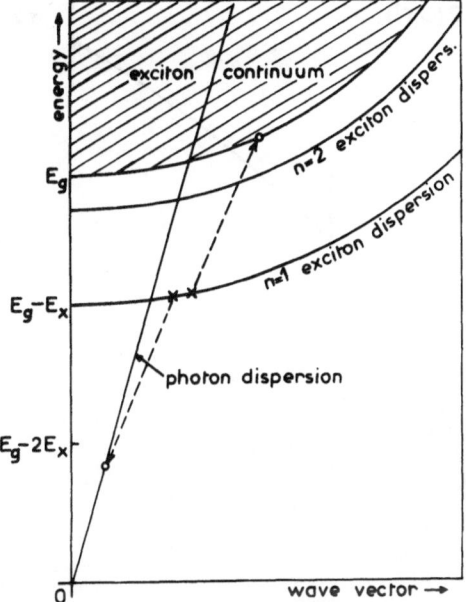

Fig. 2. Dispersion curves of the exciton and the photon

in which one of the excitons scatters into an excited excitonic state or dissociation state while the other scatters into a photon-like state, as shown in Fig. 2.

The momentum conservation in this process gives the following relation:

$$K + K' = q + K''$$

where K and K' are the momenta of the two colliding excitons, K'' is the momentum of the excited exciton or of the unbound pair of electron and hole, and q, the photon momentum, is assumed to be negligible. The remaining electron–hole pair or exciton has a momentum equal to the sum of the momenta of the colliding excitons. The energy balance in this process is as follows:

two excitons → electron–hole pair + photon

$$h\nu_P = E_g - 2E_x - E_{e,h}^k$$

where $h\nu_P$ is the energy of the emitted photon, E_g the energy gap and E_x the energy of the exciton; $E_{e,h}^k$ is the kinetic energy of the unbound pair electron–hole created during the collision. The exciton kinetic energy is neglected.

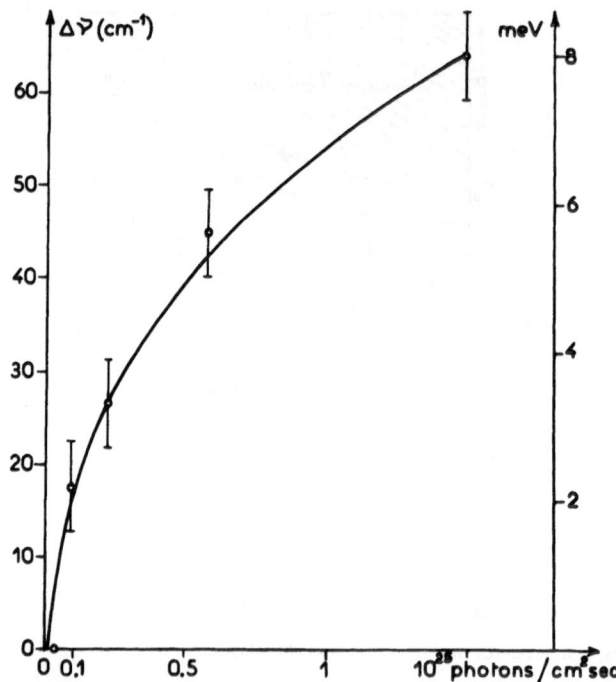

Fig. 3. Shift of the P emission line versus the intensity of UV photons at 4.2 K. We have taken the position of the P line at the lowest excitation intensity as the origin of the shift. ○ experimental points, — theoretical curve

At low excitation intensities, the bands are empty. The unbound pairs created during the process have negligible kinetic energy. At high excitation intensities, however, a high density of free carriers is first created, as shown by Goto and Langer in their absorption studies [4, 5]. The bottoms of the bands are then filled. When the exciton–exciton collision recombination takes place, the unbound pair created in the process must have higher energies. Their kinetic energy $E_{e,h}^k$ can no longer be neglected.

A shift of the emission line P towards lower energies is then expected:

$$h\Delta v = -E_{e,h}^k .$$

Such a shift has indeed been observed in CdS [6], as shown in Fig. 3 [1,2]. This shift has been measured for different excitation intensities at liquid helium temperature.

---

[1] J. Bille has observed a similar shift of the stimulated emission due to the exciton–exciton collision process in CdS [7].

[2] M. Pilkuhn has also reported a similar behavior of this line in III–V compounds (GaAs, InP) [8].

Assuming a band-filling effect with elliptical bands having only one extremum, Haken and Nikitine have calculated a line shift equal to [6]:

$$h\Delta v = [m_e^{*-1} + m_h^{*-1}] (h^2/8) (3/8\pi)^{2/3} n^{2/3}$$

where m_e^* is equal to $\sqrt[3]{m_{x,e}^* m_{y,e}^* m_{z,e}^*}$, and $m_{i,e}^*$ are the elements of the effective mass tensor, assumed diagonal, for the conduction band. m_h^* is defined in a similar way for the valence band. n is the number of free carriers per unit volume. In the steady state, n depends on the square root of the absorbed photon flux i, therefore the line shift is proportional to $i^{1/3}$. This theoretical dependence (drawn in solid line in Fig. 3) is in good agreement with the experimental points. Theoretical evaluations of this shift have also been made and give reasonable values.

In CdS, the electron and hole mass values are:

$$m_{z,e}^* = 0.15 \, m_0, \quad m_{x,e}^* = m_{y,e}^* = 0.17 \, m_0,$$
$$m_{z,h}^* = 5 \, m_0, \quad m_{x,y}^* = m_{y,h}^* = 0.7 \, m_0.$$

Let us assume that the maximum concentration of free carriers created in the crystal is given by the usual expression of the Mott concentration:

$$n^{1/3} a_x \simeq 0.2,$$

where a_x is the exciton radius ($a_x \simeq 29$ Å in CdS).

This maximum concentration is $3.3 \cdot 10^{17}$ carriers/cm^3 in CdS. A maximum shift of the exciton–exciton emission line not exceeding 3 meV should therefore be expected. This value is 2 or 3 times less than the experimental one. Two remarks are, however, necessary.

1. The value given above for the Mott concentration may be underestimated. The exciton seems still to be a stable excitation of the crystal at these high excitation levels. We have studied the anomalies of reflection corresponding to the excitonic absorption lines during ultraviolet-light excitation. No appreciable changes of these maxima and minima have been observed. This fact has been confirmed by measurements of the absorption induced by ultraviolet excitation [4, 5].

2. The study of thin samples has shown the appearance at the same high concentrations of new lines, which are stimulated by and are due to a different recombination mechanism [9]. However, no stimulated emission has been detected with thick samples.

b) Biexciton–Biexciton Collisions

Another collision process has been suggested by Saito et al. [2] to explain a new emission line, P_M, observed between the biexciton emission line, (M), and the P lines (Fig. 4). This process is an inelastic collision

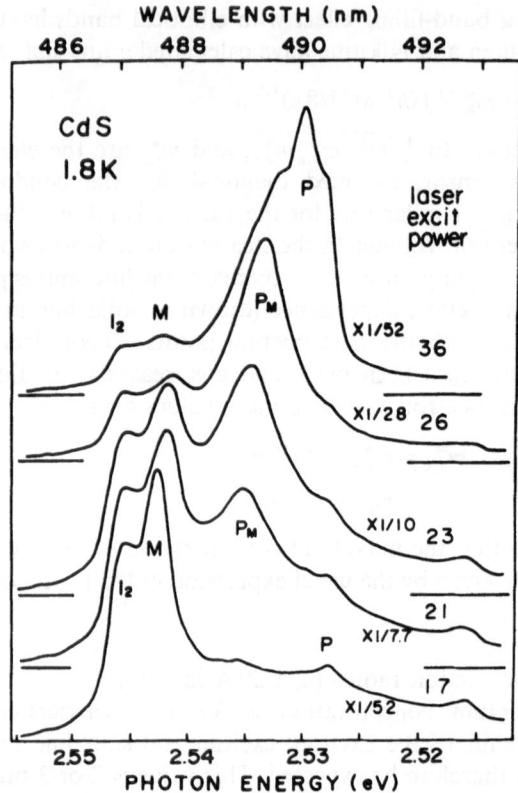

Fig. 4. Luminescence spectra of a CdS crystal at 1.8 K under various ultraviolet-light excitations (according to Saito et al.)

of two exciton molecules; one of the four excitons involved in the process is radiatively annihilated, leaving the other three as free excitons. The energy balance of the process is:

two exciton molecules → three excitons and one photon.

If we neglect the different kinetic energies, the emitted photon has the following energy:

$$h\nu_{P_M} = E_g - E_x - 2E_M.$$

Saito et al. were able to show from the position of the emission line P_M and by a study of its shape that this emission may be due to a collision of two exciton molecules, leaving three free excitons (Γ_5).

The study of the intensity variation of the line P_M with the excitation intensity i also supplies arguments for their interpretation. Figure 5

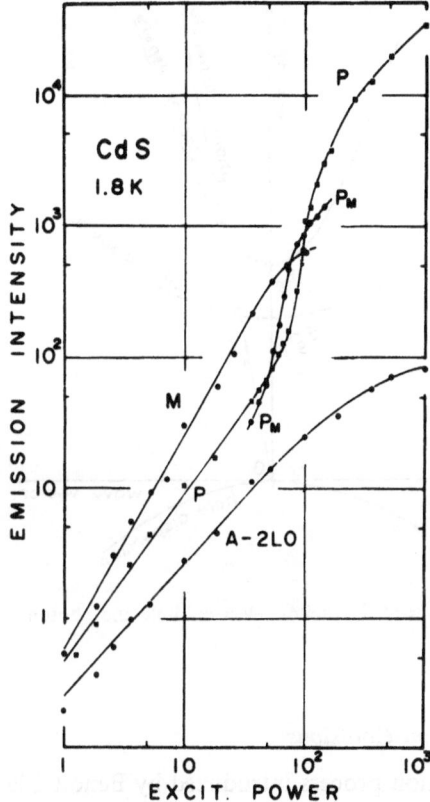

Fig. 5. Variation of the intensities of the M, P_M, P, and A-2 *LO* emission lines versus the UV-light intensity in a CdS crystal at 1.8 K (according to Shionoya et al.)

shows the variation in intensity of the M, P_M, and P lines. Just before the M line tends to saturate, the P_M line starts to grow very rapidly with a superlinear dependence of *i* (the slope is of the order of four on logarithmic scales) and becomes larger than the M line. This corresponds to the decay of exciton molecules, essentially by the collision channel. At the same time, a large number of excitons are created during the exciton molecule collision process, producing the sharp increase in the intensity of the P line. At low excitation intensities, the number of exciton molecules is larger than the number of excitons but, when the collisions of exciton molecules become important, this relation is reversed.

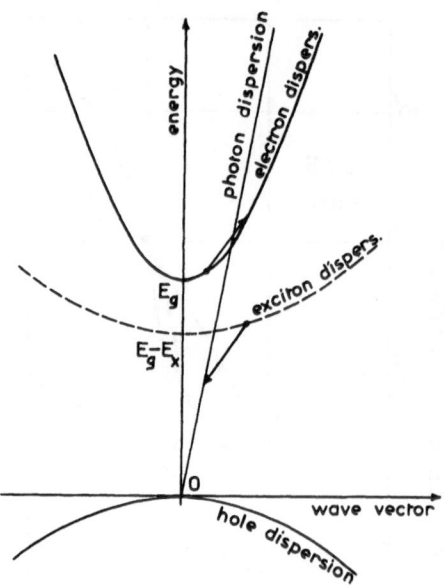

Fig. 6. Dispersion curves of the exciton, electron, hole and photon

c) Electron–Exciton Collisions

In this recombination process introduced by Benoît à la Guillaume [1], the exciton is radiatively dissociated, giving its kinetic energy to the electron, as shown in Fig. 6.

The momentum conservation gives the following relation:

$$K + k_i = k_f + q$$

where K is the exciton momentum, k_i and k_f the momenta of the electron in its initial and final states; q, the photon momentum, is negligible. Therefore, in the collision process, the exciton momentum is given to the electron. The energy balance is:

exciton + electron → photon + electron

$$E_g - E_x + \hbar^2 k^2/2M_x + E_g + \hbar^2 k_i^2/2m_e^* = h\nu + E_g + \hbar^2 k_f^2/2m_e^* .$$

If we assume, as Yu et al. [10] did, that the electron mass is much smaller than the hole mass ($m_h^* \gg m_e^*$), then the electron mass is negligible compared to the exciton mass M_x and, at thermal equilibrium, the exciton momentum K is much larger than the electron momentum k_i.

The emitted photon energy hv is simply given by:

$$hv = E_g - E_x - (\hbar^2 K^2/2M_x)(M_x/m_e^*).$$

The emission line intensity will be proportional to the product of the density of exciton states and the thermal distribution of excitons:

$$I(hv) \propto E^{1/2} \exp(-E/k_B T),$$

where $E = \hbar^2 K^2/2M_x$ is the exciton kinetic energy and k_B the Boltzmann constant.

$$E = \tfrac{1}{2} k_B T,$$

the emission band peak is therefore given by:

$$hv = E_g - E_x - \tfrac{1}{2}(M_x/m_e^*) k_B T.$$

A linear temperature shift is expected for this emission band. Such a shift has been observed by Bille et al. for an emission line in CdS [11]. Yu et al. have obtained similar results in copper halides [10].

In conclusion, inelastic collisions of particles are important radiative recombination processes when high densities of excitons, biexcitons and carriers are created in crystals.

The shift of the exciton–exciton emission line has been explained by a band-filling effect. We shall mention other effects on the exciton itself, which may be explained similarly.

3. Properties of Excitons at High Concentrations

New effects related to the creation of high concentrations of excitons and/or carriers have also been observed by the study of the excitonic absorption of some crystals (CdS, CuCl) during their excitation by a powerful light source. As we will see later, these effects may be explained by many-particle interactions.

a) Excitonic Absorption in Cadmium Sulfide

This study was made by Goto and Langer [4, 5]. The crystals of CdS were excited at 2 K by a powerful nitrogen laser. The decrease of the transmitted light of a Xe flashlamp was measured at a given wavelength as a function of time with a sampling oscilloscope. At any wavelength the induced absorption appears to have two distinct time components.

The slow component has a decay time of the order of 10 μsec for CdS platelets of about 6 μ thickness. The magnitude of this induced

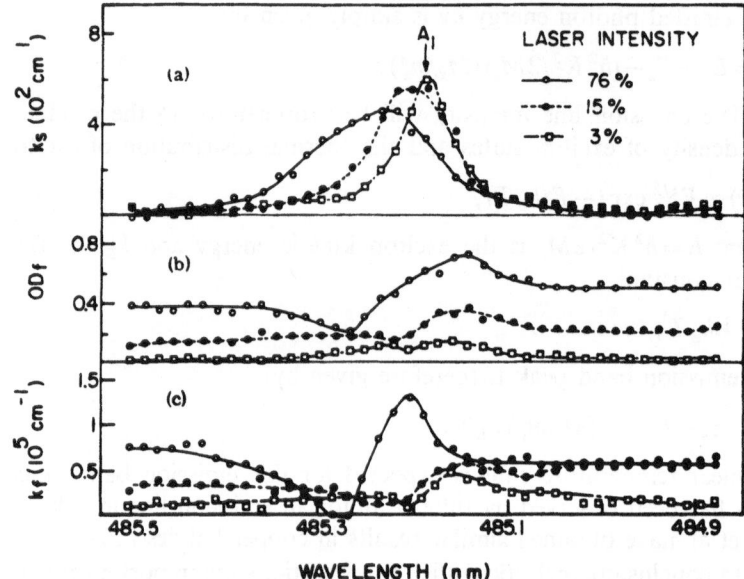

Fig. 7. In the vicinity of the A exciton, the upper section shows the induced absorption coefficient corresponding to the slow time constant effect (slow component). The center section shows the optical density associated with the fast time constant effect. The lower section of the figure shows the induced absorption constant in the excited region of 10^{-5} cm thickness (according to Goto et al.)

absorption and its time constant vary with the crystal thickness. This absorption is probably due to an increased crystal temperature. The induced absorption coefficient of the slow component is shown in the upper part of Fig. 7. When the excitation intensity increases, the A excitonic line broadens and shifts towards lower energies, as would be expected when the crystal temperature increases.

The time dependence of the fast component is indistinguishable from that of the laser pulse. The magnitude of this fast component and its time constant do not depend on the crystal thickness. The central part of the figure shows the induced optical density of the fast component in the region of the A excitonic line, and the lower part of the figure shows the induced absorption constant in the excited region of 10^{-5} cm thickness. An absorption continuum, which is larger near the absorption edge and tails off toward lower energies, can be seen. In the close vicinity of the A line the induced background can be considered as independent of wavelength. The shape of the absorption line, however, changes with increasing excitation, resembling one of the curves derived by Fano for the case of interference of an autoionized state with a continuum of states [12].

Goto and Langer have also studied the dependence of the fast component of the induced absorption on the excitation intensity for a number of fixed wavelengths. This dependence is linear, with saturation at the highest pumping intensities. The observation of the induced absorption was not restricted to measurements at 2 K; similar effects were also seen at 77 K and at room temperature.

This fast absorption continuum on the low-energy side of the A line cannot be explained by exciton–phonon transitions nor by the creation of excitons with nonzero kinetic energies. It may, however, be due to electron transitions from the conduction to the valence band, which are assisted by energy and momentum transfers from free, highly excited electrons in the conduction band. The binding energy of the exciton is not appreciably changed. We have reached similar conclusions by studying the reflection anomalies corresponding to the A exciton line. The observation of the A-2 LO emission line by Leheny et al. [13] leads to similar results. The exciton does not seem to be drastically perturbed in this high concentration of carriers and/or excitons.

b) Excitonic Absorption in Highly Excited Copper Chloride

Very different results have been obtained in the study of the absorption spectrum of copper chloride crystals under the same conditions of strong ultraviolet excitation of the crystals.

We became interested in this problem while studying the free excitonic emission line v_0 of CuCl at low temperature (Fig. 8). At low levels of UV-light excitation, this emission line is partly reabsorbed in the crystal in the $n = 1$ excitonic absorption peak. But, at high excitation levels, the minimum due to the reabsorption disappears. The possibility of a shift of the absorption line was then considered, and the absorption of highly excited crystals was therefore investigated.

α) Experimental Results on the Excitonic Absorption Spectra [14]

We have made quantitative measurements of the excitonic absorption spectra of thin samples of CuCl (0.4–1 µm) during the high-intensity UV-light excitation. These measurements have been made for a number of wavelengths at different excitation intensities. We have obtained the shape of the exciton lines and their relative intensity in the spectrum. We were also able to compare quantitatively two spectra obtained at

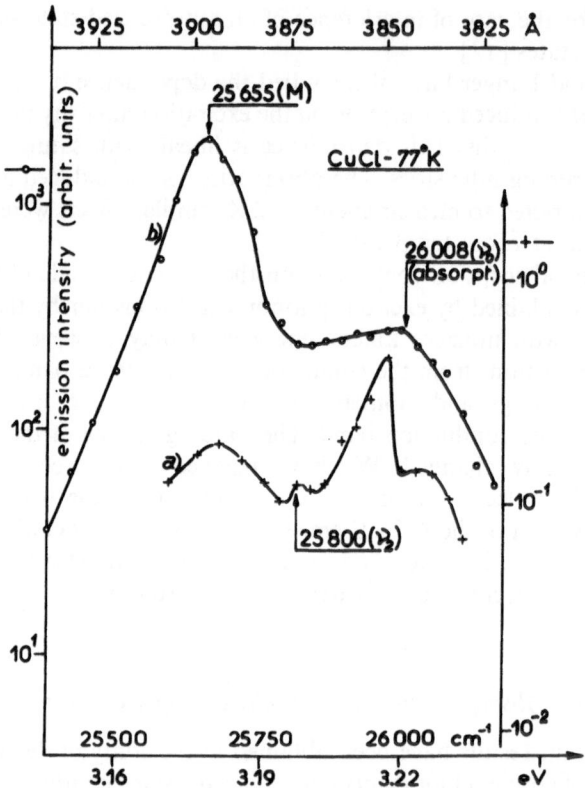

Fig. 8a and b. Luminescence spectra of CuCl at 77 K for two intensities of UV laser light excitation. (a) $7 \cdot 10^{21}$ photons/cm^2 sec, (b) 10^{25} photons/cm^2 sec

different excitation intensities. Finally, we recorded the time variation of the absorption change during and after UV excitation.

We first measured, one wavelength after the other, the intensity of a flashlamp transmitted through the sample without UV excitation, then the intensity of the same flashlamp transmitted through a quartz plate identical to the sample holder. The ratio of these two intensities gives the transmission factor of the sample. We did not, however, take into account the reflection of the light by our samples.

We then measured the transmission factor of the sample when simultaneously excited by the UV laser light. We have observed a variation in the transmission; this variation does not follow the instantaneous

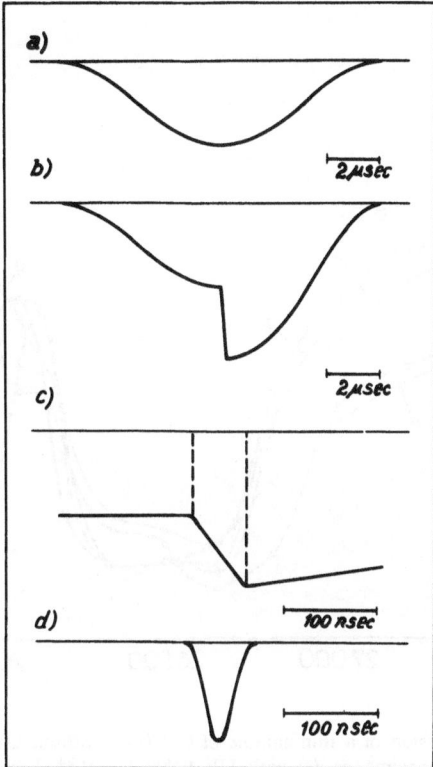

Fig. 9. (a) Transmission of the flashlamp through the sample without UV excitation (time base: 2 μsec/cm). (b) Transmission of the flashlamp through the sample wit UV laser excitation (time base: 2 μsec/cm). (c) Transmission of the flashlamp through the sample with UV laser excitation (time base: 100 nsec/cm). (d) Shape of the UV excitation pulse (time base: 100 nsec/cm)

intensity of the laser pulse but the total energy sent on the sample at a given time. The initial value of the transmission is restored only several hundreds of nanoseconds after the end of the excitation, as shown in Fig. 9.

We have also measured the transmission factor of the samples at the end of the UV excitation. The spectra obtained at 52 K are presented in Fig. 10. Curve (a) represents the "classic" absorption spectrum (without UV excitation) of our sample in the region of the two first excitonic absorption lines. The other curves represent this spectrum at different UV excitation intensities. The exciton absorption lines shift

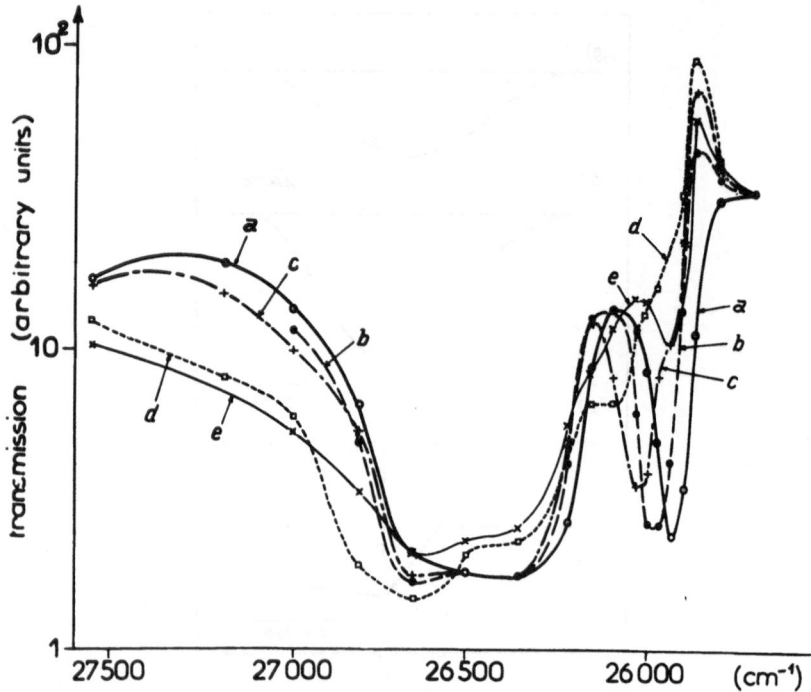

Fig. 10a–e. Transmission of a thin sample of CuCl: (a) without UV light, (b) with UV light: $4.6 \cdot 10^{23}$ photons/cm² sec, (c) with UV light: $1.3 \cdot 10^{24}$ photons/cm² sec, (d) with UV light: $5.5 \cdot 10^{24}$ photons/cm² sec, (e) with UV light: 10^{25} photons/cm² sec

towards the high-energy part of the spectrum when the excitation intensity increases and the intensity of the lines decreases, as previously noticed with qualitative measurements [15, 16]. Furthermore, a new phenomenon appears in the vicinity of the first two excitonic absorption lines.

The maximum of the first absorption line of the first excitonic series (at 25940 cm^{-1}–3855 Å without laser excitation) shifts towards the high-energy part of the spectrum when the excitation intensity increases. This shift has been measured versus the flux density of ultraviolet photons. The experimental points obtained are shown in Fig. 11. A maximum shift of 140 cm^{-1} is reached for a flux density of $5 \cdot 10^{24}$ ultraviolet photons/cm² sec. The continuous curve drawn on the same figure represents a theoretical shift of the excitonic absorption line proportional to

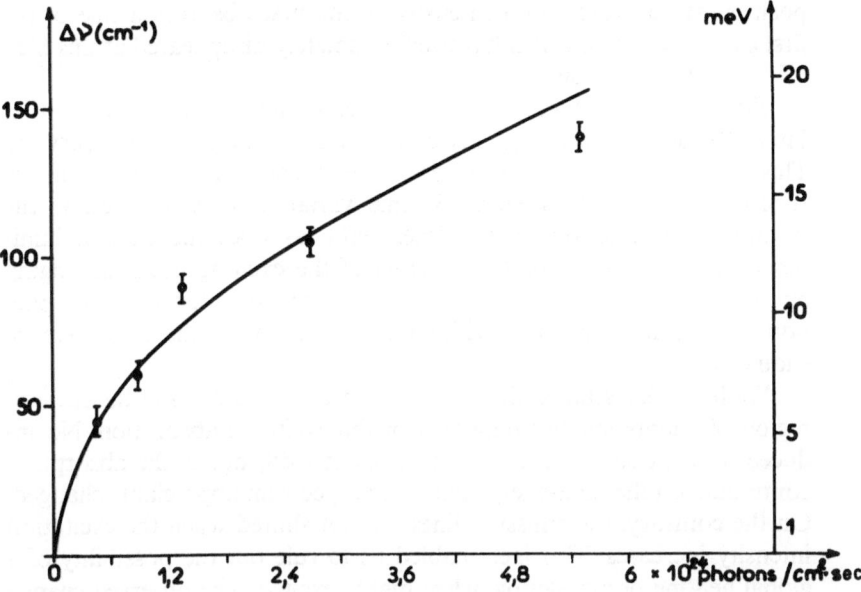

Fig. 11. Frequency shift of the first excitonic absorption line of CuCl versus the intensity of UV photons. ○ experimental points, — square-root dependence on light intensity

the square root of the excitation intensity. The agreement with the ex-perimental points is reasonably good.

The first line of the first excitonic series decreases in intensity when the excitation intensity increases. One can see that it is not a broadening of the line with reduction of its amplitude, but a global decrease in the intensity of the line. This line practically disappeared for an excitation intensity of 10^{25} photons/cm² sec.

The first line of the second excitonic series is also shifted towards the high-energy part of the spectrum when the excitation intensity in-creases. It was difficult to make a precise measurement of this effect because this line is very wide; we could nevertheless show that the intensity of this line changes only slightly.

New phenomena appear in the vicinity of the first absorption line of the first excitonic series. When the laser excitation intensity increases, although the exciton absorption line decreases in intensity, a new mini-mum of transmission appears on the low-energy side of the absorption spectrum. This phenomenon is particularly well observed for an excita-tion intensity of 10^{25} photons/cm² sec. This minimum appears at the position at zero excitation intensity of the first absorption line of the first excitonic series, which has itself completely disappeared at this high UV excitation intensity (curve e).

Furthermore, a maximum of transmission appears at low excitation intensities on the low-energy side of the exciton absorption lines (curve b). This increase of transmission does not correspond to the ordinary triplet exciton luminescence. Its time variation does not follow the instantaneous intensity of the laser pulse as does the exciton luminescence, but follows the total energy of the exciting beam impinging on the sample at a given time. It must also be noted that this phenomenon is not observed without a flashlamp, as the luminescence should be.

We have also studied the transmission of our samples in the spectral region of energy smaller than that of the excitonic absorption. No induced absorption has been observed as in CdS, nor is the absorption continuum on the high-energy part of the spectrum appreciably changed. On the contrary, the emission lines are not shifted when the excitation intensity increases. This fact enabled us to rule out the possibility of a global heating of our sample when highly excited. The observed change in the excitonic absorption spectrum is not due to an increase in the temperature of the sample.

β) Theoretical Analysis

These experimental results have not been completely explained, but some partial interpretations have been given. Due to the strong light excitation, a great number of electrons and holes are created. They will rapidly thermalize to the extrema of the bands. When these electrons and holes are not very rapidly forming excitons, a nonequilibrium situation, in which the excited electrons occupy the bottom of the conduction band, has been assumed by Haken et al. [15]. When an exciton is then created by absorption of a photon, its wave function cannot be formed out of all the conduction band states because it must be orthogonal to the occupied states (Pauli principle). The binding energy of an exciton has been calculated by a variational principle, when the states of the partly filled conduction band have been projected out. The following energy shift of the $1S$ exciton line towards the higher energies has been obtained:

$$\Delta E_{1S} = 32\pi a_x^3 E_x n$$

where E_x is the unchanged binding energy of the exciton, a_x is the radius of the exciton in the ground state, and n is the free carrier concentration.

The variation of the observed shift of the absorption peak, as a function of the intensity of the impinging light, can now be explained. This may be done by studying the rate equation for the electron density:

$$dn/dt = Ai - \gamma n^2 .$$

The first term gives the creation of electrons (or holes) by the absorbed photon flux i, A being the absorption coefficient. The second term describes the recombination of two carriers with the recombination rate γ. Using realistic parameters A, i and γ, one finds that, under the present experimental conditions, the steady state is reached very rapidly in the first phase of the pump process. Thus, one may solve the above equation under the assumption that the time variation of n may be neglected. We thus find that n depends on the square root of i:

$$n = [(A/\gamma) i]^{1/2} .$$

Inserting reasonable values for the constants, namely:

$$A = 10^5 \text{ cm}^{-1}$$

$$\gamma = 3 \cdot 10^{-7} \text{ cm}^3 \text{ sec}^{-1}$$

$$i = 10^{25} \text{ photons/cm}^2 \text{ sec} ,$$

we obtain an electron density $n = 1.7 \cdot 10^{18} \text{ cm}^{-3}$. Introducing this value into the theoretical expression of the line shift and letting $E_x = 190 \text{ meV}$ and $a_x = 7 \text{ Å}$ for CuCl, a shift of the exciton absorption line of 86 cm^{-1} is obtained, which is of the same order of magnitude as the experimental shift.

Furthermore, the shift depends linearly on the carrier concentration n, hence on the square root of the light intensity i. Such a square root dependence is drawn in Fig. 11. The fit with the experimental points is reasonably good.

The excitonic absorption spectrum observed corresponds to the creation of new excitons in this nonequilibrium situation. However, when the exciton recombines on account of the selection rules $\Delta K = 0$ and $K = 0$ (K being the exciton wave vector), only the "optical" excitons will participate in this recombination. These excitons are formed of all the states, so that their binding energy is not changed. The exciton emission line will not be shifted, as observed experimentally.

It must be emphasized that the above arguments remain valid if some of the electrons and holes, or all of them, form excitons. In this case, however, only the value of the numerical factor in the equation of the shift changes.

The elementary excitation energy necessary to create a new exciton at a high density of excitons has also been calculated by Hanamura [17, 18] and by Keldysh and Kozlov [19]. Hanamura introduces operators of electron–hole pairs obeying exact boson commutation rules. The hamiltonian operator, consisting initially of fermion operators, is then expressed in terms of these new operators in the effective mass approximation for a two-band model. He supposes that the system is in a stationary state, with conservation of the number of pairs, and that the excitons are in a Bose condensed state ($K = 0$).

Keldysh and Kozlov, with the same assumption as Hanamura, introduce quasi-electron and quasi-hole operators, which obey the fermion commutation rules and are related to the electron and hole operators by the transformation of Bogoliubov.

These two theories, almost equivalent, predict a shift of the $1S$ exciton level, which is given by the following formula:

$$\Delta E = f\, a_x^3\, E_x\, n_x$$

where n_x is the exciton concentration and f a numerical factor, of the order of 9π according to Hanamura, and of unity according to Keldysh and Kozlov.

These theories predict a shift in the exciton level of the order of magnitude of the experimental one, if all electron–hole pairs form excitons. The square root dependence of the shift on the excitation intensity i can also be deduced from exciton and biexciton rate equations. However, these theories predict that emission and absorption both involve the same energy level. The emission band should then, like the absorption band, be shifted towards higher energies. This is in contradiction with the experiment. In these theories, Bose condensation and superfluidity of excitons are assumed. One is tempted to conclude that the experiment is in contradiction with Bose condensation of excitons.

The bleaching of the first absorption line of the first excitonic series observed at these high concentrations could be explained by assuming that we cannot indefinitely create new excitons in our crystals. The highest density of excitons possible may be $1/a_x^3$, which is of the order of 10^{21} excitons/cm^3 [20–22].

The time behavior of the transmission variation may be directly related to the population of singlet excitons, which have a longer lifetime than the triplet excitons. Their direct recombination is indeed strictly forbidden, in contrast to triplet excitons [23].

The new effects observed near the $1S$ exciton absorption line have not yet been clearly explained. The minimum of transmission reappearing at the position at zero excitation of the $1S$ exciton level may be due to a decrease in exciton lifetime [24] at these high excitations, hence a decrease in exciton densities. The shift of the absorption line is no longer observed.

In conclusion, this study has produced some new observations on the behavior of excitons at these high densities. Many effects still have to be clearly explained. Some new experiments are needed, taking into account e.g. the light reflection of the crystals, which has been completely neglected in these studies.

4. General Conclusion

At high concentrations, inelastic collisions between excitons and biexcitons, and between electrons and excitons are very important radiative recombination processes. They give rise to stimulated emission with very large gain.

The exciton remains a stable entity at these levels of excitation, but the effects of high concentrations of quasi-particles have been revealed by the study of the excitonic absorption of light. A new absorption due to free carrier transitions is induced in highly excited CdS crystals. These transitions interfere with excitonic transitions. The excitonic absorption is strongly modified in highly excited CuCl samples. A shift of the exciton absorption peak is observed and explained by a nonequilibrium concentration of carriers and/or excitons.

References

1. Benoît a la Guillaume, C., Debever, J. M., Salvan, F.: Phys. Rev. **177**, 567 (1969).
2. Saito, H., Shionoya, S., Hanamura, E.: Solid State Commun. **12**, 227 (1973).
3. Magde, D., Mahr, H.: Phys. Rev. Letters **24**, 890 (1970).
4. Goto, T., Langer, D. W.: Phys. Rev. Letters **27**, 1004 (1970).
5. Langer, D. W., Goto, T.: Proc. of the 11th Conf. on Physics of Semiconductors. Warsaw (Poland) (1972).
6. Levy, R., Grun, J. B., Haken, H., Nikitine, S.: Solid State Commun. **10**, 915 (1972).
7. Bille, J.: (Private communication).
8. Pilkuhn, M.: Non-Linear Optics Conference, Titisee (Sept. 1971).
9. Levy, R., Grun, J. B.: J. Luminescence **5**, 406–412 (1972).
10. Yu, C. I., Goto, T., Ueta, M.: J. Phys. Soc. Japan **34**, 693 (1973).
11. Bille, J., Fischer, T., Liebing, H., Ruppel, W.: Proc. of the 11th Int. Conf. on Physics of Semiconductors, Warsaw (Poland) (1972).
12. Fano, U.: Phys. Rev. **124**, 1866 (1961).
13. Leheny, R. F., Nahory, R. E., Shaklee, K. L.: Phys. Rev. Letters **28**, 437 (1972).
14. Levy, R.: Thesis, Strasbourg (1973).
15. Bivas, A., Levy, R., Grun, J. B., Comte, C., Haken, H., Nikitine, S.: Optics Commun. **2**, 227 (1970).

16. Levy,R., Bivas,A., Grun,J.B.: Phys. Letters A **36**, 159 (1971).
17. Hanamura,E.: Semicond. Conf. Boston 487 (1970).
18. Hanamura,E.: J. Phys. Soc. Japan **29**, 50 (1970).
19. Keldysh,L.V., Kozlov,A.N.: Sov. Phys. JETP **27**, 521 (1968).
20. Grun,J.B., Nikitine,S., Bivas,A., Levy,R.: Intern. Conf. on Luminescence, Delaware (1969) J. Luminescence **1,2**, 241 (1970).
21. Nikitine,S., Haken,H.: Int. Conf. on Luminescence, Leningrad (1972).
22. Nikitine,S.: In Cooperative Phenomena, H. Haken and M. Wagner, eds., p. 20–42. Berlin-Heidelberg-New York: Springer 1973.
23. Hanamura,E.: Private communication.
24. Haken,H., Nikitine,S.: Private communication.

Dr. R. Levy
Dr. A. Bivas
Dr. J. B. Grun
Prof. Dr. S. Nikitine
Laboratoire de Spectroscopie
et d'Optique du Corps Solide
Université Louis Pasteur
5, rue de l'Université
F-67000 Strasbourg (France)

VI. Laser Action of Excitons

Theory of Stimulated Emission by Excitons*

H. Haken and S. Nikitine

Contents

Laser action of excitons has been found by Mysyrowicz et al. [1] and others and has since been investigated mostly experimentally by several groups. In the present article we want to give a survey about some recent theoretical approaches dealing with this problem. The subsequent article will deal with recent experimental studies by the Strasbourg group.

In order to achieve laser action of excitons the crystals are first excited by optical pumping or by electron beam bombardment. In these processes electrons are brought from the valence band to the conduction band or higher conduction bands from where the electrons and holes relax to low lying energy states and eventually form excitons. In the following we want to discuss various decay processes of excitons into photons and other elementary excitations.

1. Different Approaches

Laser theory may be quite generally developed at three different levels [2].

a) The rate equation approach, which deals with light intensities or photon numbers neglecting phase relations and usually also quantum fluctuations. This approach gives a good insight into the various laser

* This work was carried out within the frame of the agreement on French-German scientific cooperation between C.N.R.S., Paris, Université Louis Pasteur, Strasbourg, and Universität Stuttgart.

processes and into the distribution of the emitted laser power over the laser modes.

b) The semiclassical approach which still neglects quantum mechanical fluctuations but deals with field amplitudes. It thus treats phase relations correctly, and allows in particular to study phase locking phenomena.

c) The quantum mechanical approach which fully takes into account the quantum mechanical fluctuations. It is in particular capable of deriving the coherence properties and the photon statistics of laser light.

As we will see below, exciton laser action is usually a rather complex phenomenon from both the theoretical and experimental point of view. Therefore the rate equation approach seems to be most appropriate at the present stage. At some later instances of this article we will, however, also point out the fully quantum mechanical approach.

2. The Rate Equations of the Laser for Various Decay Processes

a) Recombination Radiation of Excitons without Further Processes (ex → hν)

If not otherwise noted we will treat excitons as bosons. The number of excitons will be denoted by N_{ex}, the number of photons by n. More precisely, N_{ex} and n are quantum statistical averages over the corresponding number operators and can adopt continuously varying values ≥ 0. The photons are assumed to belong to a single resonator mode. (Otherwise suitable averages must be taken.) We assume first that only one kind of excitons is present which emit light by their decay to the crystal ground state. The rate equation for n can be obtained by the following prescription (for a rigorous derivation from first principles consult the appendix):

The temporal change of n is determined by the following processes:

1. Spontaneous decay of an exciton. The corresponding production rate of n is given by

$$G N_{ex} \tag{2.1}$$

where G is the radiative decay rate of an exciton into a photon of a given mode:

$$G = 2|g|^2/\gamma$$

g is proportional to the optical matrix element, γ is the decay constant of the exciton phase.

2. The stimulated emission rate is obtained from (2.1) by multiplying it with n:

$$GN_{ex}n. \tag{2.2}$$

Thus the total production rate is

$$GN_{ex}(n+1). \tag{2.3}$$

3. Losses by reabsorption: Because both photons and excitons are treated as bosons, they play a symmetric role. The absorption of a photon is thus accompanied by "spontaneous and stimulated emission" of an exciton. Exchanging the role of N_{ex} and n in (2.3), we obtain for the loss rate

$$-G(N_{ex}+1)n. \tag{2.4}$$

4. Losses through the mirrors, by scattering, etc. These losses are proportional to n: Denoting the decay constant by $2\varkappa$, we obtain as loss rate

$$-2\varkappa n \tag{2.5}$$

$2\varkappa = 1/t_0$, where t_0 is the lifetime of the photon in the cavity. (We denote all kinds of resonators even with very high losses as a "cavity".) The lifetime of the mode is limited by

$$t_0 = (L/c)(1-R)^{-1} \tag{2.6}$$

where L is the length of the crystal, R the reflection coefficient of the endfaces and c the light velocity in the crystal. For small crystals one may also expect considerable diffraction losses. Further losses are caused by inhomogeneities in the crystal etc.

Putting (2.3)–(2.5) together, the rate equation reads

$$dn/dt = G(N_{ex}-n) - 2\varkappa n. \tag{2.7}$$

Because the photon number appears on the right hand side of (2.7) with negative sign, laser action cannot take place. The reason for this rests in the Bose statistics of excitons which are coupled to the light field which also obeys Bose statistics. As is seen from the sum of (2.3) and (2.4), the stimulated emission processes ($\propto n$) are completely compensated by absorption processes. Thus in order to obtain stimulated emission of excitons other mechanisms than the direct recombination must be considered. The essential point is that reabsorption processes at the same frequency (i.e. the same energy) must be prevented. This can be achieved in different ways, which we will now list:

b) Recombination Radiation of Excitons with Further Processes

α) While one exciton (ex) is emitting a light quantum ($h\nu$), it gives the energy to an additional elementary excitation. Such elementary excitations may be

1. one or several phonons[1]

$$\text{ex} \longrightarrow h\nu + h\nu_{\text{phonon}}$$

or

$$\text{ex} \longrightarrow h\nu + lh\nu_{\text{phonon}} \quad (l = 2, 3, \ldots) ;$$

2. a second exciton (i.e. exciton–exciton collision) including[2] the dissociation of an exciton into an electron and a hole

$$\text{ex}_1 + \text{ex}_2 \longrightarrow h\nu_{(\text{ex}-\text{ex})} + e + h ;$$

3. excitation of conduction electrons

$$\text{ex} + e \longrightarrow h\nu + e' ;$$

4. excitation of holes

$$\text{ex} + h \longrightarrow h\nu + h' ;$$

5. excitation of plasmons

$$\text{ex} \longrightarrow h\nu + h\nu_{\text{plasmon}} .$$

A further class of processes of the required sort is a radiative recombination of an exciton within a *biexciton*, because it is unprobable that an exciton is again created in the neighbourhood of another exciton already present, provided the exciton concentration is not too high.

β) Excitons belonging to complexes with internal excitation of the complex.

γ) It has been observed experimentally that at high exciton concentrations [5] there is a shift between the emission and absorption lines which also prevents the reabsorption of the emitted light, so that laser action might become possible.

We now represent the rate equations for the various processes. We first consider only excitons with one internal state and denote the number of excitons with total wave vector q by N_q. The total number of excitons is denoted by $N_{\text{ex}} = \Sigma_q N_q$.

---

[1] See e.g. Haug [3].
[2] See [4].

1. Parametric Process[3] *with Emission of a Phonon* $(ex \rightarrow h\nu + h\nu_{phonon})$. Because the phonons participating in this process are usually longitudinal optical phonons, we denote the number of photons created by this process by n_{lo}. The number of phonons with wave vector q is denoted by n_q. We present a heuristic derivation of the corresponding rate equation (for a derivation from first principles see Haug [4]).

Again we treat photons, phonons and excitons as Bose particles and use the fact that the production rate is proportional to the corresponding number plus one

$$n_{lo} + 1, \quad n_q + 1, \quad N_q + 1 \tag{2.8}$$

the annihilation rate to the number only

$$n_{lo}, \quad n_q, \quad N_q . \tag{2.9}$$

The individual terms of the rate equation for the photon number are as follows:

The *photon creation rate* by decay of an exciton with wave vector q, and the simultaneous creation of a phonon with the same number (due to momentum conservation) is given by the product of the corresponding rates (2.8) and (2.9).

$$C_q N_q (n_{lo} + 1)(n_q + 1) \tag{2.10}$$

where C_q is a q-dependent rate constant.

The loss rate due to the inverse process is given by

$$- C_q (N_q + 1) n_{lo} n_q . \tag{2.11}$$

Because all excitons q may contribute, we have to sum up (2.10) and (2.11) over all values of q. Adding finally the loss rate due to the finite photon lifetime,

$$- 2 \varkappa_{lo} n_{lo}$$

we obtain the following rate equation:

For the photon number:

$$\dot{n}_{lo} = - 2 \varkappa_{lo} n_{lo} + \sum_q C_q [N_q (n_{lo} + n_q + 1) - n_{lo} n_q] . \tag{2.12}$$

The equation for the total number of excitons $N_{ex} = \Sigma_q N_q$ can be derived in an analogous way

$$\dot{N}_{ex} = P - \sum_q 2\gamma_q N_q - \sum_q C_q [N_q (n_{lo} + n_q + 1) - n_{lo} n_q] . \tag{2.13}$$

---

[3] The process described here is basically well known in electronics, nonlinear optics, mechanics. It occurs when one oscillation (the pump) is decomposed into two other oscillations (signal and idler) so that $\nu_{pump} = \nu_{signal} + \nu_{idler}$.

We briefly discuss the different terms. P is the pump rate $=$ number of excitons generated per second. It is assumed that exciton formation by the pump process occurs fast compared to the laser process. $2\gamma_q$ is the decay constant of excitons with wave vector q which is determined by all decay processes of excitons not giving rise to laser light emission. The sum containing γ_q represents the losses. The last sum describes the difference of the annihilation and creation processes of excitons corresponding to (2.10) and (2.11).

The equation for the phonons participating in the laser process are given by

$$\dot{n}_q = 2\hat{\gamma}_q(\bar{n}_q - n_q) + C_q[N_q(n_{lo} + n_q + 1) - n_{lo}n_q] \tag{2.14}$$

$\bar{n}_q$ denotes the thermal distribution of the phonons, $2\hat{\gamma}_q$ is the inverse relaxation time of the phonons to reach thermal equilibrium. We demonstrate that the above equations can be considerably simplified under special assumptions. We assume that the excitons for different q-values are thermally distributed so that

$$N_q = N_{ex}f_q(T). \tag{2.15}$$

We further assume that we operate at low temperatures and that the phonons are strongly damped so that

$$n_q \ll 1, \quad n_q \ll N_q. \tag{2.16}$$

Introducing further the abbreviations

$$\sum_q C_q f_q = \beta, \tag{2.17}$$

$$\sum_q 2\gamma_q f_q = \gamma \tag{2.18}$$

the rate Eqs. (2.12) and (2.13) reduce to the photon rate equation

$$\dot{n}_{lo} = -2\varkappa_{lo}n_{lo} + \beta N_{ex}(n_{lo} + 1) \tag{2.19}$$

and the exciton rate equation

$$\dot{N}_{ex} = P - \gamma N_{ex} - \beta N_{ex}(n_{lo} + 1). \tag{2.20}$$

For our following discussion we will present rate equations of a similar type for the other possible processes.

2. *Rate Equations with Exciton–Exciton Collision* $(ex_1 + ex_2 \rightarrow h\nu + e + h)$. These equations, whose derivation will be indicated in Section 5 reads (n_{ex}: photon number)

Photon rate equation:

$$dn_{ex}/dt = \alpha N_{ex}^2(n_{ex} + 1) - 2\varkappa_{ex}n_{ex}. \tag{2.21}$$

The first expression on the right hand side describes spontaneous and stimulated emission of photons connected with a bimolecular exciton process $\propto N_{ex}^2$, the last term represents as usual the losses. Reabsorption is neglected here.

Exciton rate equation:

$$dN_{ex}/dt = -2\alpha N_{ex}^2 (n_{ex} + 1) - \gamma N_{ex} + P . \tag{2.22}$$

The first term on the right hand side stems from losses due to stimulated and spontaneous emission of photons during the collision process. The factor 2 occurs, because two excitons are destroyed per event. The second term represents the decay of the exciton number by all other kinds of processes; P finally is the pump rate. Here it is assumed that exciton formation occurs quickly compared to all other processes described by Eq. (2.22).

3. *Rate Equations for Excitation of Electrons (Holes)* $(ex + e \rightarrow hv + e')$. We consider the process in which an exciton decays to a photon, giving simultaneously part of its energy to an electron (or hole) with which it collides. We denote the number of photons created by this process by n_{el}, the number of electrons by N_{el}. We expect (as can be also derived from first principles) that the photon production rate is proportional to the collision events which are a prerequisite for this process, i.e. $\propto N_{el} N_{ex}$. The *photon rate equation* thus reads

$$dn_{el}/dt = \tilde{\beta} N_{el} N_{ex} (n_{el} + 1) - 2\varkappa_{el} n_{el} . \tag{2.23}$$

The electron rate equation depends on which processes will be included.

4. *Rate Equations for Biexcitons* $(biex \rightarrow hv + ex)$. We use the following notation: n_{Bi}: number of photons created by the emission from biexcitons, N_{Bi}: number of biexcitons, N_{ex}: number of excitons, w: formation rate of biexcitons. Because we may neglect the reabsorption process for not too high numbers of excitons, the *photon rate equation* reads:

$$\dot{n}_{Bi} = \delta N_{Bi}(n_{Bi} + 1) - 2\varkappa_{Bi} n_{Bi} . \tag{2.24}$$

The rate of biexciton formation and annihilation is determined by the following processes: formation from two excitons: $\propto w N_{ex}^2$, decay by stimulated and spontaneous emission of photons (n_{Bi}): $\propto -N_{Bi}(n_{Bi} + 1)$, all other losses $\propto -\gamma_{Bi} N_{Bi}$.

Thus we obtain the *biexciton rate equation*

$$\dot{N}_{Bi} = w N_{ex}^2 - \gamma_{Bi} N_{Bi} - \delta N_{Bi}(n_{Bi} + 1) . \tag{2.25}$$

The exciton rate equation contains the following terms: pump rate P, formation rate of biexcitons $-2w N_{ex}^2$, decay of biexcitons into excitons (at low temperatures rather improbable) $\tilde{\gamma} N_{Bi}$, decay of excitons $-\gamma N_{ex}$. Finally, for each biexciton decay in the laser process, one exciton is created with the rate $\delta N_{Bi}(n_{Bi}+1)$.

$$\dot{N}_{ex} = -2w N_{ex}^2 + \tilde{\gamma} N_{Bi} - \gamma N_{ex} + P + \delta N_{Bi}(n_{Bi}+1). \tag{2.26}$$

In certain crystals several or all of the above mentioned processes may occur simultaneously. The recipe for obtaining the adequate rate equations for the total system is as follows. The rate equations for the photons and biexcitons remain unchanged, but the rate equations for the excitons must be altered in the following way. The time derivative dN_{ex}/dt is given by the sum of the right hand sides of the corresponding Eqs. (2.20), (2.22), (2.26), where, however, the term $P - \gamma N_{ex}$ must be taken only once.

3. Stationary Solution

In the following theoretical discussion (Sections 3 and 4) it should be borne in mind that the experimental situation is rather complex: the pump light may be inhomogeneous with respect to space and time. Spatial imhomogeneities of the laser process may be not only in the direction of the laser emission but also perpendicular to it. In the pump process a number of relaxation processes with presumably quite different time constants may be involved.

Finally, and most important, it seems that many of the stimulated processes occur as spatial or temporal transients. Nevertheless it is desirable to start with clear cut simple models which allow to elucidate the role of the various parameters. It is in this spirit that we first seek stationary solutions of the rate equations of Section 2. With respect to the discussion of experiments described later in this volume we investigate the competition between the phonon-assisted process and the exciton–exciton collision process.

The rate equations for these processes have been listed above and read

$$dn_{ex}/dt = \alpha N_{ex}^2(n_{ex}+1) - 2\varkappa_1 n_{ex}, \tag{3.1}$$

$$dn_{lo}/dt = \beta N_{ex}(n_{lo}+1) - 2\varkappa_2 n_{lo} \tag{3.2}$$

(note a small change of the indices at the $\varkappa$'s)

$$dN_{ex}/dt = -2\alpha N_{ex}^2(n_{ex}+1) - N_{ex}\beta(n_{lo}+1) - \gamma N_{ex} + P. \tag{3.3}$$

We first investigate in detail the properties of the steady state for which we may put the left hand sides of Eqs. (3.1), (3.2), (3.3) equal 0. In order to discuss the method of approach we first give the example in which

n_{lo} only is present which is achieved by putting $\alpha = 0$ and thus $n_{ex} = 0$. From Eq. (3.2) we obtain

$$n_{lo} = \beta N_{ex}(2\varkappa_2 - \beta N_{ex})^{-1}. \tag{3.4}$$

From (3.4) together with (3.2) we obtain for the stationary state after a short calculation

$$N_{ex} = (2\hat{\gamma})^{-1} \{K + \hat{p} \pm \sqrt{K^2 + 2\hat{p}\hat{\varkappa}(1 - \hat{\gamma}) + \hat{p}^2}\} \tag{3.5}$$

where we have introduced the abbreviations

$$K = \hat{\varkappa}(1 + \hat{\gamma}); \quad \hat{\varkappa} = 2\varkappa_2/\beta; \quad \hat{\gamma} = \gamma/\beta; \quad \hat{p} = P/\beta. \tag{3.6}$$

(3.5) gives us the number of excitons N_{ex} as a function of pump power. Thus by inserting N_{ex} into (3.4) we may readily calculate the number of photons. Two limiting cases are of special interest. For P *small* we obtain from (3.5) and (3.4)

$$N_{ex} \approx P(\beta + \gamma)^{-1} \tag{3.7}$$

and

$$n_{lo} \approx \tfrac{1}{2}(\beta/\varkappa_2) P(\beta + \gamma)^{-1}. \tag{3.8}$$

On the other hand for P *big* we obtain from (3.5)

$$N_{ex} \approx \hat{\varkappa} - \hat{\varkappa}^2/\hat{p} + \hat{\varkappa}^3(1 - \hat{\gamma})/\hat{p}^2 \tag{3.9}$$

and thus for n_{lo}

$$n_{lo} \approx \hat{p}/\hat{\varkappa} - \hat{\gamma} - 1 \equiv P/2\varkappa_2 - \gamma/\beta - 1 \geqq 0. \tag{3.10}$$

The condition that $n_{lo} > 0$ yields the usual laser condition. We now observe that the special cases (3.7), (3.8), (3.9), (3.10) could have been obtained from the basic Eqs. (3.2) and (3.3) in a much simpler way avoiding the square root expression (3.5) as follows. Let us consider again the relevant equations

$$\beta N_{ex}(n_{lo} + 1) - 2\varkappa_2 n_{lo} = 0 \tag{3.11}$$

and

$$-\beta N_{ex}(n_{lo} + 1) - \gamma N_{ex} + P = 0. \tag{3.12}$$

For P *small* [i.e. $\hat{p} \ll \tfrac{1}{2}\hat{\varkappa}(1 + \hat{\gamma})$] we assume $n_{lo} \ll 1$ so that we may neglect n_{lo} compared to one. From (3.12) we readily obtain

$$N_{ex} \approx P(\beta + \gamma)^{-1} \tag{3.13}$$

whereas (3.11) now yields

$$n_{lo} \approx \beta N_{ex}/2\varkappa_2 \approx \tfrac{1}{2}(\beta/\varkappa_2) P(\beta + \gamma)^{-1}. \tag{3.14}$$

For P big (i.e. $\hat{p} \gg \hat{\varkappa}(1 + \hat{\gamma})$) we may assume $n_{lo} \gg 1$. Then we may start with Eq. (3.11), which after division by n_{lo} yields

$$N_{ex} \approx 2\varkappa_2/\beta . \tag{3.15}$$

Introducing (3.15) into (3.12) allows us to solve (3.12) for n_{lo}:

$$n_{lo} \approx P/2\varkappa_2 - \gamma/\beta - 1 . \tag{3.16}$$

We now return to the general case of (3.1)–(3.3) where we again confine our analysis to the stationary state. The results are as follows:
Solving (3.1) and (3.2) yields

$$n_{ex} = \alpha N_{ex}^2 (2\varkappa_1 - \alpha N_{ex}^2)^{-1} \tag{3.17}$$

and

$$n_{lo} = \beta N_{ex} (2\varkappa_2 - \beta N_{ex})^{-1} . \tag{3.18}$$

We now distinguish between the following cases:

1. P small, $n_{ex} \ll 1, n_{lo} \ll 1$. Neglecting terms quadratic in N_{ex} we obtain from (3.3)

$$N_{ex} \approx P(\gamma + \beta)^{-1} \tag{3.19}$$

and thus from (3.17) and (3.18)

$$n_{ex} \approx \alpha N_{ex}^2/2\varkappa_1 , \tag{3.20}$$

$$n_{lo} \approx \beta N_{ex}/2\varkappa_2 . \tag{3.21}$$

2. P big. We distinguish the following cases:
α) $n_{ex} \gg 1; n_{lo} \ll 1$.
We readily obtain from the Eqs. (3.1)–(3.3) (again stationary state)

$$N_{ex} \approx \sqrt{2\varkappa_1/\alpha} \tag{3.22}$$

and

$$n_{ex} \approx \tfrac{1}{4}\varkappa_1^{-1} (P - (\gamma + \beta)\sqrt{2\varkappa_1/\alpha}) - 1 \tag{3.23}$$

and

$$n_{lo} \approx \beta \sqrt{2\varkappa_1/\alpha}/2\varkappa_2 . \tag{3.24}$$

Thus n_{lo} is now saturated whereas n_{ex} increases linearly with the pump power.

β) $n_{ex} \ll 1$, $n_{lo} \gg 1$.

In this case we obtain

$$N_{ex} \approx 2\varkappa_2/\beta,$$ (3.25)

$$n_{ex} = \alpha N_{ex}^2 (2\varkappa_1 - \alpha N_{ex}^2)^{-1} \approx 2\alpha \varkappa_2^2/\beta^2 \varkappa_1,$$ (3.26)

$$n_{lo} \approx \tfrac{1}{2}\varkappa_2^{-1} (P - (\gamma + \beta) N_{ex} - 2\alpha N_{ex}^2)$$ (3.27)

or, using (3.25)

$$n_{lo} = P/2\varkappa_2 - \gamma/\beta - 1 - 2\alpha \cdot 2\varkappa_2/\beta^2.$$ (3.28)

γ) $n_{ex} \gg 1$, $n_{lo} \gg 1$.

In this case we obtain from (3.1)

$$N_{ex} = \sqrt{2\varkappa_1/\alpha}$$ (3.29)

whereas (3.2) yields

$$N_{ex} = 2\varkappa_2/\beta.$$ (3.30)

Obviously (3.29) and (3.30) contradict each other. Thus the case γ) which assumes laser action in both decay channels (ex–ex and phonon-assisted) is not possible. Because experimentally the coexistence of laser action has been observed, we discuss possible explanations. As may be readily deduced from conventional laser theory, simultaneous laser action becomes possible under the following circumstances:

a) stationary state:

1. laser action of different modes is supported by spatially different regions,

2. it is supported by different kinds of excitons

b) the process is transient.

4. Transient States

A most probable possibility is the explanation by transient effects. Transient processes are described by equations of the form e.g. (2.19), (2.20) which are nonlinear differential equations and which in general require a computer solution. In order to get a first insight into at least the qualitative behaviour one must therefore make additional assumptions which allow to solve the corresponding equations explicitly. We choose as an example the equations

$$dn/dt = gN(n+1) - 2\varkappa n,$$ (4.1)

$$dN/dt = P - gN(n+1) - \gamma N$$ (4.2)

(or a set of such equations) where we have dropped all indices. In the following we discuss several approximation methods.

1. We first consider *transient states without gain saturation*. In this approximation we assume that laser light is built up so quickly that N may be kept fixed. Denoting the gain coefficients by G_j, the equations for the different laser modes may be now written in the form

$$dn_j/dt = G_j(n_j + 1) - 2\varkappa_j n_j. \tag{4.3}$$

If the gain is bigger than the losses,

$$G_j - 2\varkappa_j > 0, \tag{4.4}$$

the solution of (4.3) reads

$$n_j = G_j(G_j - 2\varkappa_j)^{-1} (e^{(G_j - 2\varkappa_j)t} - 1). \tag{4.5}$$

If the gain is smaller than the losses,

$$G_j - 2\varkappa_j < 0, \tag{4.6}$$

the solution reads

$$n_j = G_j(2\varkappa_j - G_j)^{-1} (1 - e^{-|G_j - 2\varkappa_j|t}). \tag{4.7}$$

By putting $ct = x$ (c is the light velocity in the material) we may proceed from the time dependent equations (4.3) to space dependent equations which then would apply to the experimental setup discussed in the articles by Levy et al. and Shaklee [6] in this volume.

2. As our second example we treat *transient states including gain saturation* but assume the gain is so high that we may neglect the pump and the decay processes during the laser process. This means we may replace equations (4.1), (4.2) by

$$dn/dt = gN(n+1), \tag{4.8}$$

$$dN/dt = -gN(n+1). \tag{4.9}$$

Taking the sum of the Eqs. (4.8) and (4.9) we immediately verify that $N + n$ is a constant of motion. We write

$$N = N_0 - n \tag{4.10}$$

where N_0 is the initial number of excitons because initially practically no photons are present. Inserting (4.10) into (4.8) yields

$$dn/dt = g(N_0 - n)(n + 1) \tag{4.11}$$

which may be solved in an elementary manner yielding

$$n = N_0(e^{gt(N_0+1)} - 1)(N_0 + e^{gt(N_0+1)})^{-1} \tag{4.12}$$

with the initial condition $n(0) = 0$, $N(0) = N_0$.

For small times $t \lesssim 1/gN_0$ we obtain from (4.12)

$$n \approx N_0^2 gt/N_0 = N_0 gt \tag{4.13}$$

and for big times $t \gg 1/gN_0$

$$n = N_0 . \tag{4.14}$$

While (4.12) is a good approximate solution of (4.1), (4.2) for small and intermediate times, for large t it becomes worse because the gain (which is proportional to N) decreases and the decay constants $\varkappa$ and γ become important. A numerical estimate, however, shows that due to the high gain of exciton lasers the approximation (4.12) as solution of (4.1), (4.2) is a reasonable one. We now treat the equations for the exciton–exciton collision process

$$dn/dt = gN^2(n+1), \tag{4.15}$$

$$dN/dt = -2gN^2(n+1). \tag{4.16}$$

A similar method of solution may be applied, yielding the following equation for n as a function of time

$$N_0 \frac{n+1}{N_0-2n} \exp\left\{\frac{N_0+2}{N_0} \cdot \frac{2n}{N_0-2n}\right\} = \exp\left\{(N_0+2)^2 gt\right\}. \tag{4.17}$$

We discuss several special cases.
   a) $n \ll 1$ (i.e. small times)

$$n \approx gtN_0^2 ; \tag{4.18}$$

   b) $n \approx 1 \ll N_0$

$$n = N_0 e^\xi - 1 \tag{4.19}$$

where

$$\xi = N_0^2 gt - \ln N_0 \tag{4.20}$$

   c) to treat $n \gg 1$ we put $x = n(N_0 - 2n)^{-1}$ and obtain

$$2x + \ln x = \xi. \tag{4.21}$$

For $x \approx 1$, i.e. $1 \ll n \approx \frac{1}{3} N_0$ we find

$$n \approx N_0(\xi+1)(2\xi+5)^{-1} \tag{4.22}$$

and for $x \gg 1$, i.e. $n \gg 1$

$$n \approx \frac{1}{2} N_0 \xi/(1+\xi) \to \frac{1}{2} N_0 . \tag{4.23}$$

3. We finally represent an approach (used e.g. by Shaklee), which assumes that the exciton number adjusts itself adiabatically to the photon number. This corresponds mathematically to putting $dN/dt = 0$ in (4.2). It means that the decay constant γ must be bigger than the inverse time of the laser pulse. Under this assumption (4.2) yields

$$N = P[g(n+1)+\gamma]^{-1} \tag{4.24}$$

which after insertion into (4.1) leads to

$$dn/dt = gP(n+1)[g(n+1)+\gamma]^{-1} - 2\varkappa n. \tag{4.25}$$

This first order differential equation may again be solved explicitly and yields the following relation

$$\frac{(y-y_1)^\alpha (y-y_2)^\beta}{(y_0-y_1)^\alpha (y_0-y_2)^\beta} = e^{a(t_0-t)} \tag{4.26}$$

where

$$y = n+1, \tag{4.27}$$

$$a = 2\varkappa g, \tag{4.28}$$

$$\alpha = (\gamma/g + y_1)(y_1 - y_2)^{-1}, \tag{4.29}$$

$$\beta = -(\gamma/g + y_2)(y_1 - y_2)^{-1}. \tag{4.30}$$

y_1 and y_2 are defined as follows

$$y_{1,2} = b/2a \pm (1/2a)\sqrt{b^2 + 4ac} \tag{4.31}$$

where the following abbreviations have been used:

$$b = gP - 2\varkappa\gamma + 2\varkappa g, \tag{4.32}$$

$$c = 2\varkappa\gamma. \tag{4.33}$$

y_0 corresponds to the initial photon number, n_0, at time t_0. (4.26) can be considerably simplified, if $P \gg 2\varkappa$. Then $|\alpha| \gg |\beta|$ and (4.26) reduces (under the initial condition $t_0 = 0$, $n_0 = 0$) to

$$n = (P/2\varkappa)(1 - e^{-2\varkappa t}). \tag{4.34}$$

This is equivalent to replacing the Eq. (4.25) from the very beginning by

$$\dot{n} = P - 2\varkappa n. \tag{4.35}$$

The following chapter is aimed at experts in quantum theory of the laser and may be omitted without loss of understanding of the following article.

5. Derivation of the Basic Rate Equations from First Principles

In this paragraph we sketch the derivation of rate equations in the case of exciton–exciton collision from first principles. We use the method of second quantization and introduce the following creation and annihilation operators

photons of the "cavity" mode b^+, b,
excitons with wave-vector k B_k^+, B_k,
dissociated electron-hole pairs with
total wave number k and internal
state j $\hat{B}_{k,j}^+, \hat{B}_{k,j}$.

We start from the following Hamiltonian

$$\hbar^{-1} H = \omega b^+ b + \sum_k \varepsilon_k B_k^+ B_k + \sum_{kj} \varepsilon_{kj} \hat{B}_{kj}^+ \hat{B}_{kj} + g b^+ B_0 + g B_0^+ b$$

$$+ \sum_{q_1 q_2 Q j} (V_{q_1 q_2 Q j} \hat{B}_{q_1 + Q, j}^+ B_{q_2 - Q}^+ B_{q_2} B_{q_1} + \text{h.c.}). \tag{5.1}$$

ω is the frequency of the laser mode, $\hbar \varepsilon_k$ the excitation energy of an exciton with translational wave vector k, ε_{kj} is the excitation energy of an electron-hole pair with total wave vector k and internal motion j, g is the coupling constant, i.e. essentially the optical matrix element between the laser mode and excitons with wave vector $k \approx 0$. V is essentially the Coulomb interaction matrix element which occurs when 2 excitons collide leaving one exciton and one dissociated pair. In writing down the Hamiltonian (5.1) we have included only those terms which are important for the processes which we will discuss. The terms occurring under the last sum are described by a diagram of the form

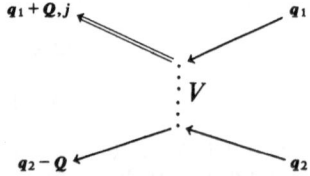

We are ultimately interested in a diagram of the form

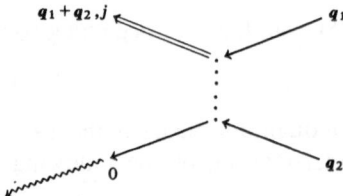

which can be lumped together in a diagram of the form

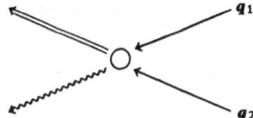

This diagram describes the required process: two colliding excitons give rise to a dissociated exciton and an emitted photon.

The method of solution has been well established in laser theory and is the following:

We proceed by the usual prescription for the time derivative of an arbitrary operator Ω

$$\dot{\Omega} = i\hbar^{-1}[H, \Omega]$$

to Heisenberg equation of motions which read, for instance for b^+, B_0^+

$$\dot{b}^+ = (i\omega - \varkappa) b^+ + igB_0^+ + F_b^+(t),\tag{5.2}$$

$$\dot{B}_0^+ = (i\varepsilon_0 - \gamma_0) B_0^+ + i \sum_{q_1 q_2 j} V_{q_1 q_2 q_2, j}^* B_{q_1}^+ B_{q_2}^+ \hat{B}_{q_1 + q_2, j}$$

$$+ 2i \sum_{q_1, Q, j} V_{q_1, 0, Q, j} \hat{B}_{q_1 + Q, j}^+ B_{-Q}^+ B_{q_1}\tag{5.3}$$

$$+ igb^+ + F_{B_0}^+(t).$$

In these Heisenberg equations of motion we have included the effect of damping and fluctuation: $\varkappa$ and γ_0 are damping constants, whereas F^+ are the fluctuating forces which are as usual assumed to have a Gaussian-Markovian distribution. For details of this method see [2].

Because our further procedure is rather lengthy we indicate only the most essential steps. We assume that the rapid time variation of b^+ is given by $e^{i\omega t}$ (with a slowly varying amplitude), and make a corresponding assumption for

$$B_q^+ \sim e^{i\varepsilon_q t},$$

$$\hat{B}_{Q,j}^+ \sim e^{i\varepsilon_{Q,j} t} \quad \text{etc.}$$

This allows (neglecting the time-dependence of the slowly varying amplitudes) the following solution of (5.3)

$$B_0^+ = \frac{ig}{i(\omega - \varepsilon_0) + \gamma_0} b^+ + i \sum_{q_1 q_2 j} \frac{V_{q_1, q_2, q_2, j}^*}{i\Delta\varepsilon_{q_1, q_2, j} + \gamma_0} B_{q_1}^+ B_{q_2}^+ \hat{B}_{q_1 + q_2, j}$$

$$+ 2i \sum_{q_1, Q, j} \frac{V_{q_1, 0, Q, j}}{i\Delta\varepsilon_{q_1, Q, j}' + \gamma_0} \hat{B}_{q_1 + Q, j}^+ B_{-Q}^+ B_{q_1} + \tilde{F}_{B_0}^+(t)\tag{5.4}$$

where $\Delta\varepsilon$ are linear combinations of ε_q and ε_{Qj}, and $\tilde{F}$ is again a fluctuating force. (5.4) allows us to eliminate B_0^+. One then readily realizes that the equations of motion could have been derived by means of an effective Hamiltonian (provided γ_0 may be neglected) which is of the form of the first two expressions in (5.1) plus the expression

$$V_{\text{eff}} \cdot b^+ \hat{B}^+_{q_1 + q_2 j} B_{q_2} B_{q_1} + V^*_{\text{eff}} \cdot B^+_{q_1} B^+_{q_2} \hat{B}_{q_1 + q_2 j} b \, . \tag{5.5}$$

The Langevin equations [i.e. the Heisenberg equations with damping and fluctuations as derived from (5.5)] have the general structure

$$\begin{aligned}
\dot{b}^+ &= \ldots B^+ B^+ \hat{B} \\
\dot{\hat{B}} &= \ldots b^+ B B \\
\dot{\hat{B}}^+ &= \ldots b B^+ B^+ \\
\dot{B}^+ &= \ldots b^+ \hat{B}^+ B
\end{aligned} \tag{5.6}$$

where the right hand sides are certain sums over the operators which still carry indices.

If one eliminates $\hat{B}^+$, $\hat{B}$ from these equations adiabatically one finds equations of the form

$$b^+ = \ldots B^+ B^+ B B b^+ + \text{damping} + \text{fluctuations} \, , \tag{5.7}$$

$$\dot{B}^+ = \ldots b^+ B^+ B^+ B b + \text{damping} + \text{fluctuations} \, . \tag{5.8}$$

Note that in (5.6), (5.7), (5.8) B^+, B still carry indices q and that the right hand sides are sums over these indices. Multiplying (5.7), (5.8) by b, B respectively, taking averages over the quantum fluctuations, using a random phase approximation with respect to the B^+'s, B's and putting

$$\overline{b^+ b} = n \, , \tag{5.9}$$

$$\overline{B^+ B} = N \tag{5.10}$$

we find the following rate equations

$$\dot{n} = -2\varkappa n + G N^2 (n+1) \, , \tag{5.11}$$

$$\dot{N} = -2\gamma N - 2G N^2 (n+1) + P \, . \tag{5.12}$$

In it, G is an effective averaged gain coefficient and P is the pump. Thus it has been possible to derive our original rate equations.

Appendix

Derivation of Eq. (2.7)

We treat a single mode (in the form of a running wave) and denote the corresponding photon creation and annihilation operators by b^+, b, respectively. Due to the k-selection rule for direct band-to-band transitions the single mode couples only to a single kind of excitons. The corresponding creation and annihilation operators are denoted by B^+, B, respectively. Using the rotating wave approximation [2], the Hamiltonian reads

$$H = \hbar\varepsilon B^+ B + \hbar\omega b^+ b + \hbar(gB^+ b + Bb^+ g^*) \tag{A.1}$$

where $\hbar\varepsilon$ is the exciton energy and $\hbar\omega$ the photon energy. g is proportional to the optical matrix element. We proceed to Heisenberg equations of motion and introduce damping constants $\gamma, \varkappa$ and fluctuating forces F_B^+, F_b^+ [2]. We thus obtain

$$\dot{B}^+ = i\varepsilon B^+ + ig^* b^+ - \gamma B^+ + F_B^+ , \tag{A.2}$$

$$b^+ = i\omega b^+ + igB^+ - \varkappa b^+ + F_b^+ . \tag{A.3}$$

We assume exact resonance, i.e. $\varepsilon = \omega$ and replace

$$B^+ \quad \text{by} \quad B^+ e^{i\omega t}, \quad b^+ \quad \text{by} \quad b^+ e^{i\omega t}. \tag{A.4}$$

This yields

$$\dot{B}^+ = -\gamma B^+ + ig^* b^+ + F_B^+ , \tag{A.5}$$

$$\dot{b}^+ = -\varkappa b^+ + igB^+ + F_b^+ . \tag{A.6}$$

The fluctuating forces have the following properties

$$\langle F^+ \rangle = 0 \tag{A.7}$$

and

$$\langle F_B^+ (t) F_B(t') \rangle = 2\gamma\delta(t - t')\,\bar{N}; \quad \langle F_b^+ (t) F_b(t') \rangle = 2\varkappa\delta(t - t')\,\bar{n} \tag{A.8}$$

where $\bar{N}$ is the equilibrium number of excitons and $\bar{n}$ the number of photons in thermal equilibrium. We assume $\gamma \gg \varkappa$, which allows us to neglect the time derivative in (A.5) so that we obtain immediately

$$B^+ = \gamma^{-1}(ig^* b^+ + F_B^+). \tag{A.9}$$

Introducing (A.9) into (A.6) yields

$$\dot{b}^+ = -\varkappa b^+ - (|g|^2/\gamma)\, b^+ + (ig/\gamma)\, F_B^+ + F_b^+ \tag{A.10}$$

which has the solution

$$b^+ = \int_0^t e^{-\bar{\varkappa}(t-\tau)} \{(ig/\gamma)\, F_B^+ (\tau) + F_b^+ (\tau)\}\, d\tau + b^+ (0)\, e^{-\bar{\varkappa}t} \tag{A.11}$$

where

$$\bar{\varkappa} = \varkappa + |g|^2/\gamma . \tag{A.12}$$

To derive a rate equation for the photon number, we form

$$n = \langle b^+ b \rangle = \int_0^t \int_0^t e^{-\bar{\varkappa}(t-\tau)} e^{-\bar{\varkappa}(t-\tau')} \langle \quad \rangle d\tau d\tau' + \langle b^+(0) b(0) \rangle e^{-2\bar{\varkappa}t}$$

$$(A.13)$$

with

$$\langle \quad \rangle = (|g|^2/\gamma^2) \langle F_B^+(\tau) F_B(\tau') \rangle + \langle F_b^+(\tau) F_b(\tau') \rangle . \qquad (A.14)$$

Using (A.8), we transform (A.14) into

$$\{2(|g|^2/\gamma) \bar{N} + \bar{n} 2\varkappa\} \delta(\tau - \tau') . \qquad (A.15)$$

Introducing (A.15) into (A.13) and performing the integration yields

$$\langle b^+ b \rangle = \bar{\varkappa}^{-1}(1 - e^{-2\bar{\varkappa}t}) \{(|g|^2/\gamma) \bar{N} + \varkappa\bar{n}\} + \langle b^+(0) b(0) \rangle e^{-2\bar{\varkappa}t} . \qquad (A.16)$$

Apparently (A.16) can be considered as the solution of the following rate equation

$$(d/dt) \langle b^+ b \rangle = 2(|g|^2/\gamma) \bar{N} - 2\bar{\varkappa} \langle b^+ b \rangle . \qquad (A.17)$$

With $n = \langle b^+ b \rangle$ and (A.12), (A.17) reads

$$dn/dt = 2(|g|^2/\gamma)(\bar{N} - n) - 2\varkappa n \qquad (A.18)$$

which is identical with (2.7).   q.e.d.

References

1. Mysyrowicz, A., Grun, J. B., Bivas, A., Levy, R., Nikitine, S.: Phys. Letters A **25**, 286 (1967).
2. Haken, H.: "Laser Theory" Encyclopedia of Physics, 25/2 c. Berlin-Heidelberg-New York: Springer 1970.
3. Haug, H.: J. Appl. Phys. **39**, 4671 (1968).
4. Bénoit à la Guilleaume, C., Debever, J. M., Salvan, F.: Phys. Rev. **177**, 567 (1969). Magde, D., Mahr, H.: Phys. Rev. Letters **24**, 890 (1970).
5. Bivas, A., Levy, R., Grun, J. B., Comte, C., Haken, H., Nikitine, S.: Optics Commun. **2**, 227 (1970).
   See also the article by R. Levy, J. B. Grun, A. Bivas and S. Nikitine: in this volume.
6. Shaklee, K. L.: In this volume where further references are given.

Prof. Dr. H. Haken
Institut für Theoretische Physik der Universität
7000 Stuttgart-Vaihingen
Federal Republic of Germany

Prof. Dr. S. Nikitine
Laboratoire de Spectroscopie
et d'Optique du Corps Solide
Université Louis Pasteur
5, rue de l'Université
F-67000 Strasbourg (France)

Experimental Investigation of the Competition of Stimulated Emissions Involving Exciton *

R. Lévy, J. B. Grun, and S. Nikitine

Contents

1. Introduction

The theoretical part of this subject is treated in the foregoing paper by Haken and Nikitine. Processes of stimulated emission involving excitons (or biexcitons) have been studied in the Strasbourg laboratory and in other laboratories. Some of these processes will be discussed in this paper. As the theoretical background was covered by Haken [1], only a short summary of the particular points directly connected with our experiments is given here.

2. Theoretical Background

a) Stimulated Emission from Excitons and Biexcitons

It has been shown [1] that no stimulated emission can be expected from excitons unless the exciton absorption line bleaches by high intensity of excitation, shifts in the spectrum, so that it is no longer super-

---

* This work was carried out within the frame of the agreement on French-German scientific cooperation between C.N.R.S., Paris, Université Louis Pasteur, Strasbourg, and Universität Stuttgart.

imposed on the resonance emission line. This situation has been observed and reported in the paper by Lévy, and some of his observations are possibly indications of such an exceptional emission. Biexcitons have been shown to give rise to stimulated emission. This effect has been observed by the Bell Laboratory group [2].

These processes, however, are beyond the scope of this paper, where interest is concentrated on two specific processes:

a) the recombination of an exciton with an emission of a photon and a phonon;

b) the recombination of one exciton in an inelastic collision with another exciton, with emission of a photon and dissociation of the second exciton into an electron and a hole;

c) competition between these two processes.

b) Stimulated Emission in the Process of Emission of a Photon and a Phonon

This effect was first observed by the Strasbourg group on CdS crystals [3]. The theory has been given by Haug [4], and we summarize the theoretical results as follows. The process suggested is a simultaneously emission of a photon and a LO phonon:

$$(ex) \rightarrow h[v(ex\text{-}LO) - v_{(LO)}].$$

The rate equation is as shown by Haken:

$$dn_{LO}/dt = \beta N_{ex}(n_{LO} + 1) - 2\kappa_1 n_{LO}.$$

Here n_{LO} is the number of photons generated in this process, N_{ex} is the number of excitons per cm^3, β is a rate coefficient, and κ_1 a loss coefficient. It can be seen that in a steady state when $\beta N_{ex} > 2\kappa_1$ the coefficient of n_{LO} is positive, which indicates a laser action.

c) Nonelastic Collisions of Excitons

In our experiments the above process competes with a second process, which is the nonelastic collision of excitons:

$$(ex) + (ex) \rightarrow h v(ex\ ex) + e + h.$$

Here $v(ex\ ex)$ is the frequency emitted in this process, and e and h are respectively a free electron and a free hole. It has been shown [1] that the rate equation for this process is

$$dn_{ex\ ex}/dt = \alpha N_{ex}^2(n_{ex\ ex} + 1) - 2\kappa_2 n_{ex\ ex},$$

where $n_{ex\ ex}$ is the number of photons emitted, and α is a rate coefficient. In the right-hand part of this equation, the first term represents stimulated emission, the second term spontaneous emission, and the last term losses. It can be seen that in this process laser action is also possible when $N_{ex}^2 > 2\kappa_2/\alpha$.

d) Competition between the Two above Effects [5]

It is clear that the first process (ex $-LO$) increases linearly with N_{ex}, but the second with N_{ex}^2. Now, the essential result of the theory [1] in this respect, is that both processes simultaneously cannot lead to stimulated emission; when stimulated emission sets in for one process the other must show saturation.

If $n_{ex\ ex} \gg 1$ but $n_{LO} \ll 1$, then $n_{ex\ ex} \simeq ap - b$, where p is the pumping rate, and a and b are constants given by Haken. However, $n_{LO} \simeq$ const. So the first effect increases with pumping, and the second is constant. For the symmetrical case, $n_{ex\ ex} \ll 1$ and $n_{LO} \gg 1$, it has been shown that $n_{ex\ ex} =$ const, and $n_{LO} = a'p - b'$, which is the reverse of the previous case[1].

The case $n_{LO} \gg 1$ and $n_{ex\ ex} \gg 1$ is impossible as the condition for one process is $N_{ex} = \sqrt{2\kappa_1/\alpha}$ and for the other process $N_{ex} = 2\kappa_2/\beta$, which cannot be compatible. Such a situation, if observed experimentally, can only happen in a transient state, not in a steady state, as shown by Haken [1].

These theoretical results enable us to understand the experiments described in the second part of this paper.

3. Experimental Results on the Stimulated Emission of CdS

The crystals of cadmium sulfide studied are vapor-grown platelets with natural surfaces[1]. No special attempts were made to have a laser cavity.

a) The Sample is Excited Perpendicular to its Surface
   and the Luminescence is Studied in the Direction Parallel
   to the Excited Surface [6]

The intensity of the emission lines has been measured versus the excitation intensity at the position of the exciton–exciton collision emission and at the position of the emission line due to the recombination of

---

[1] We thank Professor F. Raga (Cagliari), for providing the samples used in these experiments.

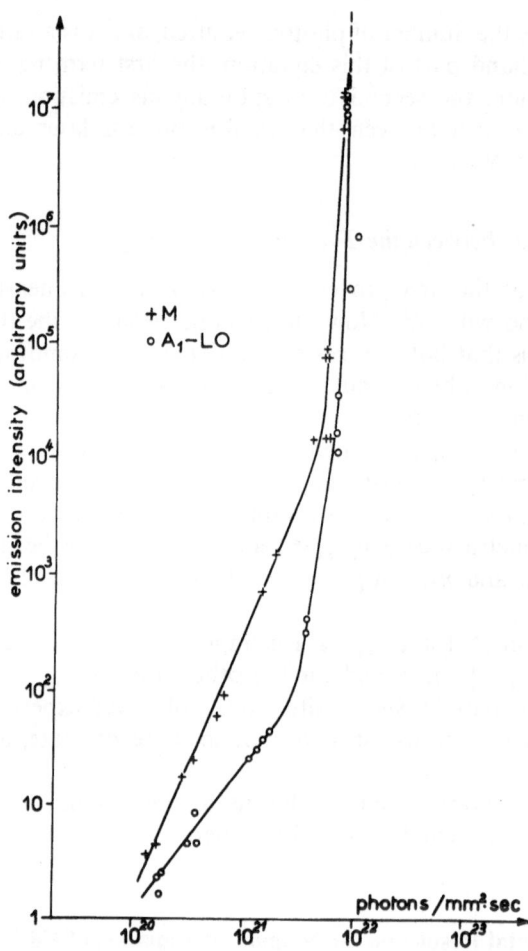

Fig. 1. Variation of the intensity of the stimulated M and A_1-*LO* emission lines with the UV laser light intensity at 4.2 K

a free exciton with simultaneous emission of a *LO* phonon (A-*LO*)[2]. These measurements are shown in Fig. 1.

A square dependence of the spontaneous emission intensity of the M line is observed at low excitation intensities. This behavior is in good agreement with the kinetic theory of the exciton–exciton recombination process given by Magde and Mahr [7]. At higher pump intensities

[2] As the first exciton line in CdS is usually named A, we have prefered the notation A-*LO* instead of the more general notation Ex-*LO*.

the stimulated emission starts to increase exponentially. The characteristic increase in intensity is observed on the curve.

The variation of the emission intensity of the A-*LO* line versus the excitation intensity is linear at low excitation intensity and becomes also superlinear for a certain threshold different from that of the M emission line.

We have obtained results similar to those of the preceding investigators [8, 10] especially Benoît à la Guillaume [8].

To understand the behavior of these stimulated emissions observed at high excitation intensities, we had to introduce the equations of variation of the number of emitted photons versus the time, given by Haken [1] for different recombination processes. As stated above, it has been shown [1] that in order to explain how both lines are stimulated, we have to suppose that the stimulated emission grows in a nonstationary state at the beginning of the laser pulse.

b) The Sample is Excited in a Direction Nearly Parallel to the Surface and the Luminescence is Observed Perpendicular to this Direction

The luminescence of all the samples studied was first excited by a low density of photons (with a high-pressure mercury lamp). The spectrum shown in Fig. 2 by a dashed line is an example of the exciton-recombination emission lines observed with our samples.

The spectra in Fig. 2 drawn in solid lines represent the emissions of the same sample under various ultraviolet laser light excitations. We observe first, at low intensities, the spontaneous emission of the sample (spectra a and b), but after a certain threshold the stimulated emission becomes so strong that we record essentially this stimulated emission (spectra c, d, and e).

Spectrum (a) is obtained with the lowest ultraviolet light intensity used in our experiments. The emission line M observed at 4907 Å (2.526 eV) is an exciton–exciton interaction. The main arguments for this interpretation are the spectral position of the line about two exciton binding energies away from the band gap, and the superlinear variation of the line intensity with the excitation intensity.

As the intensity of the ultraviolet light increases — spectra b and c — the line M shifts toward lower energies. This shift has been interpreted as a filling of the extrema of the conduction and valence bands during the exciton–exciton interaction.

When the intensity of excitation reaches a certain threshold, the emission becomes stimulated, as can be seen from the intense emission lines that appear suddenly in spectra c, d, and e. In spectrum c a stimulated line at 4919 Å (2.520 eV) is observed. In spectrum d, the line A_1-*LO*

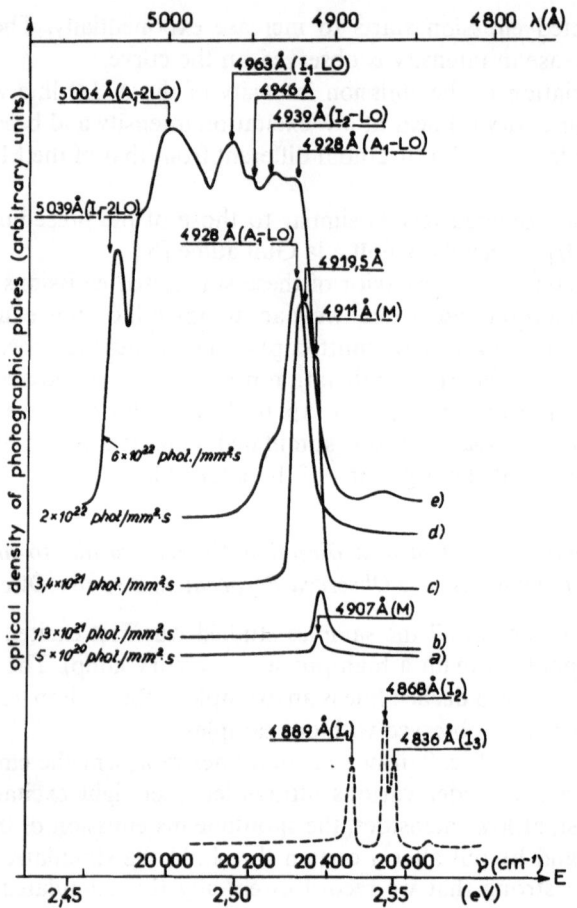

Fig. 2. Emission spectra of a sample of CdS excited by a high density of UV photons (laser) at 4.2 K. The spectrum represented by a dashed line shows the luminescence of the sample excited by a low density of UV photons (SP 500) at 4.2 K

(4928 Å – 2.515 eV) is stimulated, due to the recombination of a free exciton with simultaneous emission of an *LO* phonon. In spectrum e, finally, a group of lines corresponding essentially to phonon-assisted exciton recombinations is stimulated [3, 12].

A detailed study of the spectrum obtained at the onset of the stimulation was necessary, as we observed that a small change in the intensity of the exciting light gives different stimulated lines. Figure 3 represents the lines obtained for different excitation intensities. A simultaneous

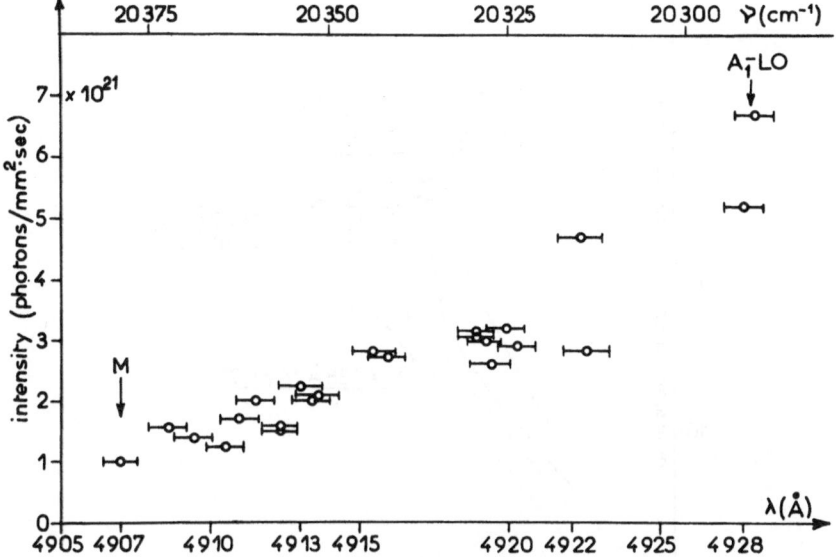

Fig. 3. Emission lines obtained with different intensities of UV light excitation at 4.2 K

study was made of the variation in the intensity of these lines with the intensity of light excitation; Fig. 4 gives some of the curves obtained.

At the lowest intensities of excitation (Fig. 3) the line observed is the M exciton–exciton recombination emission. As shown in Fig. 4, the line intensity varies with the square of the excitation intensity. The line shifts toward the lower energies of the spectrum as the excitation intensity increases. When the excitation intensity is slightly increased, emission lines appear successively, at 4913 Å (2.523 eV), 4915 Å (2.522 eV), 4919 Å (2.520 eV), 4923 Å (2.518 eV) and 4928 Å (2.515 eV), as can be seen in Fig. 3.

The dependence of the intensity of these lines on the excitation intensity is shown in Fig. 4. At low intensities of ultraviolet light excitation, the intensity of the lines varies in proportion to the ultraviolet light intensity. The lines are thus due to recombination processes that are different from the exciton–exciton interaction recombination, which varies quadratically with the intensity of the ultraviolet light.

At a certain threshold, the emission line (4913 Å – 2.523 eV) becomes stimulated and grows stronger than the M line, which becomes saturated. As the excitation increases, the intensity of this line is saturated too, and a new line (4915 Å – 2.522 eV) appears and becomes stimulated.

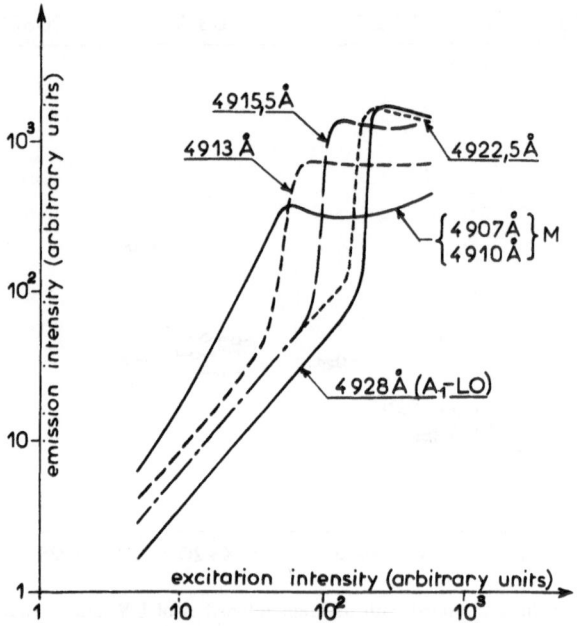

Fig. 4. Variation of the intensity of different emission lines with the UV laser light intensity at 4.2 K

Several emission lines of lower and lower energies in the spectrum become stimulated and eventually saturated, as the intensity of the ultraviolet light increases.

The processes involved in the lasing are assumed to be recombination of bound excitons to excited states of donors (proposed by Era and Langer), an exciton–bound exciton interaction recombination, and finally, the simultaneous emission of photons and *LO* phonons.

References

1. Haken, H.: Int. Symposium on Excitons at high density and polaritons, Tonbach (RFA), Oct. (1973).
   Nikitine, S., Haken, H.: Int. Conference on Luminescence, Leningrad (1972).
2. Shaklee, K. L., Leheny, R. F., Nahory, R. E.: Phys. Rev. Letters **26**, 888 (1971).
3. Mysyrowicz, A., Grun, J. B., Bivas, A., Levy, R., Nikitine, S.: Phys. Letters **25** A, 286 (1967).
   Levy, R., Bivas, A., Grun, J. B.: J. Phys. **31**, 507 (1970).
4. Haug, H.: J. Appl. Phys. **39**, 4681 (1968).
5. Nikitine, S.: Cooperative Phenomena. Berlin-Heidelberg-New York: Springer 1973.

6. Levy, R.: Thesis, Strasbourg (1973).
7. Magde, O., Mahr, H.: Phys. Rev. **2**, 4098 (1970).
8. Bénoît à la Guillaume, C., Debever, J. M., Salvan, F.: Phys. Rev. **177**, 567 (1969).
9. Gross, E., Permogorov, S., Razbirin, B.: J. Phys. Chem. Solids **27**, 1647 (1966).
10. Era, K., Langer, D. W.: J. Appl. Phys. **42**, 3 (1971).
11. Levy, R., Grun, J. B., Haken, H., Nikitine, S.: Solid State Commun. **10**, 915 (1972).
12. Mysyrowicz, A., Grun, J. B., Raga, F., Nikitine, S.: Phys. Letters A **24**, 335 (1967).

Dr. R. Levy, or J. B. Grun and Prof. S. Nikitine
Laboratoire de Spectroscopie
et d'Optique du Corps Solide
Université Louis Pasteur
5, rue de l'Université
F-67000 Strasbourg (France)

Experimental Studies of Excitons at High Densities

K. L. Shaklee

Contents

Studies of excitons at high densities pose interesting, and often difficult new experimental [1–3] and theoretical [4–6] questions. It is the purpose of this lecture to attempt to tie together, within the spirit of the scientific method, some aspects of these theoretical and experimental questions. Although the study of excitons at high densities is a relatively new field, there is already a wealth of experimental and theoretical information available. I cannot hope to cover the entire field in the time allotted, so I will discuss some selected topics which I feel illustrate some of the problems now facing us in this field. This talk will be heavily prejudiced toward the experimentalist's view of the present state of the art.

In the first section we will give a short discussion of some of the observable properties of high density exciton (HDE) systems, and review the experimental techniques used for studying such systems. In Section 2 we will consider, in some detail, two different experiments. The first experiment we will describe is the measurement of the reflectivity of an intense light beam by excitons in CdS. The results are analyzed using a harmonic oscillator model with spatial dispersion. It is found that the exciton damping parameter is linear in the intensity of the probe light beam. The second experiment we describe is an optical gain measurement [8] on highly excited CuCl [9]. Our discussion here will be devoted to features in the stimulated luminescence spectrum which have been attributed to excitonic molecule decay [9–11]. We attempt to interpret the observations in terms of a simple 3-level model for the excitonic molecule [12], and conclude that the observed kinetics and saturation behavior are not in good agreement with the model. In the last section we give the conclusions and some suggestions for definitive experiments which might shed some light on the validity of the excitonic molecule model.

1. Techniques

There exists a wealth of experimental observations, models for these observations and predictions for new observations in HDE systems. It is not possible in the short time available to discuss these points in any detail. We give a list of some of the kinds of information that we would like to have to check these models and predictions in Table 1. The references given are examples of work pertinent to these points, but are not intended to be exhaustive bibliographies. The various experimental tools that we have at our disposal for studying these systems are listed in Table 2. Most of the entries in Table 2 are self-explanatory. However, the method of ellipsometry has not been reported for use in studies of HDE systems. Such measurements are highly desirable since they give simultaneously the values of both the real and imaginary parts of the dielectric constant of the HDE system. Such information could, in principle, be obtained from reflectivity measurements through a Kramers-Kronig transformation, but on a HDE system it is difficult to obtain a reliable reflectivity spectrum over a wide enough spectral range to insure convergence with the desired sensitivity. Aspnes [13] has recently reported the design of a sensitive automatic ellipsometer which could be easily adapted to measurements on HDE systems. The reported system has sufficient sensitivity to measure changes in the complex dielectric function of one part in 10^4. This is of significance since it indicates a sensitivity in determining optical gain of $10 \, \text{cm}^{-1}$ which is well below the measured gains of $100–10000 \, \text{cm}^{-1}$ in most semiconductors.

In the remainder of this section we will describe briefly the experimental techniques for measuring the reflectivity of intense light beams, and for measuring optical gain spectra of HDE systems.

Table 1. Kinds of information

Types of excitations present

Coexistence
Spatial Homogeneity
Binding Energies of Bound States
Energy Transfer Mechanisms
Energy Bottlenecks
Particle-Particle Interactions
Lifetimes
   radiative
   nonradiative
Particle Densities

Table 2. Coupling methods

Interaction	Experiments which used
Magnetic	Stimulated Emission in GaAs[a]
	Cyclotron Resonance[b]
Electric	$p-n$ Junction Noise[c]
	Photoconductivity[d]
Electromagnetic fields	Absorption[e]
	Reflectivity[f]
	Light Scattering
	Elastic[g]
	Inelastic[h]
	Luminescence
	Spontaneous[i]
	Stimulated[j]
	Ellipsometry[k]
	$\Delta\varepsilon \sim 10^{-4}$
	$\Delta\alpha \sim 10\,\mathrm{cm}^{-1}$

[a] Shay, J. L., Johnston, W. D., Jr.: Phys. Rev. B**6**, 1605 (1972).
[b] Hensel, J. C., Phillips, T. G., Rice, T. M.: Phys. Rev. Letters **30**, 227 (1973).
[c] Asnin, V. M., Rogachev, A. A., Sablina, N. I.: JETP Letters **11**, 99 (1970).
[d] Asnin, V. M., Rogachev, A. A.: JETP Letters **9**, 248 (1969).
[e] Goto, T., Langer, D. W.: Phys. Rev. Letters **27**, 1004 (1971).
[f] See Ref. [7].
[g] Pokrovsky, Y. E., Svistunova, K. I.: JETP Letters **13**, 212 (1971).
[h] Worlock, J. M., Shaklee, K. L.: To be pub.
[i] See Ref. [10].
[j] See Ref. [1].

The experimental [7] arrangement for making reflectivity measurements of an intense light beam is shown in Fig. 1. The source of high intensity monochromatic light is a tunable pulsed dye laser. The laser, which has a maximum output power of 10 kW in a 10^{-8} sec pulse, consists of the dye methylumbelliferone dissolved in methanol and excited by a pulsed N_2 laser [14]. The dye laser has a typical band width of 1 Å and a repetition rate of 30 pulses/sec. The laser output passes through a variable attenuator and is focused to a 20 μ diameter spot on the surface of a CdS platelet which has been suspended in liquid He. The intensity of the incident beam is monitored by a silicon photodiode D_1 which detects a small fraction of the light reflected from a beam splitter, and the reflected beam is detected by another photodiode D_2. The relative reflectivity spectrum can be easily determined from the ratio to the output signals of detectors D_1 and D_2 for several different wavelength settings of the tunable dye laser.

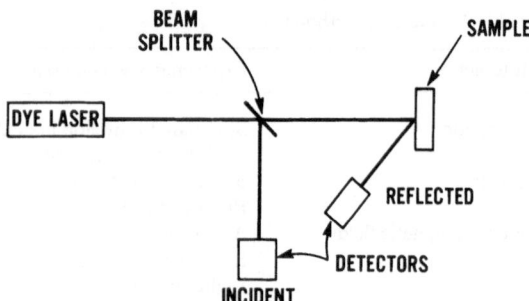

Fig. 1. Experimental arrangement for making reflectivity measurements using an intense light source

Fig. 2. Excitation geometry for gain measurements

The technique for measuring optical gain in highly excited semi-conductors has been described in detail elsewhere [8]. We will point out only the important features here. The rectangular beam of a pulsed N_2 laser with an output power of 10^5 W in a 10^{-8} sec pulse is focused on the sample surface to create a rectangular region of excitation as indicated in Fig. 2. The width w of the excited region is typically 20–50 μ while the length l is continuously varied from 0–5 mm by a slit assembly placed in the laser beam. The super-luminescent (stimulated) light passing out the end of the sample is dispersed by a monochromator and detected by a fast photomultiplier.

The output intensity can be monitored as a function of the excitation intensity, the excitation length, or the spectrometer wavelength setting. For determining optical gain it is most convenient to use the excitation length as the continuously variable parameter. The output intensity I then has the form [15]

$$I = 8(J/g)(e^{gl} - 1)$$

where J is the spontaneous emission intensity per unit length, g is the optical gain factor in units of inverse length, and l is the length of the excited region. The equation is valid for values of $gl < 10$, and under these conditions g is refered to as the small signal gain. For larger values of $g\,l$, the gain begins to saturate, and more careful treatment of the output intensity must be given. This will be discussed in detail in a later section.

2. Examples

Figure 3 shows the normal incidence reflectivity spectra of CdS at 2 K for three different incident light intensities. The structures correspond

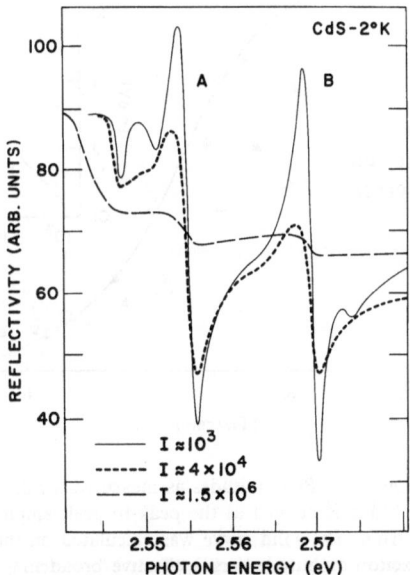

Fig. 3. Reflectivity spectra of CdS at 2 K for three different incident light intensities

to the *A* and *B* excitons of CdS [16]. The solid curve, obtained at a relatively low incident intensity of $\sim 10^3$ watts/cm^2, is quite similar to spectra obtained in conventional low-intensity reflectivity measurements [16]. The unusual structures at the lower energies near 2.55 eV are caused by back surface reflections and absorption corresponding to I_2 and I_3 bound exciton complexes. The small irregularity just below 2.58 eV corresponds to the $n = 2$ state of the *A* exciton.

With higher intensity illumination we obtain the dashed curves in Fig. 3. The peak-to-peak amplitudes of both *A* and *B* are seen to decrease and for intensities greater than 10^6 watts/cm^2 they are barely seen. Note in the short-dash curve that the amplitude of *B* is roughly half the size of *A*, indicating that *B* exciton structures are decreasing faster than *A* exciton structures. This might be expected since *B* excitons have the possibility of decaying into *A* excitons. Figure 3 also shows that for both *A* and *B* excitons, the minima in the spectra apparently do not shift with an increase in light intensity. The maximum preceding each

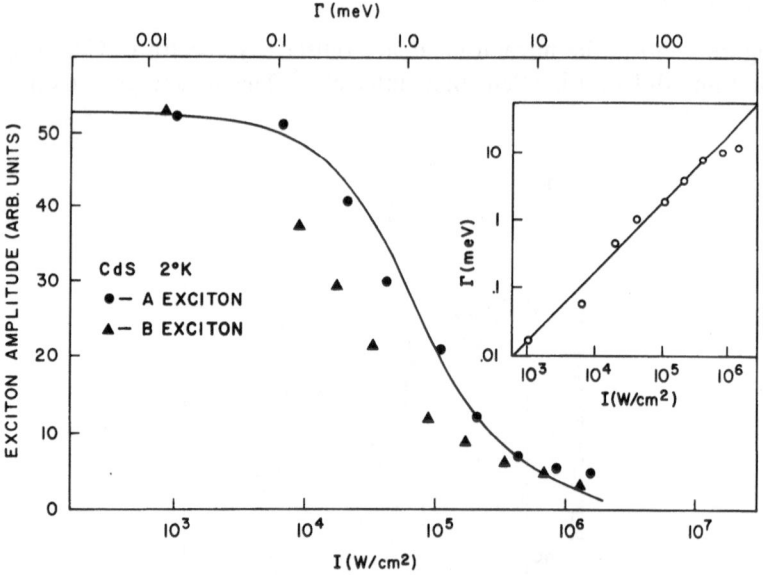

Fig. 4. Points give exciton (*A* or *B*) amplitudes as measured for different light intensities (lower scale). Here amplitude is defined as the peak-to-peak amplitude observed in the exciton reflectivity spectrum. The solid curve was calculated on the basis of oscillatory theory and gives the exciton amplitude versus effective broadening parameter Γ (upper scale) and is adjusted to the *A* exciton data. The insert shows the broadening parameter vs. light intensity *I*

minimum does shift slightly to lower energies. Since there is no shift of the structure as a whole, however, we rule out sample heating as the principal cause of the shrinking of the reflectivity structure [9].

In Fig. 4 the points show the measured peak-to-peak amplitudes of both the A and B exciton reflection structures plotted versus incident intensity (lower horizontal scale). The more rapid decrease of B exciton amplitude mentioned above is clearly seen. The solid curve was calculated on the basis of oscillator theory, and is discussed in the next paragraph.

The low intensity exciton reflectivity spectrum of CdS can be represented very well in terms of a classical oscillator if spatial dispersion is included [17]. In this model, the reflectivity $R = f(\omega, \Gamma, F, M, z)$, where Γ is the damping parameter, F is the oscillator strength, M is the exciton effective mass, ω is the photon frequency, and z is the depth of a surface barrier layer for excitons. No explicit dependence on intensity has been included. Hence, if the model is to fit the data in Fig. 4, one or more of the parameters of f must depend on the incident light intensity. Figure 5 shows the calculated reflectivity spectra for the oscillator model for different values of the damping parameter Γ using material parameters appropriate to CdS. The barrier depth z has been neglected since we have not, at present, observed the sharp spike associated with the longitudinal exciton [17]. The line shapes and amplitudes at different values of Γ in Fig. 5 are strikingly similar to the line shapes and amplitudes at different intensities in Fig. 3. In fact, we find that in order to fit the data, Γ must be made a linear function of the light intensity,

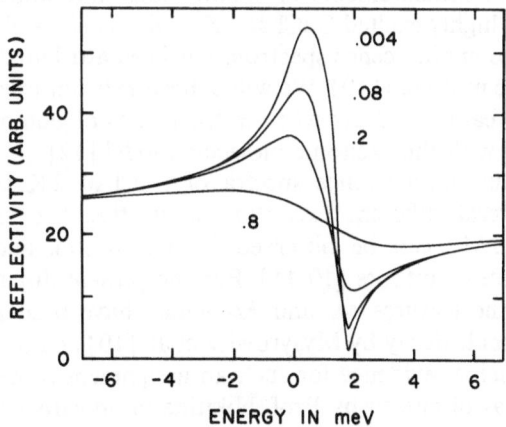

Fig. 5. Exciton reflectivity spectra calculated using a classical oscillator model for four different values of the broadening parameter Γ

I. This can be seen in Fig. 4 where the curve represents the calculated *A* exciton amplitude (peak-to-peak) as a function of Γ (upper horizontal scale). An excellent agreement with the data points for the *A* exciton was obtained by translating the upper horizontal scale, Γ, parallel to the lower scale *I*, which is equivalent to multiplying *I* by a constant, *K*. Hence, in this intensity regime, one should replace Γ by $\Gamma = KI$ in the oscillator model. It should be noted that, according to the data, the oscillator strength is not changing with light intensity in this mode. Changes in strength give energy shifts in the spectrum which are not observed (except possibly at the very highest intensities where the structures are already greatly obscured).

The oscillator model gives a good representation of the experimental points over a range of more than three orders of magnitude in incident intensity. However, the model fails at the highest intensities ($I > 10^6$ watts/cm^2) where the calculated curve lies below the experimental points. The failure is more dramatic when one realizes that the predicted energy separation of the reflectivity maxima and minima for an oscillator increases very rapidly for large Γ and becomes too large to be consistent with the observed spectrum at the highest intensities. This breakdown of the oscillator model at very high intensities might be expected on the basis of strong exciton-exciton interaction.

Although the origin of the linear relationship between Γ and *I* is not clear, these experimental results indicate that interesting new effects occur when high intensity light beams are used for probes such as in a reflectivity measurement.

As a second example of an experimental study of a HDE system, we will discuss measurements of spontaneous and stimulated luminescence [9] of highly excited CuCl at 2 K. This system is of interest since a feature in the luminescence spectrum has been attributed to the decay of an excitonic molecule [10]. We will concentrate our attention on this luminescence feature and ask whether the results of gain measurements are consistent with the excitonic molecule model [12].

Spontaneous luminescence spectra of CuCl at 2 K are shown in Fig. 6 for several different excitation intensities. Six main features (A, B, C, D, M_1, M_2) can be observed in the spectra, and have been discussed by other authors [10, 11]. For the present discussion we are interested in the features M_1 and M_2 which have been attributed to excitonic molecule decay by Mysyrowicz et al. [10] and by Souma et al. [11]. The principal evidence for such an assignment comes from three observations, as discussed by Prof. Nikitine in an earlier lecture. They are: (a) the kinetics, (b) the spectral position, and (c) the lineshape.

The stimulated luminescence spectra of this system are shown in Fig. 7 using the excitation intensity (a) and the excitation length (b) as

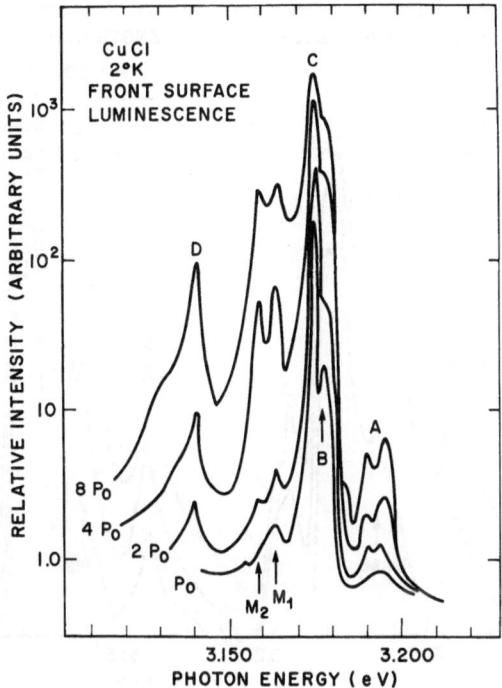

Fig. 6. Spontaneous emission spectra of CuCl at 2 K for several different excitation levels

parameters. These spectra were taken using an optical gain measurement technique which has been described elsewhere [1, 7]. The principal feature of this technique is that the length of the excited volume is carefully controlled. This is important since, in stimulated emission measurements, we consider the excited volume of the sample as a superluminescent optical amplifier [19] where the output intensity is some superlinear function of the length of the excited region. The spectra in Fig. 7a indicate that the stimulated emission intensity has a nonlinear dependence on excitation level as would be expected from measurements of spontaneous luminescence shown in Fig. 6. The spectra in Fig. 7b indicate that the molecule line also has a nonlinear dependence on excitation length. This is demonstrated more completely in Fig. 8 where we plot the stimulated luminescence intensity as a function of excitation length as is normally done for the gain measurement. We can see that there is a well defined exponential region for each excitation intensity, and the gain can be determined from a measurement of the slope of the exponential region. The measured gains are indicated in the

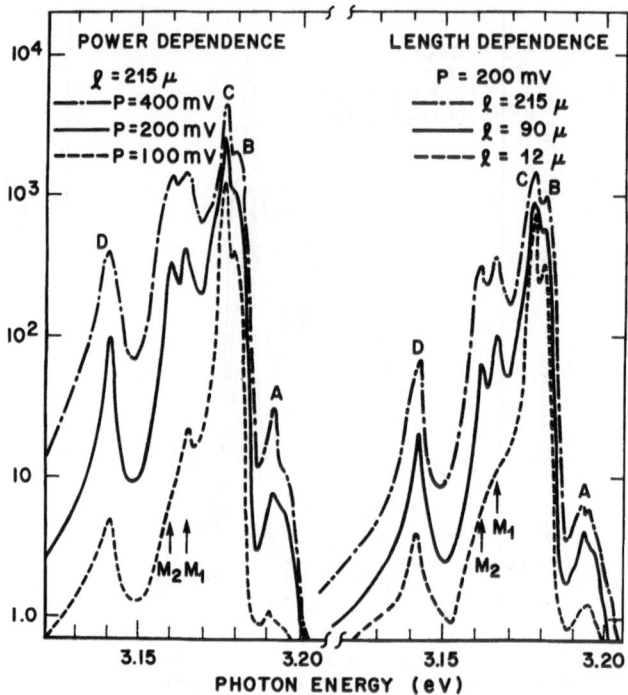

Fig. 7. Stimulated emission spectra of CuCl at 2 K using excitation intensity (a) and excitation length (b) as parameters. A power setting of 400 mV corresponds to an incident intensity of 5 MW/cm²

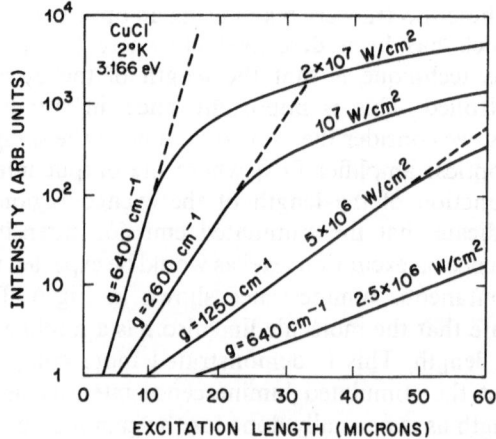

Fig. 8. Variation of the stimulated luminescence intensity of the excitonic molecule line M_1 as a function of excitation length for several different excitation intensities. The measured gain is indicated for each curve

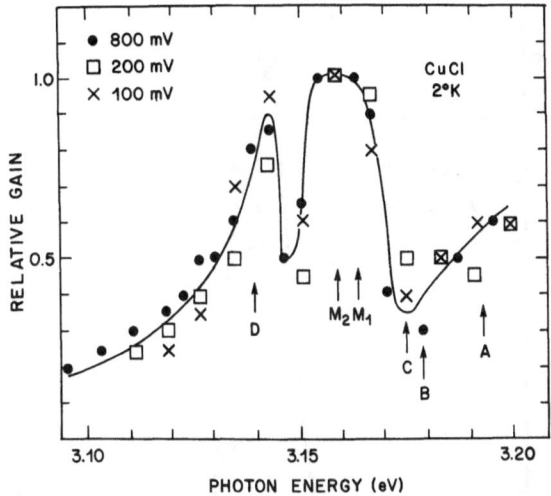

Fig. 9. Plot of the spectrum of the optical gain parameter of CuCl at 2 K for several different excitation intensities. In each spectrum the observed gain has been divided by the excitation intensity to obtain a normalized gain spectrum. The peak gain at 10 MW/cm² corresponds to a value of 6400 cm⁻¹

figure. It is possible then, by performing this measurement at several wavelengths, to obtain the spectrum of the optical gain factor as shown in Fig. 9. In this figure the observed gains have been divided by the excitation intensities used to obtain a normalized gain spectrum. Notice that the points scaled in this way all fall on the same curve. The positions of the various features in the stimulated luminescence spectra have been indicated in Fig. 9 for reference.

There are three regions of interest in a gain measurement [1] depending on the intensity in the stimulated beam. These regions are spontaneous emission, gain, and saturated stimulated emission. We have shown data for gain measurements, and for spontaneous luminescence measurements which have also been discussed in the literature. Alternatively, spontaneous emission spectra can be obtained from stimulated emission spectra taken for short excitation lengths where the gain-length product gl is much less than one. We can obtain saturated stimulated emission spectra by allowing the excitation to become long enough such that the stimulated light flux builds up to the saturation level. The spectrum obtained under these conditions is shown in Fig. 10.

The saturated stimulated emission spectrum in Fig. 10 has five peaks (labeled a, b, c, d, e) in the vicinity of the excitonic molecule lines M_1 and M_2. There is some indication of a shoulder on the high energy

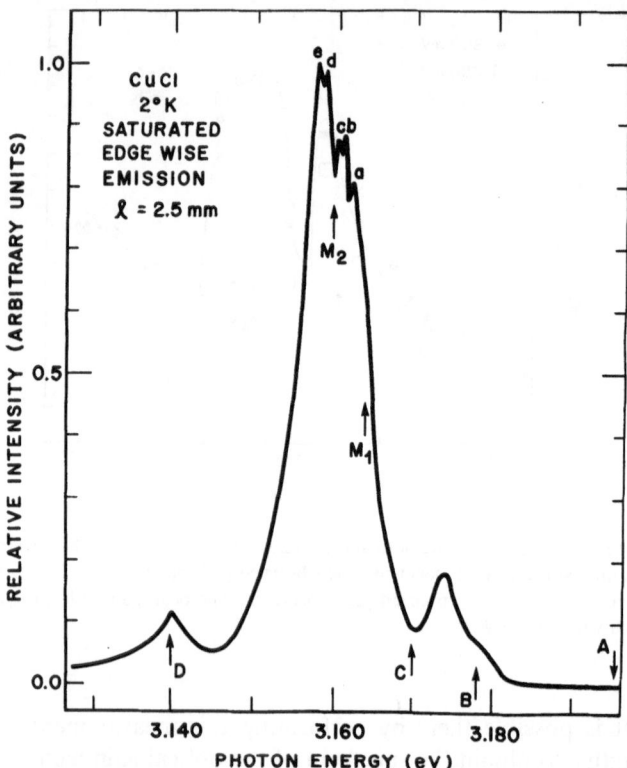

Fig. 10. Saturated stimulated emission spectrum of CuCl at 2 K for an excitation length of 2.5 mm and an excitation intensity of 10 MW/cm²

Table 3. Splitting of fine structure in saturated stimulated emission spectrum (Fig. 10)

a–b	1.2 meV
b–c	0.9 meV
c–d	1.4 meV
d–e	0.9 meV

side of the large peak in Fig. 10 which may correspond to the molecule peak M_1 in the spontaneous luminescence. The energies of the sharp peaks are listed in Table 3.

So far we have only presented the data on stimulated luminescence in the vicinity of the excitonic molecule in CuCl. It remains to ask whether this data is consistent with the excitonic molecule model.

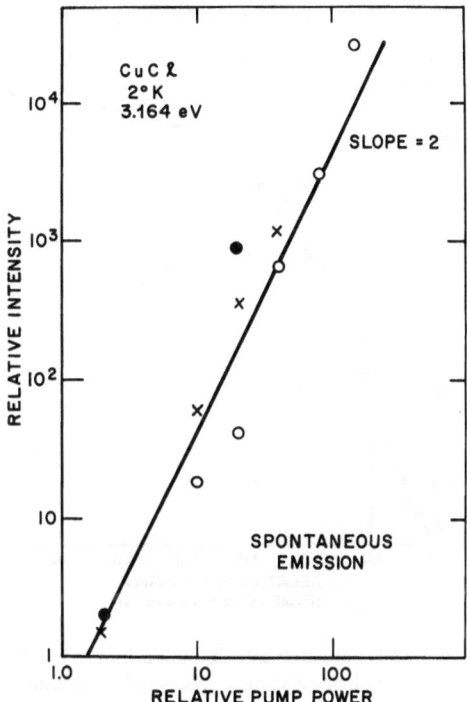

Fig. 11. Data points show the luminescence intensity of the M_l line in CuCl at 2 K as a function of excitation intensity. For comparison, the solid line is drawn with a slope of two to indicate a quadratic dependence of luminescence intensity on excitation level

Let us first consider the kinetics of the excitonic molecule lumin-escence. Knox et al. [12] have shown that the molecule luminescence intensity should be quadratic in excitation intensity, becoming linear only at high excitation levels. Such a dependence has been observed experimentally by Mysyrowicz et al. [10], Souma et al. [11], and in the present experiments as indicated by Fig. 11 where we show the excitation intensity dependence of the strength of the molecule luminescence line at 2 K. Note that the points agree with a quadratic dependence of the luminescence intensity on excitation level. The satura-tion to a linear dependence is not seen, presumably because at low temperatures [10, 11] the bound excitons provide a fast decay process inhibiting the buildup of exciton density.

If we now investigate the kinetics of the gain as measured in Fig. 8, we find a quite different result. Over the excitation intensities indicated in Fig. 8, the gain is nearly linear in excitation intensity and is indicated

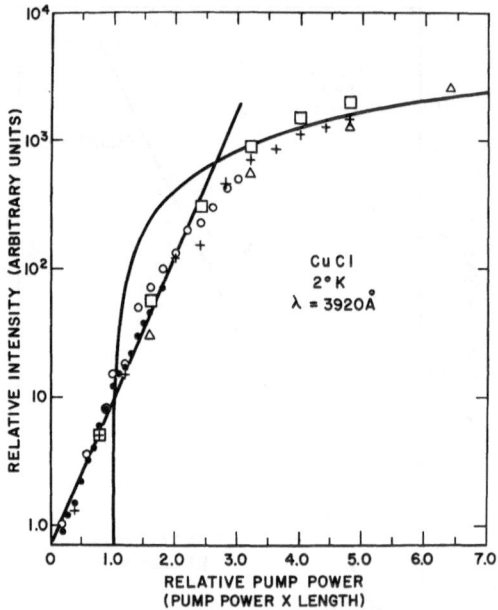

Fig. 12. Data points show the stimulated emission intensities of CuCl at 2 K as a function of the product of excitation length and excitation intensity for excitation intensities (different symbols) covering a range of factor of ten

in Fig. 12 where we show that when the length axis is scaled by the excitation intensity, the stimulated emission intensities fall on a single exponential. Further, however, if we measure the kinetics of the saturated stimulated emission by using long excitation lengths we find that this quantity is quadratic in excitation intensity as shown by Fig. 13.

To summarize the kinetics, we find the spontaneous emission intensity to be quadratic, the gain to be linear and the saturated stimulated emission intensity to be quadratic in the excitation intensity. This seems like a rather contradictory set of observations to be reconciled with the excitonic molecule model. However, it has been pointed out that the observed gain value is a net gain, hence the losses of the system play an important role in determining this quantity. Therefore, it is necessary to make a careful mathematical model for gain measurements on systems of excitonic molecules in order to determine whether these observed kinetics are indeed reasonable in terms of excitonic molecules.

In constructing such a model, we proceed as in the case of a 4-level model discussed elsewhere [18], except that we must now consider a 3-level model as shown in Fig. 14, and we must pay careful attention

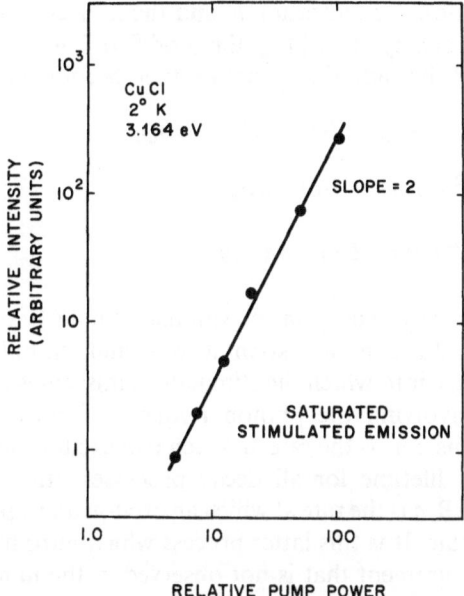

Fig. 13. Data points show the saturated stimulated emission intensity of CuCl at 2 K for an excitation length of 500 μ and several different excitation intensities. The solid line is drawn to indicate a quadratic dependence of the emission intensity on the excitation level

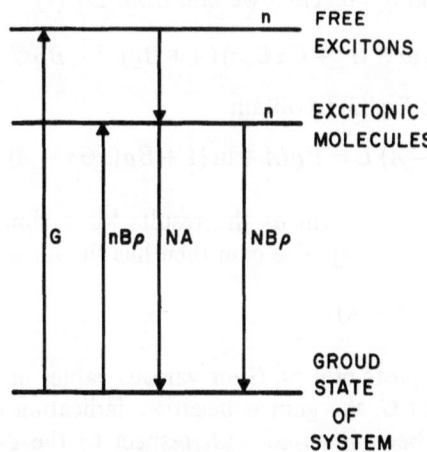

Fig. 14. Schematic diagram of a 3-level model for representing excitonic molecule decay

to both the exciton concentration n, and the molecule concentration N [12]. Assuming steady state [18], the modified Einstein rate equations for n and N and the radiation equation then become [19]

$$\dot{n} = G - 2\alpha n^2 - n/\tau + N(A + B\varrho) - nB_a\varrho$$

$$\dot{N} = 0 = \alpha n^2 - N(A + B\varrho) + nB_a\varrho \tag{1}$$

$$d\varrho/dx = \varrho' = NK(A\Omega + B\varrho) - nkB_a\varrho$$

where ϱ is the energy density in the stimulated light beam, A and B are the Einstein coefficients for spontaneous and stimulated emission, Ω is the solid angle into which the stimulated emission is confined, and K is a constant involving the photon frequency, Planck's constant etc. In these equations, αn^2 is the rate at which two excitons form a molecule, τ is the exciton lifetime for all decay processes other than molecule formation, and $nB_a\varrho$ is the rate at which an exciton and a photon combine to form a molecule. It is this latter process which introduces a loss term in the gain measurement that is not observed in the luminescence.

Using Eq. (1) we find

$$N = [\alpha\tau^2/(A + B\varrho)] G^2 - G\tau B_a\varrho \tag{2}$$

subject to the condition that $G \ll 1/4\alpha\tau^2$. Since the spontaneous emission intensity is proportional to N, we have established the range of excitation intensities over which we would expect to find a quadratic dependence of luminescence on excitation level, subject to given values of τ and α and the condition $\varrho = 0$. Then we find from Eq. (1)

$$\varrho' = k(A\Omega + B\varrho)(\alpha\tau^2 G^2 + G\tau B_a\varrho)(A + B\varrho)^{-1} - B\varrho G\tau \tag{3}$$

We can integrate Eq. (3) to obtain

$$(kG\tau B/A)(\alpha G\tau - A) L = B\varrho/A + \ln\{1 + B\varrho(\alpha G\tau - A)/A\Omega\alpha G\tau\} \tag{4}$$

which is of the same form as the results for a simple 4-level system discussed elsewhere [19]. The gain then has the form

$$g = kG\tau(B/A)(\alpha G\tau - A). \tag{5}$$

Figure 15 shows plots of g vs. G for various values of $A/\alpha\tau$. Notice that at small values of G, the gain is negative, indicating that the molecule system has not been inverted with respect to the excitons. At larger values of G, g becomes positive, and for $G \gg A/\alpha\tau$ the gain becomes quadratic in G. For comparison with the experimental results presented

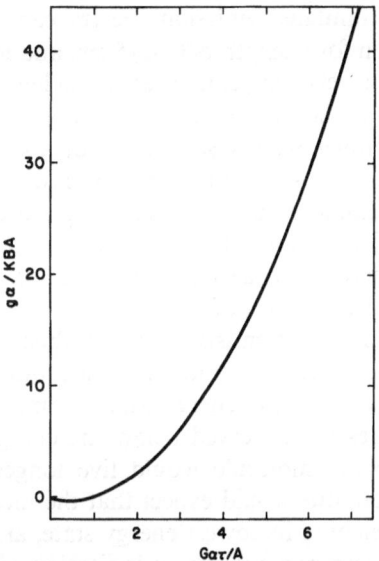

Fig. 15. Plot of excitonic molecule gain as a function of excitation level using 3-level model. The units are as given in Eq. (4)

earlier, it is important to note that the present model does not predict a gain region which is linear in G.

This example indicates that the kinetics of the gain measurement technique are not in agreement with the predictions of the excitonic molecule model. It remains to determine whether it is the experiment or the model that is suspect in this lack of agreement. As was pointed out earlier, kinetics was only one of three points listed as evidence for the existence of excitonic molecules. The other two points are the lineshape, and the molecule binding energy.

The lineshape analysis has been discussed by Souma et al. who describe the molecule luminescence lineshape in terms of a Maxwellian distribution with a temperature considerably higher than the lattice temperature. Such a temperature rise is reasonable [20] since the photons used to create the excitons contributed an excess energy of about 0.3 eV/photon to the crystal.

On the other hand, the gain spectrum shown in Fig. 9 is quite broad (~ 20 meV) in the neighborhood of the molecule emission and appears to depend linearly on excitation level. The gain spectrum is considerably broader than the molecule luminescence spectrum of Souma et al. or than the peak observed in the stimulated emission spectrum at short lengths (see Fig. 7b).

The saturated stimulated emission spectra can be seen in Fig. 7b in the spectrum taken for a length of 0.0215 cm, and in Fig. 10 where the length has been made very long compared to the length at which saturation begins. In the initial stages of saturation the molecule structure appears to be a doublet with a separation of about 4 meV, while in complete saturation the spectrum contains 5 distinct peaks with a separation of approximately 0.9 meV. The peaks are all placed to the low energy side of the molecule structure M_1 seen in spontaneous luminescence, and there is only a slight shoulder at M_1 while the spectrum appears to peak at the position of M_2.

The saturated-stimulated emission spectra then pose an interesting problem. As saturation sets in, the intensity distribution of the stimulated emission spectra moves to lower energy, and in complete saturation a set of closely spaced lines are observed below the energy of the molecule luminescence. Since the molecule would live longest for the case of spontaneous emission, one would expect that the molecule would have the best chance of reaching its lowest energy state, and hence the spontaneous emission spectrum would be an indication of the lowest energy state of the molecule. Therefore, we must conclude that the structure seen in the saturated stimulated emission spectra is an indication of the final states of the transition rather than of initial states. This conclusion is not readily understood in terms of the molecule model.

We can also examine these observations in terms of the density of excitonic molecules created by the excitation. The maximum power density used was 10^7 W/cm^2 which corresponds to a photon flux of $2 \cdot 10^{26}$ photons/cm^2 sec. If we assume a diffusion length of 2 µ and a conversion efficiency of 1 % then we would expect an excitation density of 10^{28} photons/cm^3 sec which is roughly two orders of magnitude larger than that used by Souma et al. However, no departure from square dependence was observed in the luminescence intensity data shown in Fig. 11. It is possible that the actual exciton densities achieved in the present work might be restricted by some process which limits the free exciton lifetime such as, for instance, bound exciton formation. However, even if we assume that the excitation levels are equal, or that Souma et al. achieved higher exciton densities than were achieved in the present experiment, there is still a problem in that Souma et al. used volume excitation through crystals 50 µ thick. It can be seen from Fig. 8 that at a power density of $5 \cdot 10^6$ W/cm^2 the stimulated emission has saturated at an excitation length of 50 µ. It is then possible that the change in slope from 1.7 to 1.1 reported by Souma et al. [11] is actually caused by a saturation of the stimulated beam traveling along the axis of the excited volume rather than by a limitation in the kinetics of the molecule formation process.

3. Conclusions

We have given a discussion of some of the types of experiments that can be conducted on HDE systems. We have discussed, in detail, reflectivity of intense light beams and optical gain measurements on highly excited CuCl. We find that the reflectivity spectra are modified when a high intensity probe is used, and that this modification can be phenomenologically explained in terms of an exciton damping parameter which varies linearly with incident light intensity. We are not able, at this time, to give a physical explanation of the phenomenon. From gain measurements in CuCl we find that the observed kinetics of the optical gain are not consistent with the predictions based on a simple 3-level model of an excitonic molecule. The origin of this discrepancy is not clear, but it points up the need for a definitive investigation of the existence of excitonic molecules in CuCl.

Such an investigation would require new experimental approaches to studying HDE systems. Careful reflectivity, absorption and ellipsiometric measurements on such systems could provide important information. Further infrared absorption, or possibly Raman scattering experiments could obtain information about internal structure of the molecular state (if such a state exists in CuCl).

The author is indebted to R. E. Nahory and R. F. Leheny for helpful discussions and for their collaboration in some of the work discussed here. W. F. Brinkman has made many useful comments and suggestions. The author also wishes to thank Prof. C. Benoît à la Guillaume for his helpful comments on the 3-level model discussed here.

References

1. Shaklee, K. L., Nahory, R. E., Leheny, R. F.: J. Luminescence **7**, 2841 (1973).
2. Benoît à la Guillaume, C., Debever, J. M., Salvan, F.: Phys. Rev. **177**, 567 (1969).
3. Pokrovsky, Y., Svistunova, K.: In: Keller, S. P., Hensel, J. C., Stern, F. (Eds.): Proceedings of the Tenth International Conference on the Physics of Semiconductors, CONF-700501, p. 504. (U.S. AEC Division of Tech. Information, Springfield, Va. 1970).
4. Brinkman, W. F., Lee, P. A.: Phys. Rev. Letters **31**, 237 (1973).
5. Büttner, H.: Phys. Stat. Sol. **42**, 775 (1970).
6. Hanamura, E., Inoue, M.: Proceedings of the Eleventh International Conference on the Physics of Semiconductors, p. 711. (PWN Warsaw, 1972).
7. Shaklee, K. L., Nahory, R. E.: To be published.
8. Shaklee, K. L., Leheny, R. F.: Appl. Phys. Letters **18**, 475 (1971).
9. Shaklee, K. L., Leheny, R. F., Nahory, R. E.: Phys. Rev. Letters **26**, 15 (1971).
10. Mysyrowicz, A., Grun, J. B., Levy, R., Bivas, A., Nikitine, S.: Phys. Letters **26**A, 615 (1968).
11. Souma, H., Goto, T., Ohta, T., Ueta, M.: J. Phys. Soc. Jap. **29**, 697 (1970).
12. Knox, R. S., Nikitine, S., Mysyrowicz, A.: Optics Commun. **1**, 19 (1969).
13. Aspnes, D. E.: Optics Commun. **8**, 222 (1973).

14. Shank,C.V., Dienes,A., Trozzolo,A.M., Meyer,J.A.: Appl. Phys. Letters **16**, 405 (1970).
15. Silfvast,W.T., Deech,J.S.: Appl. Phys. Letters **11**, 97 (1967).
16. Thomas,D.G., Hopfield,J.J.: Phys. Rev. **116**, 573 (1959).
17. Hopfield,J.J., Thomas,D.G.: Phys. Rev. **132**, 563 (1963).
18. Magde,D., Mahr,H.: Phys. Rev. Letters **24**, 890 (1970).
19. Shaklee,K.L., Nahory,R.E., Leheny,R.F.: Eleventh International Conference on the Physics of Semiconductors (PNW Warsaw, 1972).
20. Leheny,R.F., Nahory,R.E., Shaklee,K.L.: Phys. Rev. Letters **28**, 437 (1972).

Dr. Kerry L. Shaklee
Bell Telephone Laboratories
Holmdel, N.J. 07733, USA

VII. Excitonic Polaritons at Higher Densities

Tests of Validity of Spatial Dispersion Theories on Lead Iodide Crystal Spectra*

M. Grosmann, J. Biellmann, and S. Nikitine

Contents

---

* This work was carried out within the frame of the agreement on French-German scientific cooperation between C.N.R.S., Paris, Université Louis Pasteur, Strasbourg, and Universität Stuttgart.

1. Introduction

It is the task of physics to analyze experimental data and from the huge quantity of information thus obtained to deduce a small number of fundamental concepts related by laws from which the results of any experiment can be predicted without the necessity of carrying it out.

We try in this article to analyze briefly the experimental data collected over the last 20 years on the dispersion of dielectric media containing excitons. We discuss the new analyses worked out to explain the data, and how the predictions of these analyses fit the experiments performed to test their validity. As a complete discussion would be too long, we confine ourselves to a brief description of the problem, with a detailed analysis of the tests performed in our laboratory, essentially on lead iodide crystals in which the effect is large enough to allow accurate measurements.

2. Brief History of the Problems of Dispersion

The fundamental problem of optics in dielectrics consists in the investigation of the propagation of plane monochromatic waves in the dielectric. These waves are characterized by certain values for the frequency ω and for the wave vector k.

In the simple case of "normal homogeneous" waves the electric field is given by

$$E = E_0 \exp i(k \cdot r - \omega t), \quad k = (\omega/c)\,\tilde{n}(\omega,s)\,s,$$

where $\tilde{n}$ is the complex refraction index and $s = k/|k|$. The electric and magnetic inductions D and B can be represented in the same form.

For the purpose of experiments it is, however, impossible to put an observer in the dielectric. It is therefore necessary to have a detailed analysis of the way in which the electromagnetic waves cross the boundary between the dielectric and the vacuum in which they can be measured. This analysis was first made more than a century ago for the case of transparent dielectrics and is well known to all physicists. However, truly "transparent" media do not exist and this approximation is not universally valid. It is therefore necessary to have a theory that can predict what will happen in the case of absorbent media.

A very important and frequently encountered case is that of absorbing crystals in which absorption is due to the creation of excitons.

An extensive analysis of the theoretical problem was made during the period 1950–1960. This made it possible to predict such interesting phenomena as the optical anisotropy of nongyrotropic cubic crystals.

These predictions stimulated experimental studies of the problem in the period 1960–1970. Many of the predicted effects were indeed observed, but there was little quantitative agreement between the theoretical predictions and the experimental measurements. New studies have therefore been undertaken in the hope of developing better theories which could give quantitative predictions.

We briefly summarize the current theoretical models and the main experimental methods available for studying the problem.

3. Theoretical Analysis

a) Spatial Dispersion

Crystal optics which takes account of spatial dispersion has superseded classic crystal optics where frequency dispersion was taken into account only as a special case.

The components D_i of the electric induction vector D in a continuous medium are linked to the components E_j of the electric vector E by Maxwell equation

$$D_i(r, t) = \int\limits_{-\infty}^{t} dt' \int \hat{\varepsilon}_{ij}(t, t', r, r') \, E_j(r', t') \, d^3 r'$$

D at time t depends only on the values of E at $t' < t$; D at point r depends on the values of E at r' in the vicinity of r. In a homogeneous medium whose properties do not vary with time

$$D_i(r, t) = \int\limits_{-\infty}^{t} dt \int \hat{\varepsilon}_{ij}(t - t', r - r') \, E_j(r', t') \, d^3 r' \, .$$

If the electric field is the field of a "normal homogeneous" wave

$$E_i(R, t) = E_i(\omega, k) \exp i(k \cdot R - \omega \tau) \, .$$

If we define $\tau = t - t'$, $\quad R = r - r'$

$$\varepsilon_{ij}(\omega, k) = \int\limits_{0}^{\infty} d\tau \int \hat{\varepsilon}_{ij}(\tau, R) \exp - i(k \cdot R - \omega \tau) \, d^3 R \, ,$$

we obtain

$$D_i(\omega, k) = \varepsilon_{ij}(\omega, k) \, E_j(\omega, k) \, .$$

The variation of ε_{ij} with ω defines the classic dispersion in which $D(\omega)$ is independent of the field in the vicinity and depends only on the field $E(\omega)$.

The variation of ε_{ij} with k defines the "spatial" dispersion. Let us note that the material can be considered as a continuum only if the wavelength λ is large compared to the interatomic distances. In a crystal of lattice parameter a this condition can be rewritten

$$\lambda \gg a \quad \text{or} \quad |k| = 2\pi n/\lambda \ll 2\pi/a.$$

b) Dispersion Equation in the Case of Crystals Containing Excitons

There are several types of crystals as well as several types of excitons. We consider in detail here a convenient case in which accurate experimental measurements can be made. Analogous calculations can be performed in other cases. We give full details on the dispersion relations in the vicinity of two polarized exciton transitions in a uniaxial crystal with symmetry center, and summarize the important results in other cases.

Pekar [1] showed in detail in 1957 that when light is propagating in a crystal at frequencies close to some exciton absorption band, spatial dispersion substantially changes the dielectric constant even when $\lambda \gg a$. As a result, there is a substantial change in the relationship between the frequency and the refractive index of light, and the relations describing the interaction between matter and light are considerably changed. The most important result is the presence of additional light waves in the neighborhood of exciton absorption bands besides the birefringence waves which exist in noncubic crystals. It is particularly interesting that two light waves having the same polarization but different velocities may be present in a cubic crystal.

If there is more than one wave propagating in the material medium, the boundary conditions used to solve the classic Maxwell's equations are not sufficient to solve the problem. Pekar has therefore proposed a new "additional" boundary condition (a.b.c.). This condition is obtained phenomenologically by assuming that exciton polarization cannot exist on the crystal surface because the exciton cannot exist there.

With this "a.b.c." the problem can be solved.

c) Surface Potential Barrier

Thomas and Hopfield [2], however, analyzed experiments on CdS and showed that the solutions thus obtained were not consistent with the experimental results. In their experiments they obtained a small peak of reflectivity in the region of the "missing ray"; this peak could not be obtained by calculations using Pekar's formulas. They proposed to reconcile theory and experiment by taking into account the fact that the excitons do not disappear suddenly at the surface. According to their model, the exciton interacts with its "electrical image" through the surface of the crystal. As the exciton comes near the surface, it feels a repulsive potential going to infinity near the surface. This potential is difficult to calculate accurately but it can be conveniently approximated by a surface layer of constant thickness in which the exciton cannot exist. In this layer the dielectric constant is equal to the constant of the crystal without the exciton, whereas in the bulk the presence of excitons must be taken into account in calculating the dielectric constant.

With this correction good qualitative agreement was obtained between theory and experiment. The slight quantitative discrepancies were attributed to the rough approximations made on the shape of the surface potential.

d) Recent Improvements

In search of better quantitative agreement between experiments and theory, numerous workers have recently adopted a new approach of the problem. Agarwal, Pattanayak and Wolf [3], Maradudin and Mills [4]; Sein and Birman [5], and some others have independently explored the properties of models of semi-infinite nonlocal dielectrics to assess the effect of spatial dispersion on the reflectivity and refractivity of the material and on the properties of surface polaritons. Their additional boundary conditions follow from Maxwell's equations and they do not need to introduce microscopic considerations to complete their theories. This seems to be a great improvement on the previous analysis.

Unfortunately, the comparison of the results of this improved theoretical analysis with experimental data is not always satisfactory. In several cases the results are much worse than those of the previous analysis. Zeyher [6] has analysed in some detail the analogies and differences of the two types of theories. He has tried to develop criteria which enable one to predict which of these theories will best apply to a given crystal.

4. Experimental Methods of Investigating Spatial Dispersion of Crystals

As already pointed out, the experimental conditions do not allow direct measurement of the "spatial" dependence of the dielectric constant. This dependence appears to produce certain physical phenomena, measurements of which allow us to calculate the influence of spatial dispersion on the interaction of light with matter.

The principal experimental observations are:

a) Oscillatory Behavior of the Intensity of the Light Transmitted through a Plane Parallel Plate of Crystal as a Function of Plate Thickness

When a monochromatic light wave impinges normally from the vacuum on to the plane parallel plate of a crystal it generates in the plate two waves with two different complex wave vectors. On emerging from the plate these two waves interfere, and as a result the intensity of the transmitted light should be an oscillating function of the plate thickness. However, this function is completely different from the oscillations, due to conventional interferences.

Such oscillating curves have been observed experimentally for anthracene [7], cuprous oxide [8], and lead iodide [9]. These experiments are, however, not easy to perform: In order to obtain a large effect, the exciton absorption must be strong enough to produce large dispersion effects; but strong absorption also means low and perhaps very weak transmission. In most crystals, investigations using this method were unsuccessful for this reason.

b) Birefringence Induced in Cubic Semi-Conductors by Spatial Dispersion

Several workers [10, 11] have recently reported birefringence attributable to spatial dispersion in Si, Ge and GaAs. A strong argument in favor of this origin is the strong enhancement of birefringence near the absorption edge in GaAs and Ge. This can be interpreted as the effect of vanishing energy denominators in the contribution of the E_0 gap to the spatial dispersion birefringence (SDB). Other measurements in GaP, ZnTe, ZnSe, ZnS, have also allowed observations of this effect.

The measurements are not very easy to make as the effect is rather small (of the order of 10^{-6}). When a small birefringence is detected, it is necessary to be sure that it is not produced by residual strains in the crystal.

c) Resonant Inelastic Scattering of Polaritons

Although it is difficult to observe transmission in absorbing crystals containing excitons, it is frequently easy to observe light re-emitted by luminescence. For instance, in CdS crystals and in other II–VI compounds, the polaritons are efficiently scattered with emission of a longitudinal optical phonon. The energy of the polariton initially excited in the resonance region drops in such a way that the scattered polaritons appear in a region where the crystal is much less absorbing and can therefore re-emit an important quantity of radiation. Measurements are therefore much easier in this case than in the transmission case.

We will not discuss this problem extensively, since several communications at this meeting deal with it [12–14].

d) Effects on Reflection Spectra Induced by Spatial Dispersion

In the vicinity of an exciton absorption there is a strong reflection peak (residual ray) where it is easy to make measurements on a strong signal. On the high-energy side of this peak a low reflection region is predicted by classic dispersion theory. Hopfield and Thomas "showed" that spatial dispersion theory predicts a peak in this region. The peaks experimentally observed in this region on CdS, ZnS, and other crystals have been considered both as confirmation of spatial dispersion theories and as a very good tool for quantitative measurements of spatial dispersion in those crystals.

5. Experimental Results in Lead Iodide

a) Experimental Procedure

Lead iodide crystals have been grown from aqueous solutions. With sufficiently slow evaporation one can obtain hexagonal platelets ranging in thickness from $0.5\,\mu$ to $15\,\mu$ and in diameter from 0.1 mm to 1 mm. Such crystals present faces (0001) and $(10\bar{1}1)$ (these $30°$ of axis), and their spectra have been recorded with a microspectrophotometer having a band pass of $0.4\,\text{Å}$, a spot dimension of $15\,\mu \times 40\,\mu$, and a beam divergence of $2.5°$. The crystals were fixed to the support by an edge or a corner so as to avoid strains induced by differential dilatations of crystal and support [15].

Reflection spectra only could be studied on the $(10\bar{1}1)$ face. On the (0001) face, for thin crystals, both reflection and transmission could be studied.

b) Ordinary Reflection Spectrum with Wave Vector Normal to a (0001) Face

The ordinary $(E \perp c)$ reflection spectrum in the vicinity of the fundamental absorption edge at 20 K is shown in Fig. 1 a. Two types of samples give two types of spectra. Gähwiller and Harbeke [16] have shown that the dotted curve can be attributed to $4\,H$ polytypes, and the full-line curve to $2\,H$ polytypes. In this discussion, we consider only the case of the $2\,H$ polytypes.

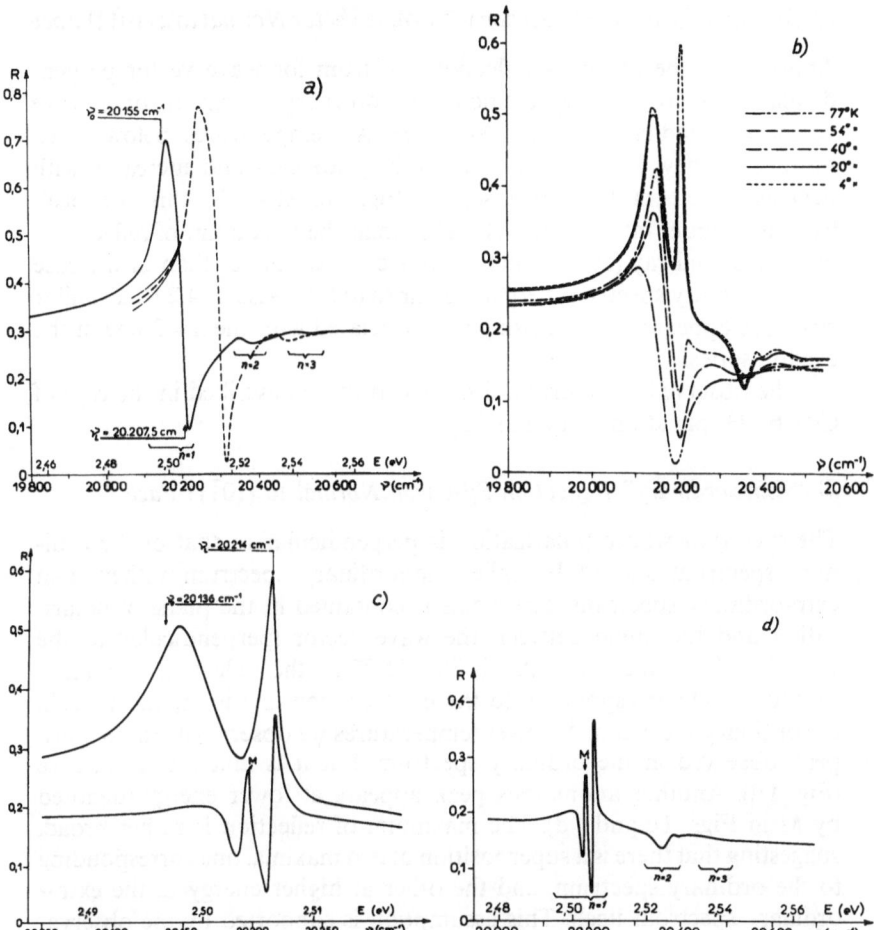

Fig. 1 a–d. Reflectivity of PbI$_2$ in the vicinity of the fundamental absorption edge: (a) ordinary spectrum normal to (0001) face at 20 K; full line: crystal grown from aqueous solution (this spectrum is analyzed and fitted in Figs. 2, 3 and 4); dotted line: crystal cleaved from zone-melted ingot; (b) ordinary spectra normal to (10$\bar{1}$1) face at various temperatures; (d) extraordinary spectrum normal to (10$\bar{1}$1) face at 4.2 K; (c) spectra b) (upper curve) and d) (lower curve) at 4.2 K in the first exciton line region

The reflection spectrum has a classic shape. A sharp maximum at 4960 Å, and a sharp minimum at 4917 Å of reflectivity are associated with the first line of the exciton series. These are the well-known "residual" and "missing" rays. To the second and the third lines of the exciton series there correspond analogous but broader and less intense structures at higher energies.

c) Ordinary Reflection Spectrum with Wave Vector Normal to a (10$\bar{1}$1) Face

Above 77 K the ordinary reflection spectrum for wave vector perpendicular to a (10$\bar{1}$1) face resembles the ordinary spectrum for a wave vector perpendicular to a (0001) face. At temperatures below 77 K, however, a new sharp peak of reflectivity appears and increases with decreasing temperature, as is seen in Fig. 1 b. At 4.2 K, this new peak has a maximum intensity (60 %) higher than the maximum of reflectivity of the residual ray (50 %) which, in this case, is smaller than in the case of the ordinary spectrum perpendicular to (0001). Also at 4.2 K, a similar structure appears in the missing ray associated with the $n = 2$ line of the exciton series.

The observed structures are similar to those described in the case of CdS by Hopfield and Thomas [2].

d) "Extraordinary" Reflection Spectrum Normal to (10$\bar{1}$1) Face

The spectrum whose polarization is perpendicular to that of the ordinary spectrum should be called nonordinary spectrum rather than extraordinary spectrum. The c axis is contained in the plane of polarization and the angle between the wave vector (perpendicular to the crystal surface) and the c axis is 30°. At 77 K, the reflection maximum (residual ray) corresponding to the $n = 1$ exciton line is smaller than in the ordinary spectrum. At lower temperatures we observe the same sharp peak observed in the ordinary spectrum, but it is much more intense (Fig. 1 d). Another anomalous peak appears at lower energy (denoted by M in Figs. 1 c and 1 d). The maximum of reflection is rather broad, suggesting that there is a superposition of two maxima, one corresponding to the ordinary spectrum, and the other at higher energy to the extraordinary spectrum itself. This assumption is supported by the observations of Harbeke et al. [17], who separated the ordinary and extraordinary spectra in incidence perpendicular to a face containing the c axis. The anomalous peaks correspond to the reflection minima observed by these authors in the two spectra, respectively.

6. Spatial Dispersion Theory and Boundary Conditions in Uniaxial Crystals

In analyzing the observed spectra, we have tried to use existing dispersion theories adapted to the particular case of a uniaxial crystal with symmetry center, such as lead iodide. Several possibilities have been proposed for dispersion theories. The boundary conditions that should be imposed for the different modes of propagation in the crystal struck by an incident wave, are not established without ambiguity. We briefly discuss the different theoretical possibilities and how they apply in our case.

a) Dispersion Equation in the Vicinity of Two Polarized Exciton Transitions in a Uniaxial Crystal with Symmetry Center

We assume that in a given frequency region only two isolated allowed exciton transitions exist. They are polarized one perpendicular and the other parallel to the crystal axis. In the case of a centrosymmetric crystal, the exciton energies are quadratic functions of the wave vector in first approximation. In the case of normal incidence of light on the crystal, homogeneous plane waves are the only modes propagating inside the crystal. The wave vector of a given propagating mode can be written:

$$k = (\omega/c)\, \tilde{n}(\omega, s)\, s \tag{1}$$

where $\tilde{n} = n + ik$ is the complex refraction index, s a real unit vector perpendicular to the plane wave, ω the circular frequency, and c the velocity of light in vacuum.

With these assumptions, the dielectric constant tensor $\varepsilon_{ij}(\omega, k)$ reduces to two components:

$$\varepsilon_p[\omega, \tilde{n}(\omega, s)\, s] = \varepsilon_{0p} + \frac{4\pi\alpha_{0p}\omega_{0p}^2}{\omega_{0p}^2 + (h\omega_{0p}\omega^2/M_{sp}^*c^2)\,\tilde{n}^2 - \omega^2 - i\omega\gamma_p} \tag{2}$$

$$(p = \perp, \parallel)$$

where ε_{0p} represents the contributions of all the other transitions outside the considered frequency region, ω_{0p} are the transverse circular resonant frequencies at $k = 0$: $\alpha_{0p}\omega_{0p}^2 = f_p e^2/m V_0$ with f_p the oscillator strengths by the unit cell volume V_0; γ_p are phenomenological damping factors, and M_{sp}^* are the exciton masses in the direction of s.

Using the auxiliary parameters A_p, M_p, Z_p $(p = \perp, \parallel)$,

$$A_p = 4\pi\alpha_{0p},$$

$$M_p = M_{sp}^*\omega_{0p}c^2/h\omega^2, \tag{3}$$

$$Z_p = (\omega^2 + i\omega\gamma_p)/\omega_{0p}^2 - 1,$$

the expressions (2) are simplified:

$$\varepsilon_p = \varepsilon_{0p} + M_p A_p/(\tilde{n}^2 - M_p Z_p). \tag{4}$$

The general wave equation $\varepsilon E = \tilde{n}^2 [E - (s \cdot E) s]$ leads to a dispersion equation, which is subdivided into two equations:

$$\tilde{n}^2 = \varepsilon_\perp (\omega, k), \tag{5}$$

$$\tilde{n}^2 = \frac{\varepsilon_\perp (\omega, k) \varepsilon_{||}(\omega, k)}{\varepsilon_\perp (\omega, k) \sin^2 \phi + \varepsilon_{||}(\omega, k) \cos^2 \phi} \tag{6}$$

where ϕ is the angle between s and the c axis.

Equation (5) has two solutions for a given direction of s corresponding to ordinary transverse waves:

$$\tilde{n}^2 = \tfrac{1}{2} \{\varepsilon_{0\perp} + M_\perp Z_\perp \pm [(\varepsilon_{0\perp} - M_\perp Z_\perp)^2 + 4 M_\perp A_\perp]^{1/2}\}. \tag{7}$$

Equation (6) has for $\phi \neq 0$ and $\phi \neq \pi/2$ three solutions corresponding to nonordinary mixed waves $(E \cdot (s \wedge c) = 0)$. It can be written explicitly as

$$(\tilde{n}^2)^3 - a_2 (\tilde{n}^2)^2 + a_1 (\tilde{n}^2) - a_0 = 0 \tag{8}$$

where

$$a_2 = (\varepsilon_0 + \varepsilon_{t_\perp} + \varepsilon_{t_{||}})$$

$$a_1 = \varepsilon_0 (\varepsilon_{t_\perp} + \varepsilon_{t_{||}}) + \varepsilon_{t_\perp} M_{||} Z_{||} + \varepsilon_{t_{||}} M_\perp Z_\perp - M_\perp M_{||} Z_\perp Z_{||}$$

$$a_0 = \varepsilon_0 \varepsilon_{t_\perp} \varepsilon_{t_{||}}$$

and

$$\varepsilon_0 = \frac{\varepsilon_{0_\perp} \varepsilon_{0_{||}}}{\varepsilon_{0_\perp} \sin^2 \phi + \varepsilon_{0_{||}} \cos^2 \phi} = \varepsilon_{0_\perp} \varepsilon_{0_{||}} / \varepsilon_\phi$$

$$\varepsilon_{t_\perp} = M_\perp (Z_\perp - A_\perp \sin^2 \phi / \varepsilon_\phi)$$

$$\varepsilon_{t_{||}} = M_{||} (Z_{||} - A_{||} \cos^2 \phi / \varepsilon_\phi)$$

$$\varepsilon_{t_\perp} = M_\perp (Z_\perp - A_\perp / \varepsilon_{0_\perp}) = \varepsilon_{t_\perp} (\phi = \pi/2)$$

$$\varepsilon_{t_{||}} = M_{||} (Z_{||} - A_{||} / \varepsilon_{0_{||}}) = \varepsilon_{t_{||}} (\phi = 0).$$

b) Boundary Conditions

The two ordinary wave amplitudes $E_{m_\perp}$ ($m = 1, 2$) on the crystal surface are related to E_0 and E_R, the incident and reflected amplitudes in vacuum, by Maxwell's equations:

$$E_0 + E_R = \sum_{m=1}^{2} E_{m_\perp}$$

$$E_0 - E_R = \sum_{m=1}^{2} \tilde{n}_m E_{m_\perp} \tag{9}$$

where $\tilde{n}_m$ are the solutions of Eq. (7).

Pekar's a.b.c., $P_{ex} = 0$ on crystal surface, provides a third relation:

$$4\pi P_{ex_\perp} = \sum_{m=1}^{2} (\tilde{n}_m^2 - \varepsilon_{0_\perp}) E_{m_\perp} = 0 \,. \tag{10}$$

This condition allows the reflection coefficient $\varrho = E_R/E_0$ to be calculated in terms of a "fictitious" index $\tilde{n}_f$:

$$\tilde{n}_f = (\varepsilon_{0_\perp} + \tilde{n}_1 \tilde{n}_2)/(\tilde{n}_1 + \tilde{n}_2) \,, \tag{11}$$

$$\varrho = (1 - \tilde{n}_f)/(1 + \tilde{n}_f) \,. \tag{12}$$

In the case of nonordinary polarization, the boundary conditions deduced from Maxwell's equations are:

$$E_0 + E_R = \sum_{m=1}^{3} (E_{\perp m} \cos\phi + E_{\|m} \sin\phi) \,,$$
$$E_0 - E_R = \sum_{m=1}^{3} \tilde{n}_m (E_{\perp m} \cos\phi + E_{\|m} \sin\phi) \,, \tag{13}$$
and

$$\varepsilon_{\perp m} E_{\perp m} \sin\phi = \varepsilon_{\|m} E_{\|m} \cos\phi \quad (m = 1, 2, 3) \,, \tag{14}$$

where $\tilde{n}_m$ are the three solutions of Eq. (8), $\varepsilon_{pm} = \varepsilon_p(\tilde{n}_m)$, given by (4) $(p = \perp, \|)$, E_{pm} are the projections of the electric field vectors E_m on a plane perpendicular to the c axis and on the c axis, respectively.

With the two following relations deduced from Pekar's a.b.c., we can calculate the reflection coefficient:

$$4\pi P_{exp} = \sum_{m=1}^{3} (\varepsilon_{pm} - \varepsilon_{0p}) E_{pm} = 0 \quad (p = \perp, \|) \,. \tag{15}$$

After some cumbersome algebra, we can express the "fictitious" index as

$$\tilde{n}_f = \frac{\varepsilon_0(\tilde{n}_2 \tilde{n}_3 + \tilde{n}_3 \tilde{n}_1 + \tilde{n}_1 \tilde{n}_2 + \varepsilon_{\perp} + \varepsilon_{l_\|}) + \tilde{n}_1 \tilde{n}_2 \tilde{n}_3 (\tilde{n}_1 + \tilde{n}_2 + \tilde{n}_3)}{(\tilde{n}_2 + \tilde{n}_3)(\tilde{n}_3 + \tilde{n}_1)(\tilde{n}_1 + \tilde{n}_2)} \,. \tag{16}$$

Agarwal et al. obtained the following boundary condition on a cubic crystal surface $(z = 0)$:

$$P_{ex} + (c/i\tilde{n}_\mu \omega)(\partial P_{ex}/\partial z)|_{z=0} = 0 \,, \tag{17}$$

where $\tilde{n}_\mu^2 = MZ$, M and Z as defined by relations (3) for an isotropic medium.

This procedure can be extended to a uniaxial medium [18] by writing relation (17) for $P_{ex_\perp}(0)$ and $P_{ex_\|}(0)$, and defining

$$\tilde{n}_{\mu p}^2 = M_p Z_p \quad (p = \perp, \|) \,. \tag{18}$$

Thus, for the two ordinary wave amplitudes we obtain:

$$\sum_{m=1}^{2} E_m/(\tilde{n}_m - \tilde{n}_{\mu_\perp}) = 0 \tag{19}$$

and for the "fictitious" index:

$$\tilde{n}_g = \tilde{n}_1 + \tilde{n}_2 - \tilde{n}_{\mu_\perp} \, . \tag{20}$$

To obtain conditions similar to relations (15) for the nonordinary waves, we must multiply each term of the sums in (15) by $(1 + \tilde{n}_m/\tilde{n}_{\mu p})$.

The "fictitious" index is expressed by

$$\tilde{n}_g = \sum_{m=1}^{3} \tilde{n}_m \delta_m \Big/ \sum_{m=1}^{3} \delta_m \tag{21}$$

where

$$\delta_i = \varepsilon_{\perp j}\,\varepsilon_{||k}\Delta_{\perp k}\Delta_{||j} - \varepsilon_{\perp k}\varepsilon_{||j}\Delta_{\perp j}\Delta_{||k} \quad (i,j,k=1,2,3)$$

and

$$\Delta_{pm} = (\varepsilon_{pm} - \varepsilon_{0p})(1 + \tilde{n}_m/\tilde{n}_{\mu p}) \, .$$

c) Potential Barrier

Using the preceding formulae, it is, however, still impossible to account for the peak in the reflectivity near the missing ray around the longitudinal frequency ω_l for the $(10\bar{1}1)$ face.

Hopfield and Thomas [2] have proposed a so-called "potential barrier" near the surface of the crystal. This is brought about by the interaction between the exciton and its "electrical image" through the surface, which can be considered as producing its reflection at the surface. Thus it is also prevented from coming too near the surface, and the repulsive potential $f(1/r)$ can be approximated as infinity in the vicinity of the surface. This is not entirely satisfactory. Sein [5] thought of having a damping factor γ increasing to infinity near the surface, but has not developed the idea so far as we know.

We have studied both cases extensively and will report the results later. Whichever model is used, it amounts to a layer of thickness l on the surface without excitons and, according to Hopfield and Thomas, with refractive index $n_0 = \sqrt{\varepsilon_{0\perp}}$ for ordinary waves and

$$n_0 = \sqrt{\varepsilon_{0\perp} \sin^2 \phi + \varepsilon_{0||} \cos^2 \phi}$$

for nonordinary waves. In the generalized case, n_0 has any value accounting for a surface layer.

Accordingly, the reflection coefficient is given by:

$$\varrho_b = (\varrho_0 + \varrho_l\, e^{2i\omega n_0 l})/(1 + \varrho_0 \varrho_l\, e^{2i\omega n_0 l})$$

where

$$\varrho_0 = (1 - n_0)/(1 + n_0)$$

and

$$\varrho_l = (\varrho - \varrho_0)/(1 - \varrho\varrho_0)$$

where ϱ is the reflection coefficient of the bulk crystal without potential barrier, calculated in Sub-Section b).

7. Analysis of Spectra

a) Ordinary Spectrum for $k\perp$ (0001) (Table 1)

We first analyzed the ordinary spectrum of reflection normal to the basal plane (0001) by a classic Kramers-Kronig analysis. Figures 2a and 2b show the index dispersion curves and the absorption coefficient curve thus obtained. The crosses show the points obtained for the extinction coefficient and the absorption coefficient from transmission measurements. There is a slight disagreement. This can be explained by an increased damping factor of the waves near the surface of the crystal.

Table 1. Parameters obtained by fitting the ordinary spectrum normal to the (0001) face

Fig.		δ %	v_0 cm^{-1}	v_l cm^{-1}	γ cm^{-1}	$4\pi\alpha_0 \cdot 10^2$	ε_0	M/m_0	l Å
	K.K.		20154.9	20207.5	25.4	6.60	12.60	∞	0
2c	Cl.	0.70	20157.1	20209.0	23.2	6.70	13.03	∞	0
2d	Cl.bp	0.96	20156.2	20227.5	16.1	9.43	13.30	∞	90.9
3.1A	P.	0.55	20156.8	20209.7	19.4	6.85	13.03	15.74	0
3.1B	A.	0.54	20156.8	20209.8	19.1	6.87	13.04	3.18	0
3.2A	P.	1.87	20153.1	20210.9	5.4	7.33	12.75	0.675 fixed	0
3.2B	A.	0.73	20156.1	20210.6	14.6	7.06	13.04	0.675 fixed	0
4.1A	P.bp	0.36	20152.9	20211.3	0.37	7.37	12.69	1.085	42.2
4.1B	A.bp	0.63	20155.5	20220.4	7.4	8.49	13.17	1.000	69.3
4.2A	P.bp	1.22	20152.0	20211.8	$\rightarrow 0$	7.54	13.70	0.675 fixed	29.0
4.2B	A.bp	0.64	20155.4	20219.1	5.9	8.32	13.15	0.675 fixed	64.3

K.K. Kramers-Kronig analysis.
Cl. Classic dispersion theory.
P. Spatial dispersion theory with a.b.c. of Pekar.
A. Spatial dispersion theory with a.b.c. of Agarwal et al.
bp Potential barrier.
δ Mean deviation value.
N.B. The four parameters v_0, v_l, $4\pi\alpha_0$ and ε_0 are not independent, but verify the relation:
$v_l^2/v_0^2 - 1 = 4\pi\alpha_0/\varepsilon_0$.

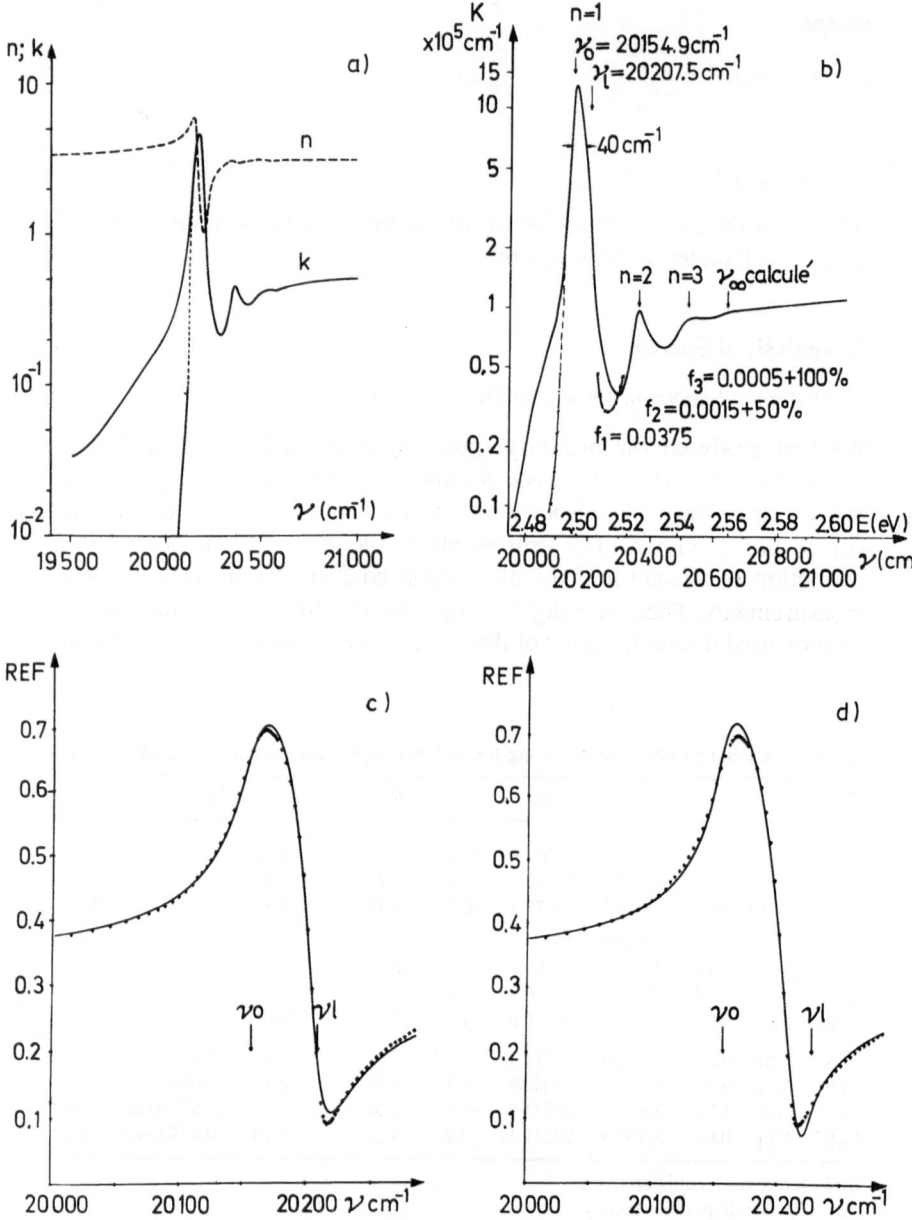

Fig. 2. (a) Dispersion and (b) absorption coefficient obtained from a Kramers-Kronig analysis of the ordinary spectrum normal to (0001) face. Fits (full line) of ordinary spectrum (crosses) normal to (0001) face according to classic dispersion theory (c) without, and (d) with potential barrier. Adjusted parameters in Table 1

This disagreement does not much influence the evaluation of the oscillator strength: $f_1 = 3.75 \cdot 10^{-2} \pm 3\%$, which gives a polarizability for zero frequency of $4\pi\alpha_0 = 0.066$. Gähwiller and Harbeke [16] seem to have obtained 50% more on a similar cleaved basal plane, but Harbeke et al. [17] about half that for the ordinary spectrum normal to the c axis.

For ε_0 we get 12.6, and the damping factor Γ, which is classically defined as the half-width of $\varepsilon'' v$, is 25 cm^{-1}. The half-width of the absorption curve is 40 cm^{-1}. This is reasonable in classic analysis for large oscillator strength and relatively small damping factor.

We have also tried to fit the experimental spectrum in the vicinity of the first exciton transition to a theoretical reflection curve for classic dispersion, shown in Fig. 2c. The constants obtained are in good agreement with those deduced from Kramers-Kronig analysis. Figure 2d shows the fit of the experimental spectrum with a theoretical reflection curve for classic dispersion including a surface layer. This does not improve the result. We tried to see if spatial dispersion curves would give a better fit.

Without potential barriers the results are quite similar in classic and spatial dispersion, as is apparent from Fig. 3.1 A and 3.1 B. There is an improvement in the fit when we use spatial dispersion with both a.b.c. Γ changes to 19 cm^{-1}.

The effective masses are $16.0\,m_0$ (a.b.c. of Pekar) and $3.2\,m_0$ (a.b.c. of Agarwal et al.). Figures 3.2A and 3.2B show similar curves when we fix the effective mass at the value $0.675\,m_0$ measured by Baldini and Franchi [19]. The oscillator strengths are the same at 3%. This is easily understood: in a spatially dispersive crystal the Kramers-Kronig analysis of the normal reflection spectrum gives a fictitious index whose integration gives the oscillator strength.

We have also fitted the ordinary reflection spectra perpendicular to the basal plane in the case of a so-called potential barrier. Using the a.b.c. of Pekar, we have a reasonably good fit, but there is a small spike at v_l. Using the a.b.c. of Agarwal (Fig. 4.1 B), the fit is not as good, but no spike appears at v_l. With the effective mass of $0.675\,m_0$, the a.b.c. of Pekar gives a reasonable fit even with $\gamma = 0$ on Figure 4.2B, whereas with the a.b.c. of Agarwal $\gamma = 5.90$ cm^{-1} and the residual ray is a little asymmetrical (Fig. 4.2B).

b) Ordinary Spectrum of Reflection Normal to a Face (10$\bar{1}$1) (Table 2)

With the polarization the same as it was previously, we have a particular feature for k (10$\bar{1}$1): a peak around ω_2. Any dispersion model without a so-called Hopfield potential barrier of generalized surface index layer cannot account for that structure.

Table 2. Parameters obtained by fitting the ordinary spectrum normal to (10$\bar{1}$1) face

Fig.		δ %	v_0 cm^{-1}	v_l cm^{-1}	γ cm^{-1}	$4\pi\alpha_0$	ε_0	$\varepsilon_{surf. layer}$	l Å
5a	Cl.sl	1.65	20138.6	20214.6	4.23	$\underline{37.0}$	$\underline{4890}$	$138 + i\ 2.5$	230.0
						$4\pi\alpha_0 \cdot 10^2$		M/m_0	
5b	Cl.bp	3.55	20136.3	20214.8	15.6	6.59	8.42	∞	189.8
5.1A	P.bp	0.64	20136.3	20214.0	1.30	6.50	8.41	0.835	154.8
5.1B	A.bp	1.20	20138.3	20213.9	1.25	6.35	8.44	0.200	156.8
5.2A	P.bp	0.84	20135.7	20213.8	0.22	6.48	8.34	0.675 *fixed*	151.1
5.2B	A.bp	2.18	20137.6	20214.6	6.91	6.52	8.51	0.675 *fixed*	171.6

Captions: See Table 1.
sl: Absorbing surface layer of dielectric constant ε.
N.B. In Cl.sl, the underlined parameters are beyond physical meaning.

Classic dispersion with a different index layer near the surface fits such a structure, as shown in Fig. 5b with ε_0. If we try for a better fit with $\varepsilon_s \neq \varepsilon_0$, we see from Fig. 5a that the physical values used are clearly outside a reasonable range. If we take the case of Fig. 5b, where the fit is not so good, we have reasonable values: f is the same as in the (0001) case; ε_0 is 9 instead of 13 in the (0001) case. Figure 5.1A shows the result of a fit with the a.b.c. of Pekar. The fit is very good: m is near $0.675\,m_0$; Γ is small. Figure 5.1B shows the fit with the a.b.c. of Agarwal. The fit is much worse than in the preceding case. Figures 5.2A and 5.2B show the same analysis with $m = 0.675\,m_0$. The results are similar with the a.b.c. of Pekar; in the case of the a.b.c. of Agarwal et al. the fit is not very good. It is not much better then, and similar to the case of classic dispersion.

c) Nonordinary Spectrum of Reflection Normal to a Face (10$\bar{1}$1) (Table 3)

When $E \cdot (k \wedge c) = 0$, we observe a structure with two peaks of "anomalous" reflection at ω_{l_1} and ω_{l_2}. Classic dispersion with a layer of index given by

$$n_0 = \sqrt{\varepsilon_{0\perp} \sin^2 \phi + \varepsilon_{0\parallel} \cos^2 \phi}$$

gives rise to a structure similar to that shown in Fig. 6a. The fit, however, is not very good and it is necessary to take spatial dispersion into account to improve it.

Figure 6.1A shows the result of a fit with the a.b.c. of Pekar. The fit is much better but some discrepancies remain. We notice that the values of $4\pi\alpha_{0\perp}$ and $\varepsilon_{0\perp}$ are higher than the values obtained on the ordinary spectra. But these values are closely correlated to those of $4\pi\alpha_{0\parallel}$ and $\varepsilon_{0\parallel}$. If we fix $4\pi\alpha_{0\perp}$ at 0.0652 we obtain practically the same

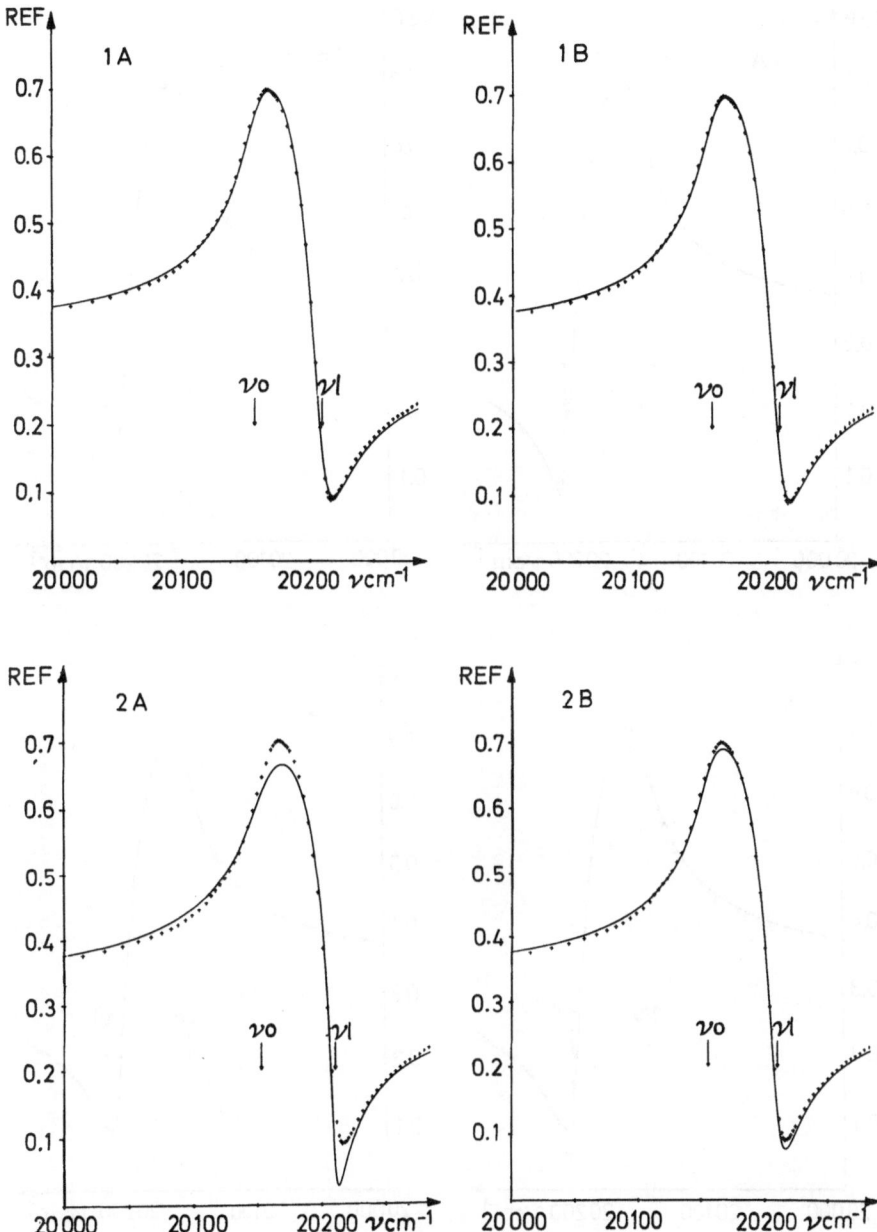

Fig. 3. Fits (full line) of ordinary spectrum (crosses) normal to (0001) face according to spatial dispersion theory without potential barrier: A: a.b.c. of Pekar; B: a.b.c. of Agarwal et al.; 1: effective mass adjusted; 2: effective mass fixed at $0.675 \, m_0$. Adjusted parameters in Table 1

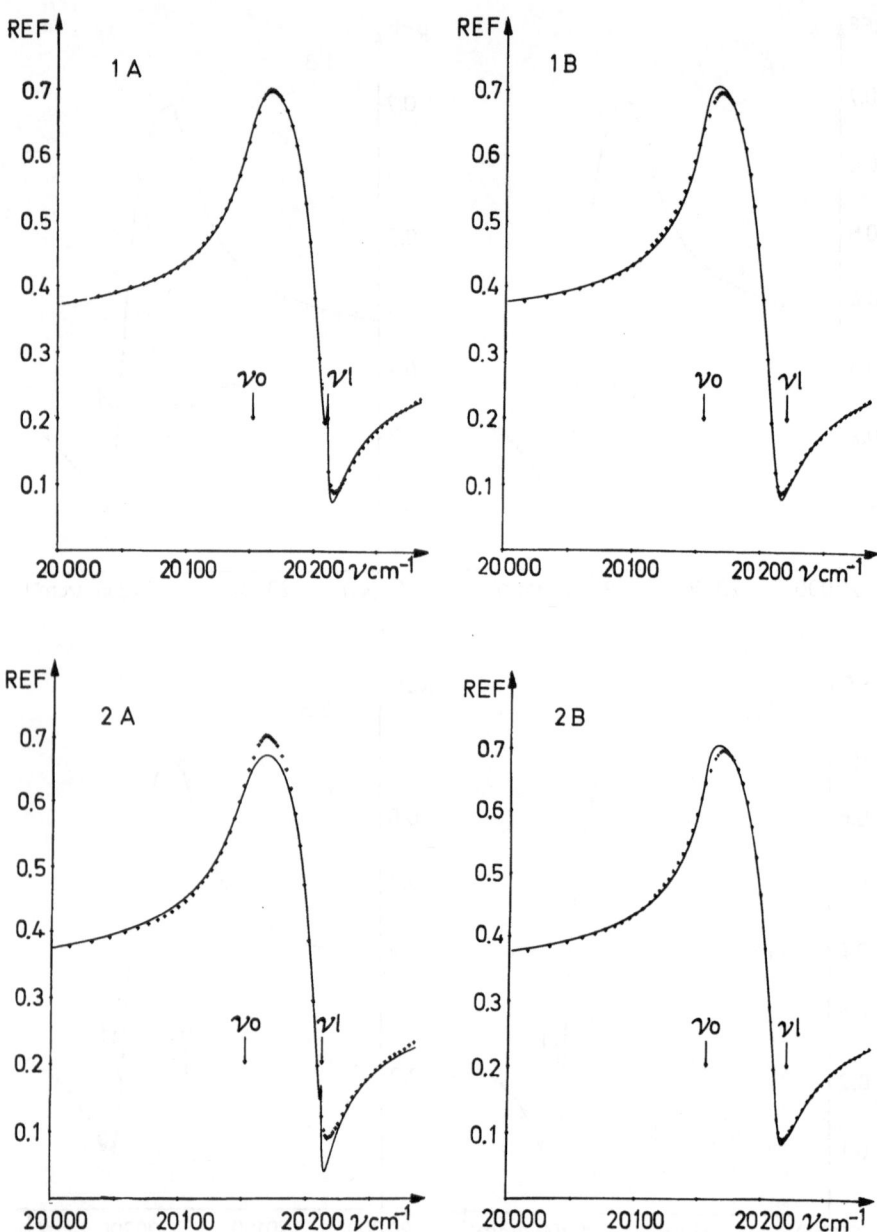

Fig. 4. Fits (full line) of ordinary spectrum (crosses) normal to (0001) face according to spatial dispersion theory with potential barrier: A: a.b.c. of Pekar; B: a.b.c. of Agarwal et al.; 1: effective mass adjusted; 2: effective mass fixed at 0.675 m_0. Adjusted parameters in Table 1

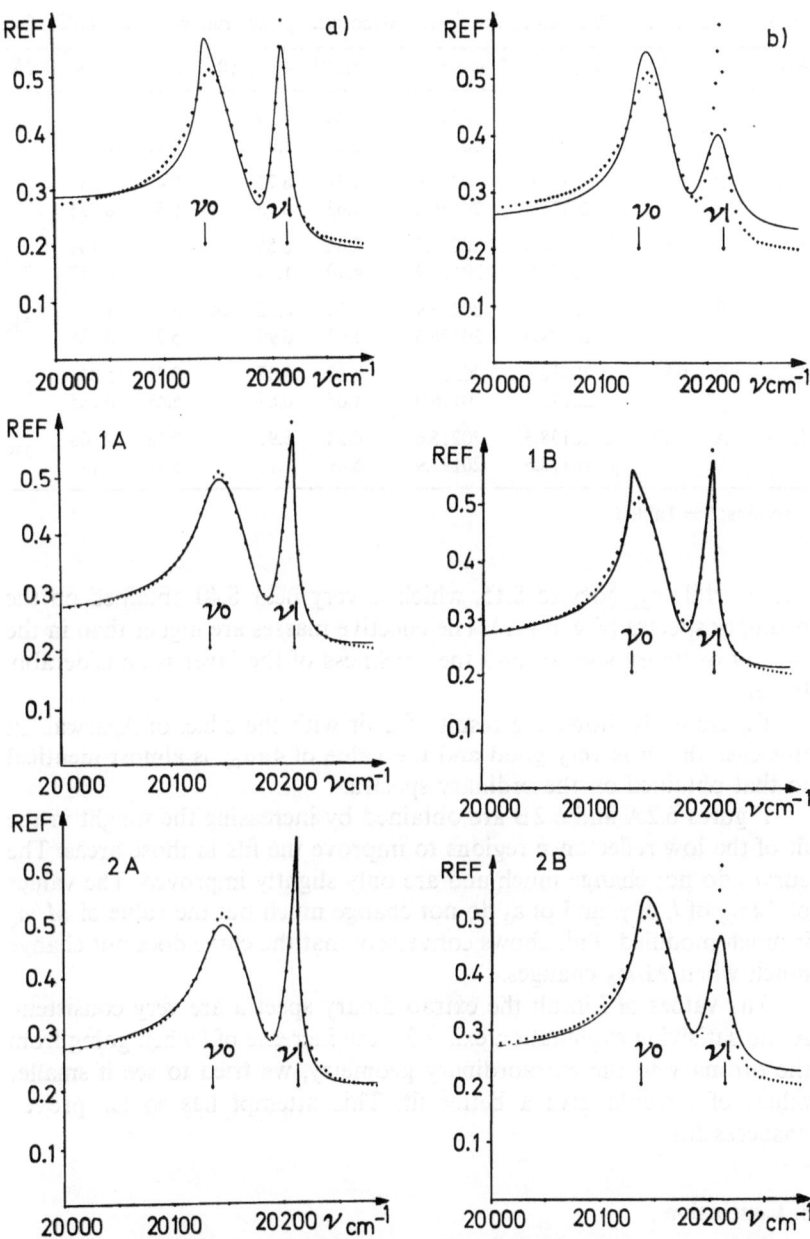

Fig. 5. Fits (full line) of ordinary spectrum (crosses) normal to (10$\bar{1}$1) face according to classical dispersion theory: a) with adjusted absorbing surface layer, b) with potential barrier; according to spatial dispersion theory with potential barrier: A: a.b.c. of Pekar; B: a.b.c. of Agarwal et al.; 1: effective mass adjusted; 2: effective mass fixed at 0.675 m_0. Adjusted parameters in Table 2

Table 3. Parameters obtained by fitting the extraordinary spectrum normal to $(10\bar{1}1)$ face

Fig.	δ %	v_0 cm^{-1}	v_l cm^{-1}	γ cm^{-1}	$4\pi\alpha_0 \cdot 10^2$	ε_0	M/m_0	lÅ
6a	Cl. 1.41	⊥ 20145.9	20214.6	3.44	7.79	11.41	∞	265
		‖ 20176.5	20196.1	10.1	1.05	5.41	∞	
6.1 A	P. 0.79	⊥ 20123.0	20215.9	1.74	8.72	9.42	1.66	236
		‖ 20179.6	20196.6	1.63	0.93	5.53	0.735	
6.1 B	A. 0.66	⊥ 20129.9	20215.7	0.84	6.55	7.67	0.141	237
		‖ 20177.9	20197.9	4.19	1.14	5.78	0.337	
6.1 A′	P. 0.79	⊥ 20135.5	20215.9	1.74	6.52 fixed	8.15	1.66	236
		‖ 20179.0	20196.6	1.63	0.99	5.71	0.735	
6.2 A	P. 0.88	⊥ 20132.8	20216.0	1.67	8.07	9.75	5.04	237
		‖ 20179.7	20196.0	1.05	0.88	5.48	0.785	
6.2 B	A. 0.72	⊥ 20138.3	20215.6	0.54	5.91	7.68	0.108	239
		‖ 20178.6	20197.9	5.49	1.11	5.81	1.14	

Captions: See Table 1.

curve, while $\varepsilon_{0\perp}$ goes to 8.15, which is very near 8.40 obtained on the ordinary spectra (Fig. 6.1 A′). The effective masses are higher than in the case of ordinary spectra and the thickness of the layer is considerably larger.

Figure 6.1 B shows the result of a fit with the a.b.c. of Agarwal. In this case the fit is very good and the value of $4\pi\alpha_{0\perp}$ is almost identical to that obtained on the ordinary spectra.

Figures 6.2 A and 6.2 B are obtained by increasing the weight in the fit of the low reflectance regions to improve the fits in those areas. The curves do not change much and are only slightly improved. The values of $4\pi\alpha_0$, of l, of γ, and of ε_0 do not change much but the value of M/m_0 is much modified. This shows conversely that the curve does not change much when M/m_0 changes.

The values of l in all the extraordinary spectra are very consistent. As no satisfying explanation exists for the increase of l when going from the ordinary to the extraordinary geometry, we tried to see if smaller values of l would give a better fit. This attempt has so far proved unsuccessful.

8. Conclusion

We have compared the validity of application of the spatial dispersion theories of Pekar and Agarwal in the case of the reflection spectra of PbI_2, in which a big effect was supposed to allow accurate measurements. Our analysis shows in fact that the big effect is essentially due to

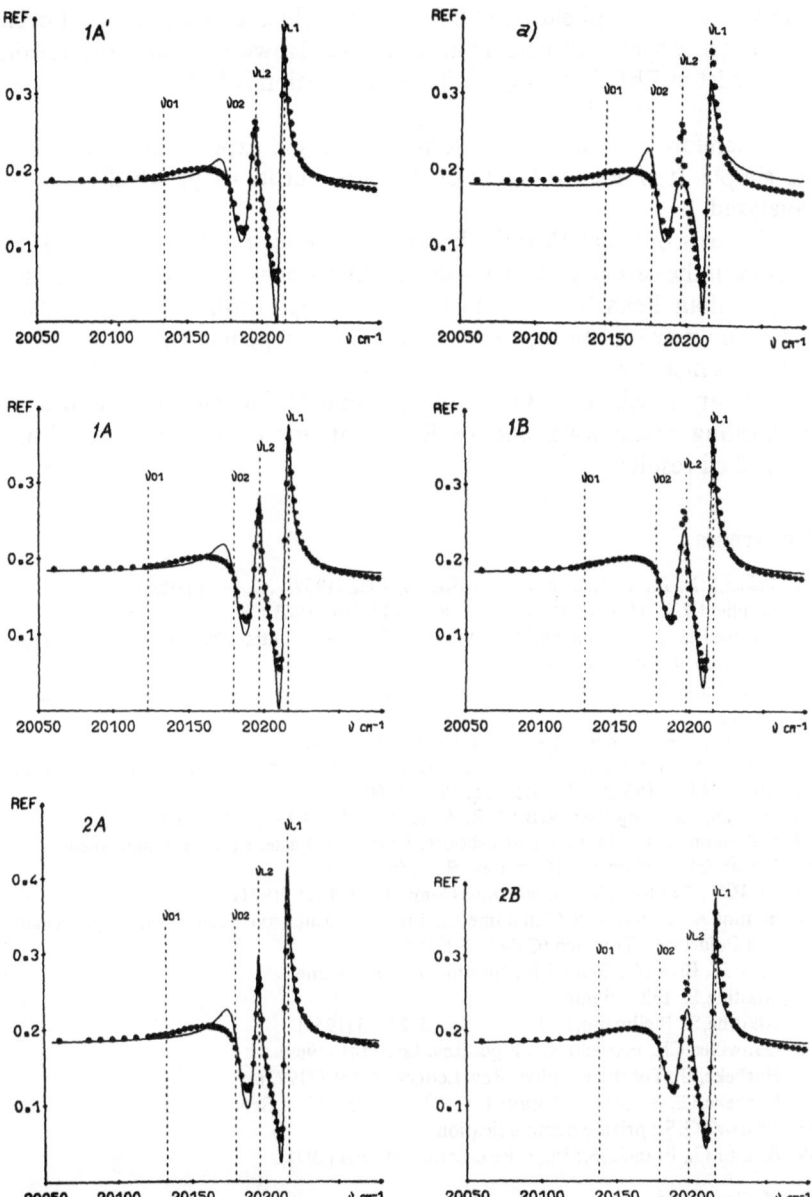

Fig. 6. Fits (full line) of nonordinary spectrum (crosses) normal to (10$\overline{1}$1) face according to classical dispersion theory with potential barrier: a) according to spatial dispersion theory with potential barrier: A: a.b.c. of Pekar; B: a.b.c. of Agarwal et al.; 1: normal weight; 2: increased weight in low reflectance region; 1 A': $4\pi\alpha_{0_1}$ fixed at 0.0652. Adjusted parameters in Table 3

the so-called "Hopfield potential barrier," which could produce it even without any spatial dispersion in the crystal. However, spatial dispersion does exist in PbI_2 but its quantitative evaluation is less simple than was previously expected.

Therefore comparison between theories and experiments is also not so simple. It is not the height of the peak but its shape which must be analyzed.

It turns out that Pekar's theory gives better results than Agarwal's theory in the case of ordinary spectra while Agarwal's theory gives better results than Pekar's theory in the case of extraordinary spectra. This opens up new questions, which we hope many people will study before our next meeting.

We are indebted to G. S. Agarwal and H. Haken for a number of interesting discussions, and to E. Tosatti for communication of unpublished results.

References

1. Pekar, S. I.: Zh. Eksperim. i Teor. Fiz. **33**, 1022 (1957); **34**, 1176 (1958).
2. Hopfield, J. J., Thomas, D. G.: Phys. Rev. **132**, 561 (1963).
3. Agarwal, G. S., Pattanayak, D. N., Wolf, E.: Phys. Rev. Letters **27**, 1022 (1971); — Opt. Commun. **4**, 255, 260 (1971).
4. Maradudin, A. A., Mills, D. L.: Taormina Research Conf. on the Structure of Matter, (1972).
5. Sein, J. J.: Ph. D. thesis, New York University (1969).
6. Zeyher, R., Birman, J. L.: Taormina Research Conf. on the Structure of Matter (1972).
7. Brodin, M. S., Pekar, S. I.: JETP **38**, 1910 (1960).
8. Gorban, I. S., Timofeiev, V. B.: C.R. Akad. Sci. URSS **140**, 791 (1961).
9. Biellmann, J.: Ph. D. thesis, Strasbourg, Université Pasteur (1972) unpublished.
10. Pasterniak, Vedam. K.: Phys. Rev. B **3**, 2567 (1971).
11. Yu, P. Y., Cardona, M.: Solid State Commun. **9**, 1421 (1971).
12. Bonnot, A., Benoît à la Guillaume, C.: Int. Symposium on Excitons at High Density and Polaritons, Tonbach (GER), Oct. 1973.
13. Levy, R., Bivas, A., Grun, J. B., Nikitine, S.: this volume.
14. Nikitine, S.: this volume.
15. Nikitine, S., Biellmann, J.: J. Phys. Suppl. **27**, 95 (1966).
16. Gähwiller, Ch., Harbeke, G.: Phys. Rev. **185**, 1141 (1969).
17. Harbeke, G., Tosatti, E.: Phys. Rev. Letters **28**, 1567 (1972).
    Harbeke, G., Bassani F., Tosatti, E.: XIth Conf. Int. Phys. Semicond. p. 163 (1972)
18. Agarwal, G. S.: private communication.
19. Baldini, G., Franchi, S.: Phys. Rev. Letters **26**, 503 (1971).

Dr. M. Grosmann
Dr. J. Biellmann
Prof. S. Nikitine
Laboratoire de Spectroscopie et d'Optique du Corps Solide
Université Louis Pasteur
5, rue de l'Université
F-67000 Strasbourg (France)

Medium and High Polariton Densities*

H. Mahr

An amazingly rich and complex set of facts on CdS-like semiconductors at He-temperatures has been presented at this Conference. As shown in Fig. 1 in a very schematic way, a lot of the effort to date has centered on producing excitations by absorption of light or electrons in a region near A which in this schematic two-particle diagram represents the free electron-free hole pair continuum state produced by the light. For very low light intensities (photons/cm² sec) the electron-hole pair created at A relaxes somehow to a bottleneck region near C, a region that I would like to call a polariton region because photon-like influences become very important there to an extent that they cause the very existence of a bottleneck region. Recent measurements of Heim and Wiesner [1] and earlier data by Benoît et al. [2] and Gross et al. [3] confirmed the bottleneck existence as predicted theoretically by Toyozawa [4].

As shown in Fig. 2, luminescence emitted after flash excitation lasts as long as 2.3 nano-seconds in the middle of the bottleneck region

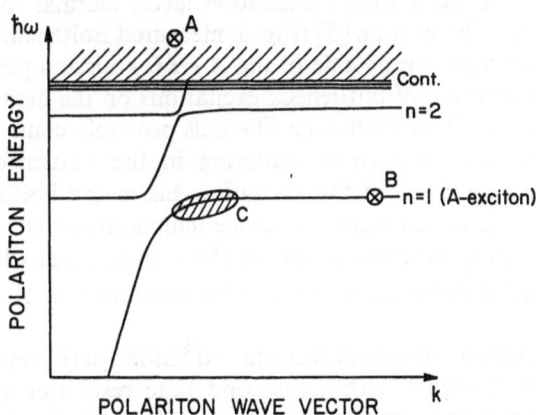

Fig. 1. Schematic (one-dimensional) Polariton dispersion curves

* Work supported in part by the U.S. Office of Naval Research under contract N 00014-67-A-0077-0019, Technical Report # 19, and by the National Science Foundation under Grant # GH-33637.

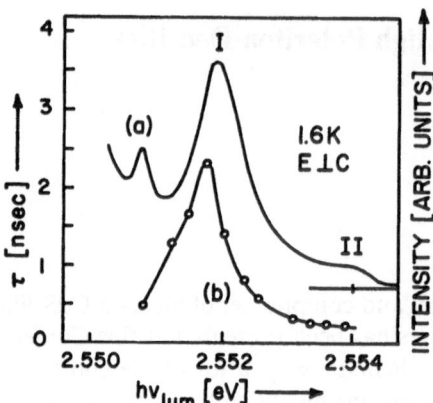

Fig. 2. (a) Luminescence spectrum of a high quality CdS crystal at 1.6 K. Peaks I and II originate from the bottleneck region. (b) Luminescence decay times after flash excitation for different polariton energies in the bottleneck region (After Heim and Wiesner, Ref. [1])

whereas excitations above and below the bottleneck exist for much shorter times. This piling up of excitations produces a distribution which is non-thermal at He-temperature. For slightly less pure samples even bottleneck excitations live only a short time, so that particle distributions over energy in the C-region differ from sample to sample. Unless measurements like those depicted are made each time for a sample no quantitative information about the initial state of excitations will be available. At slightly higher excitation levels thermal equilibrium may come about. Leheny et al. [5] (Fig. 3) measured Boltzmann-like distributions on the high energy side of the $A + LO$ emission peak, reflecting a similar distribution of bottleneck excitations on the high energy side of the bottleneck. These Boltzman-like tails probably come about through near elastic exciton-exciton scattering in the bottleneck region and indicate temperatures of the excitation gas in the bottleneck of 5.8 K to 29 K, much higher than the lattice temperature that seemed to stay well below 10 K. Nothing is known about the time at which such equilibrium distributions come about after excitation at A and how long they exist.

Recent measurements of Kuroda and Shionoya [6] reveal an amazing complexity of events with picosecond time resolution experiments in CdSe at very high excitation densities. As shown in Fig. 4, luminescence excited by a 10 psec pulse at $t = 0$ occurs first, between 50 and 150 psec of excitation, exclusively centered at the location of the M-band, found earlier by Shionoya, Saito, Hanamura and Akimoto [7]. Between 150 and 250 psec luminescence is exclusively centered around the position

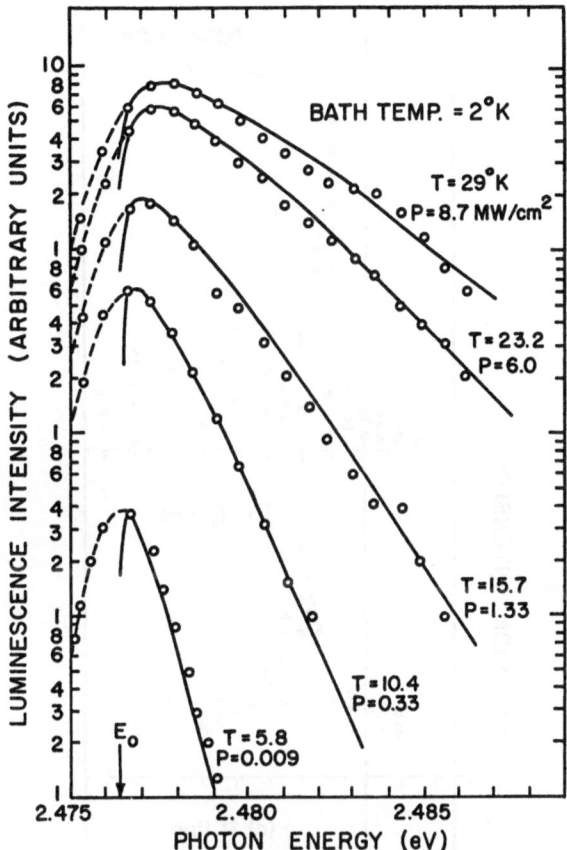

Fig. 3. Emission line shape for $2LO$-phonon assisted recombination of free excitons in CdS. The open circles correspond to original data obtained by subtracting out background luminescence. The solid curves correspond to a theoretical fit to the data using a Maxwellian distribution in energy of exciton-polaritons and a density of states factor proportional to $(E - E_0)^{\frac{1}{2}}$. The dashed curve indicates how the measured emission falls off for $E < E_0$. (After Leheny, Nahory and Shaklee, Ref. [5])

of the P_M band, found earlier by Saito, Shionoya and Hanamura [8]. The M-band was thought to be luminescence emitted during the decay of biexcitons (excitonic molecules), whereas the P_M band was thought to originate from the scattering process of biexcitons with each other. Finally, after about 250–350 psec the P-peak [9] appears which is supposedly due to bottleneck exciton-exciton scattering. These beautiful new results which confirm and extend some of the earlier preliminary data obtained by Figueira and Mahr [10], raise many questions as to

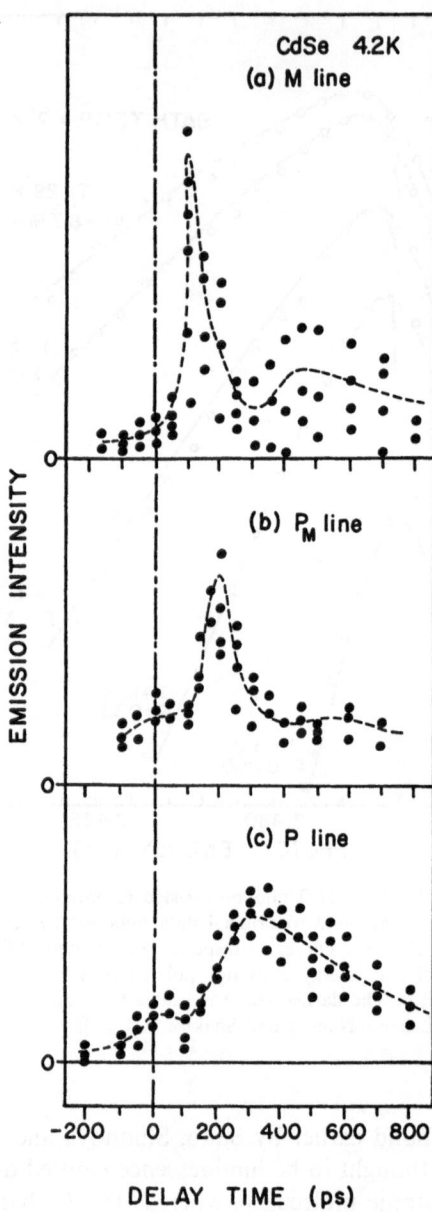

Fig. 4a–c. Time behavior of luminescence after pico second flash excitation at the peak position of the *M*-line (a), the P_M-line (b) and the *P*-line (c). (After Kuroda and Shionoya, Ref. [6])

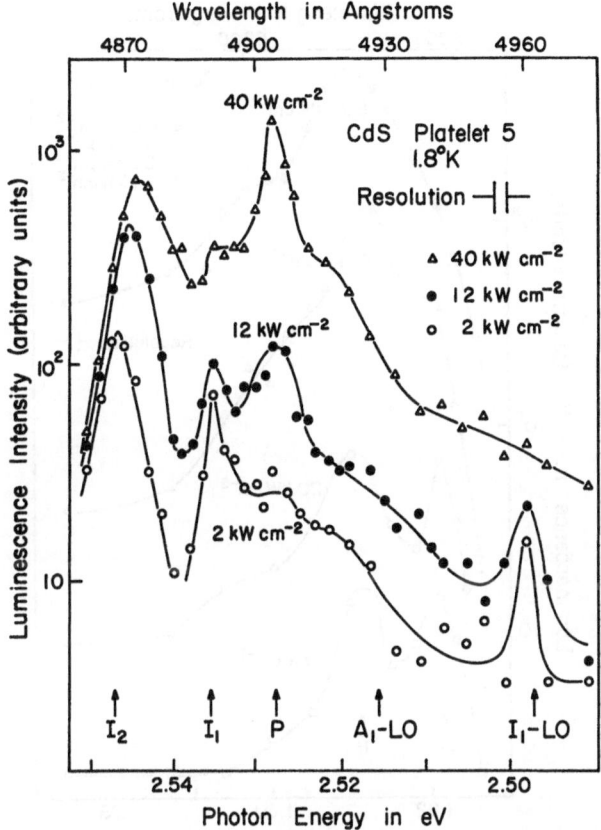

Fig. 5a–c. Luminescence spectra under nano second excitation at 1.8 K of CdS (a), CdSe (b) and ZnO (c) in the medium excitation density range. Notice the superlinear growth of the *P*-peak in all cases. (After Magde and Mahr, Ref. [12])

the sequence of events, cross sections for formation and decay and, finally, why bottleneck excitons appear so late in the time sequence and last for only 350 psec.

It is from time-resolved spectra of the kind presented by Kuroda and Shionoya [6] that much information about the dynamics of excitations is to come. More traditional methods with nanosecond pulse excitations cause the luminescence region to become horribly crowded. They raise questions as to when a particular peak appears in the sequence of events. Because of the complexity of results in the high excitation region with nanosecond excitation I will not review results in that area (see the review by Levy *et al.*, p. 171 of this volume).

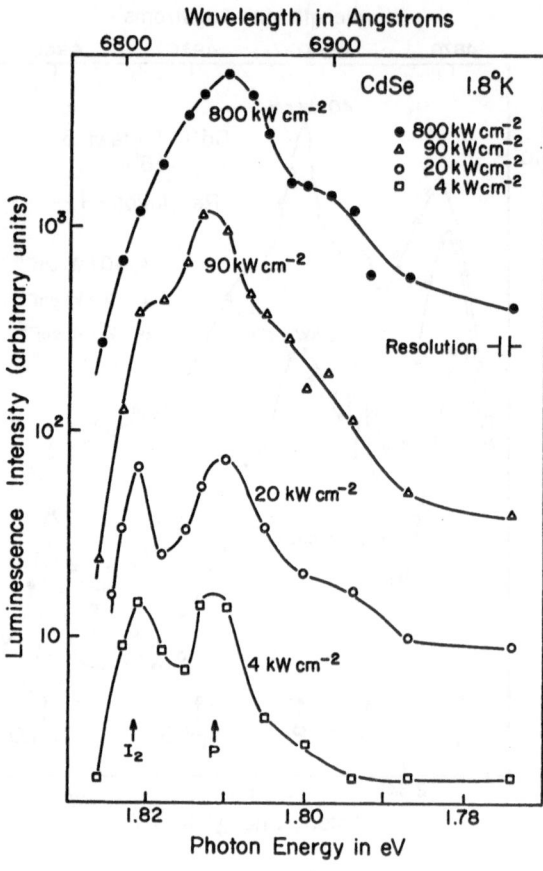

Fig. 5 b

Instead I would now like to review some of the medium density work with nanosecond excitation into the excitation continuum.

Figure 5a shows an old example of luminescence at intermediate level excitation of 10–50 kW/cm². By comparing the *P*-peak intensity with other luminescence peaks and by combining all the best known facts about high-purity CdS, Magde and Mahr [11] came up with a tentative value of the cross section for the bottleneck polariton-polariton scattering cross section. The bimolecular rate constant k is of the order of 10^{-13} cm³ sec⁻¹ for estimated bottleneck polariton concentrations of 10^{16}–10^{18} cm⁻³. We have assumed here that the polaritons that scatter are all in the bottleneck region although their temperature, if any, might be higher than the lattice temperature.

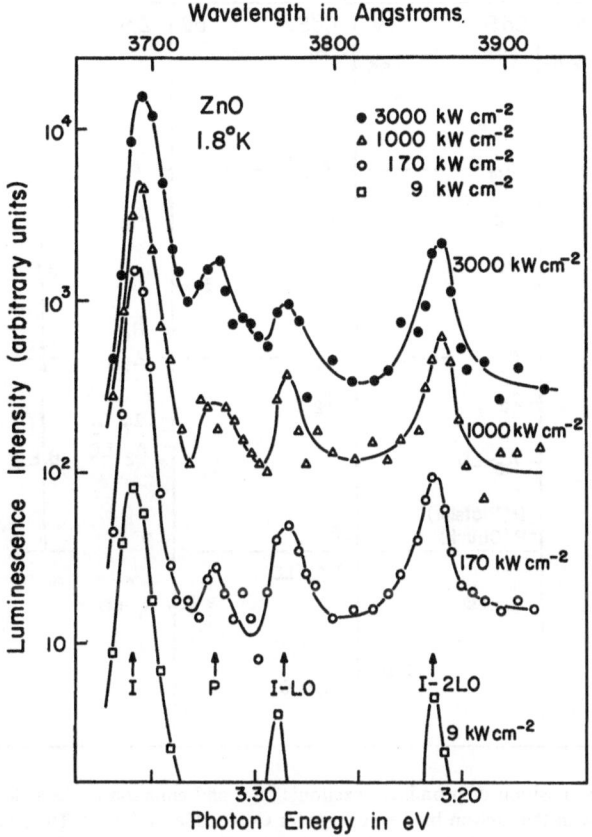

Fig. 5 c

Based on the facts [12] presented in Figs. 5a, 5b, and 5c for CdS, CdSe, and ZnO it is now well known that the position of the *P*-peak for polariton-polariton scattering in the *A*-exciton range, as shown in Fig. 6, lies at approximately twice the binding energy of the *A*-exciton, thus suggesting the following interpretation (Benoît et al. [9]; Keldysh et al. [13]; Magde and Mahr [12]) (Fig. 7): Of two excitonic polaritons in the bottleneck region, situated near *C* but not necessarily possessing the same *k*, one is scattered to higher energies into a $n = 2, 3, ..., \infty$ bound state. The other polariton is scattered into a more photon-like state near *E*. The reason for repeating these well-known facts is that a process very much like the one outlined is well known in non-linear optics. As shown in Fig. 7, below, two photons in a beam with frequency near ω_p may decay spontaneously into two other photon-like polaritons

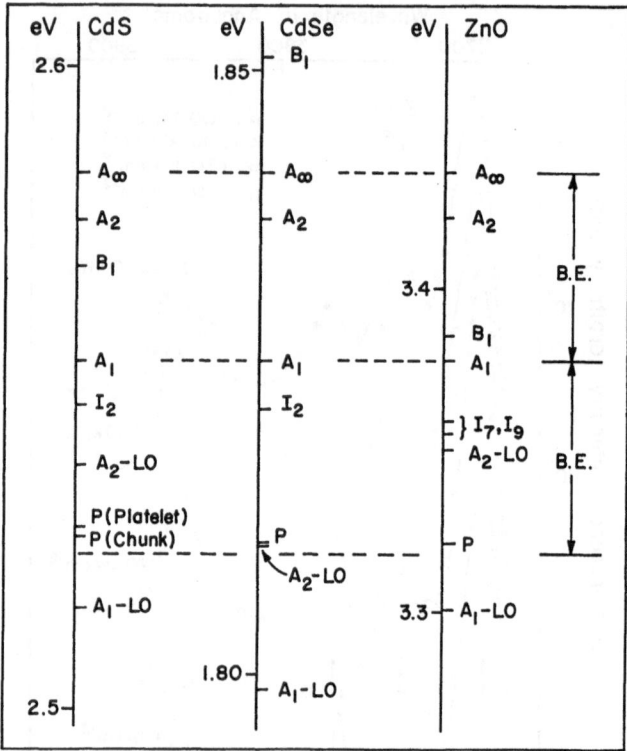

Fig. 6. Energy position of prominent excitonic lines and emission peaks at low temperatures scaled with the exciton binding energy for CdS, CdSe, and ZnO. The position of the *P*-peak is shown to be approximately another binding energy from the position of the *A*-exciton in each case

situated near *G* and *H*, with momentum and energy conserved overall. This process is called 4-photon parametric-scattering or light-by-light scattering. The two different processes shown in Fig. 7 are linked together into one and the same kind of process, namely polariton-polariton scattering, by the polariton concept as shown in Fig. 8. Two polaritons incident in a laser beam at ω_p near the absorption edge, for example, decay because of interactions between each other into a exciton-like polariton state $n = 2, 3, \ldots, \infty$, whereas the other polariton escapes as a more photon-like particle at ω_4. Other scattering processes are possible. For example, an elastic or near elastic scattering process may occur in which ω_3 and ω_4 are very near ω_p. In particular if the incident beam stays just below the intrinsic absorption region, i.e., just below

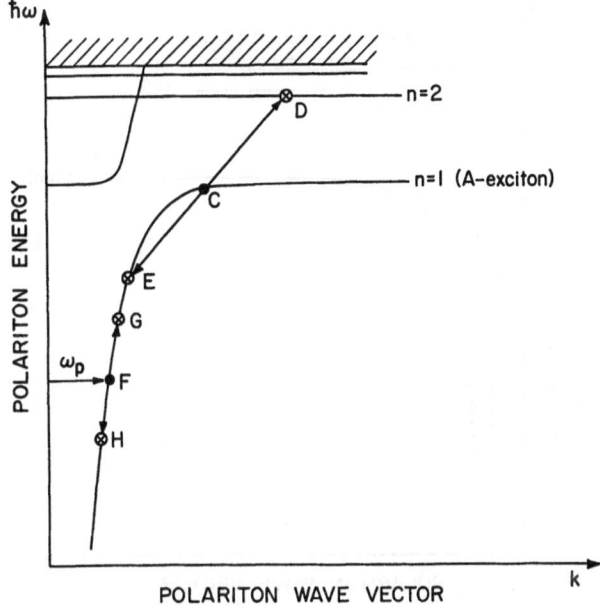

Fig. 7. Schematic polariton dispersion curve indicating exciton-exciton scattering of bottleneck excitons near C into a more exciton-like state near D and a more photon-like state near E (the P-peak location). Four-photon parametric scattering of two "photons" inside a solid may result in scattering one "photon" to higher energy (at G) and one to lower energy (at H)

the A-exciton peak, detailed information might be obtained that is otherwise, i.e., in the traditional continuum absorption of light at A of Fig. 1, difficult to get at. Whereas in absorbing near A of Fig. 1 a multitude of processes are triggered and proceed in time, with light-by-light scattering only one process takes place; in addition, angular dependences can be measured because incident excitations are only produced with one k-direction. Whereas with traditional excitation electrons, holes, excitons might influence the process and shift luminescence bands, no other particles will be created here. They can be created, however, at will by additional illumination in order to observe such effects. Finally whereas the luminescence spectrum gets horribly crowded with peaks due to molecules, molecule-molecule scattering, exciton-exciton, impurity-exciton luminescence, etc., only the process under study (Fig. 8) will give rise to a luminescence band. Furthermore, in a situation as depicted in Fig. 8, where k-vector conservation dictates that only states at $n = 2, 3, \ldots \infty$ may be occupied, the "shape" of the scattered radiation

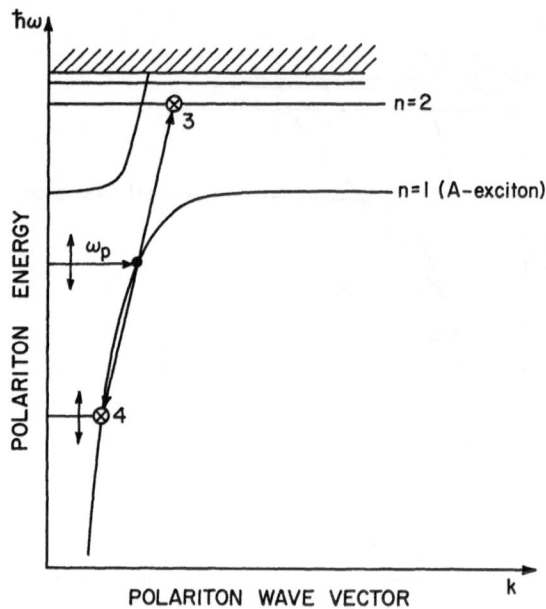

Fig. 8. Polariton-polariton scattering of two polaritons near $\hbar\omega_p$ into an exciton-like polariton state near 3 and a more photon-like state near 4

near ω_4 will reflect the density of states, scatter cross section, etc. of the region near ω_3. We might thus get information on these states, also under illumination that might shift such bands. The great advantage is the observation in the uncrowded transparent region of crystals. A great difficulty in the interpretation of photo-luminescence with traditional excitation methods is the nonuniform distribution of excitations along the sample normal because of the strong absorption constant. There is no such problem with polariton-polariton interaction if the primary excitation is below the edge. Finally, measurements of polariton-polariton interactions made as a function of the incident photon energy ω_p will give us the resonant behavior of interaction cross sections as we go from the photon-like into the exciton-like region of excitations.

All these effects are mediated by the second order non-linear susceptibility $\chi^{(3)}$ as will be shown later. Before we discuss $\chi^{(3)}$ let us review experiments on the resonant behavior of another optical non-linearity, that of $\chi^{(2)}$. Although processes involving $\chi^{(2)}$ do not have a direct relationship to high density exciton problems, they do constitute high-density polariton effects and taught us much about the resonance behavior of non-linear optical constants in general, a lesson that should

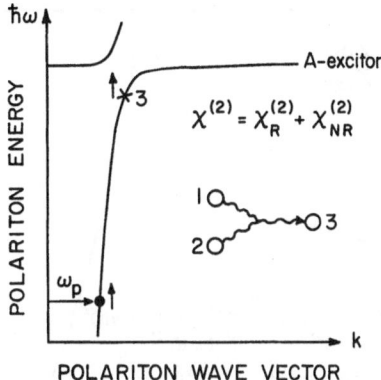

Fig. 9. Polariton fusion: Two (photon-like) polaritons near $\hbar\omega_p$ fuse to form a polariton near 3. The efficiency of fusion is related to the second order, nonlinear optical susceptibility $\chi^{(2)}$ which displays a nonresonant term and a term $\chi_R^{(2)}$, resonant in the region of the A-exciton near ω_3

help us with the polariton-polariton problem. As shown in Fig. 9, two photon-like polaritons near ω_p can join together, or "fuse" as Professor Fröhlich's group calls it, to form a polariton near ω_3. By varying ω_p we can map out the fusion probability throughout the exciton region. The particle created in the process of fusion (or second harmonic generation (SHG) as it is called in non-linear optics) may also be called the bipolariton (bi-photon). The non-linear property of the crystal that governs polariton fusion is $\chi^{(2)}$, the lowest non-linear susceptibility. It consists of a resonant and a nonresonant term. The experiment measures only the relative change of the resonant term. From a comparison of absolute values, it can be shown that the nonresonant contribution to $\chi^{(2)}$ below the resonance is one hundred times larger than the resonant contribution. The results [14] for the resonant contribution are shown in Fig. 10. $\chi_{zzz}^{(2)}$ is shown as a function of the fusion polariton-energy in the vicinity of the C-exciton in ZnO. The A and B excitons do not participate in SHG. Two resonances are clearly seen, one at the C-exciton energy at 3.42 eV and the other one, a subsidiary resonance, at 3.465 eV.

A theoretical analysis was made of these results for $|\chi|$ in terms of the simplest model possible, the anharmonic oscillator model. A term kvx^2 is added to the spring force, kx, of a harmonic oscillator, with v giving the amount of anharmonicity. As shown in detail elsewhere [14] the anharmonic model employing two nonlinear oscillators centered at two resonances, ω_0 and ω_0', predicts for the total linear dielectric con-

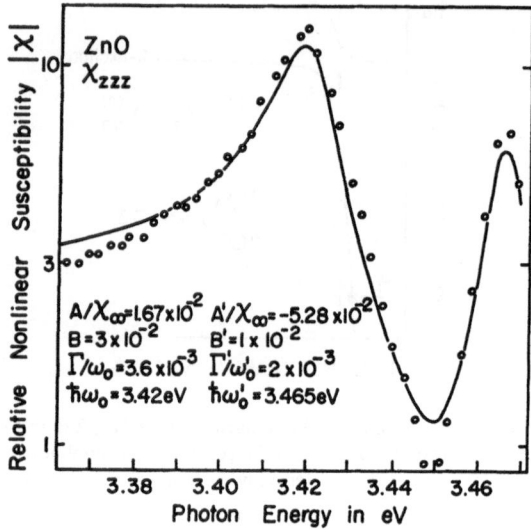

Fig. 10. Relative nonlinear susceptibility $\chi_{zzz}^{(2)}$ of ZnO in the region of the C-exciton. The circles are from measurements of second harmonic generation (polariton fusion) as a function of photon (polariton) energy. The solid curve is a theoretical fit of a two oscillator model with the constants shown. (From Haueisen and Mahr, Ref. [14])

stant (from which n and $\varkappa$ can be derived) and for the non-linear optical susceptibility χ the following results:

$$\varepsilon = \varepsilon_\infty + B\omega_0^2 \left[\omega_0^2 - \omega^2 - i\Gamma\omega\right]^{-1} + B'\omega_0'^2 \left[\omega_0'^2 - \omega^2 - i\Gamma'\omega\right]^{-1}, \qquad (1)$$

$$\chi = \chi_\infty + A\omega_0^2 \left[\omega_0^2 - \omega^2 - i\Gamma\omega\right]^{-1} + A'\omega_0'^2 \left[\omega_0'^2 - \omega^2 - i\Gamma'\omega\right]^{-1} \qquad (2)$$

where B, B' are quantities proportional to the oscillator strength of the transition, A, A' among other things are proportional to the non-linear constants v and v'. χ_∞ and ε_∞, two constants, are admittedly poor substitutes for the influence of yet higher resonances on values of ε and χ at lower photon energies. The quantity of interest in SHG is $|\chi|$ which results in a complicated expression [14], involving both resonances and a "mixed term":

$$|\chi| = \left\{ \chi_\infty^2 + \frac{A^2\omega_0^4 + 2A\chi_\infty\omega_0^2(\omega_0^2 - \omega^2)}{(\omega_0^2 - \omega^2)^2 + (\omega\Gamma)^2} \right.$$

$$+ \frac{A'^2\omega_0'^2 + 2A'\chi_\infty\omega_0'^2(\omega_0'^2 - \omega^2)}{(\omega_0'^2 - \omega^2) + (\omega\Gamma')^2}$$

$$\left. + \frac{2AA'\omega_0^2\omega_0'^2\left[(\omega_0^2\omega^2)(\omega_0'^2 - \omega^2) + \omega^2\Gamma\Gamma'\right]}{\left[(\omega_0^2 - \omega^2)^2 + (\omega\Gamma)^2\right]\left[(\omega_0'^2 - \omega^2)^2 + (\omega\Gamma')^2\right]} \right\}^{\frac{1}{2}}. \qquad (3)$$

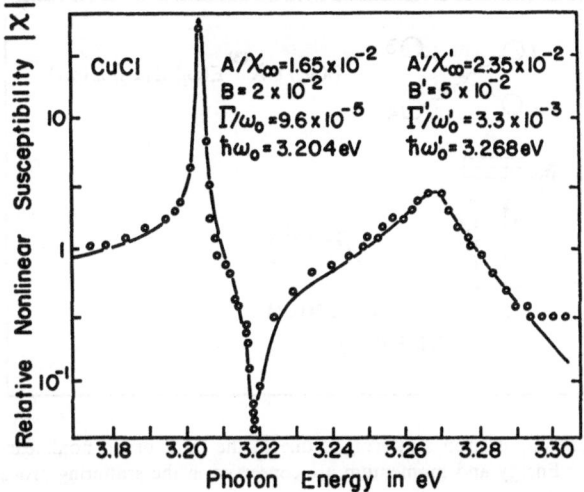

Fig. 11. Relative nonlinear susceptibility in the region of the two 1s excitons of CuCl. The circles are from measurements of second harmonic generation (polariton fusion) as a function of photon (polariton) energy. The solid curve is a theoretical fit of a two oscillator model with the constants shown. (From Haueisen and Mahr, Ref. [14])

The solid line in Fig. 10 is a best fit to the experimental data. As seen a remarkably good fit is obtained with the six parameters quoted. Remarkably, A'/χ_∞ is negative and that points out a major problem. The phase factors, $\pm$, of the various contributions to χ are not known, but are also not directly measurable in experiments.

Figure 11 shows results [14] for the region of the first $n = 1$ exciton-polariton in CuCl. A much sharper resonance is clearly displayed in the vicinity of the exciton position at 3.204 eV. The second exciton, near 3.268 eV, is much wider. Again a good fit was obtained overall with a two oscillator model.

Let me digress a moment to point out a problem. In principle and because we have circumstantial evidence otherwise, spatial dispersion should be included in a model that describes narrow resonances like that of the first exciton in CuCl. This is because coupling of one oscillator to its neighbors is more important than dissipating its energy through damping. The easiest way of incorporating this coupling is by including a kinetic energy term $\hbar^2 k^2/m^*$ into the polariton energy $\hbar\omega_0$. m^* is the effective mass of motion of the polariton-exciton. In the case of the linear dielectric constant ε this procedure leads to two complex solutions ε_1 and ε_2. If we adopt the same procedure for introducing

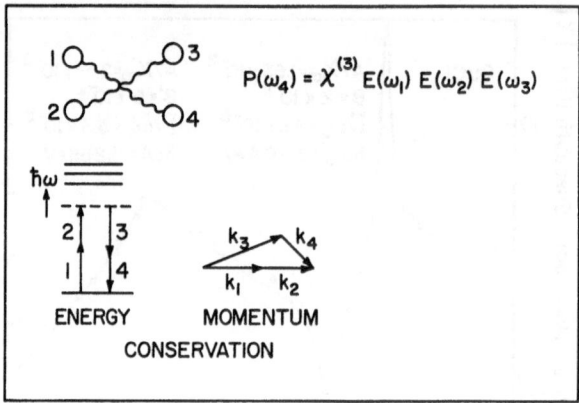

Fig. 12. Polariton-polariton scattering is due to the third order, nonlinear optical susceptibility $\chi^{(3)}$. Energy and momentum are conserved in the scattering process as shown

spatial dispersion into the non-linear constant [15]

$$|\chi| = \left\{ \chi_\infty^2 + \frac{A^2\omega_0^4 + 2A\chi_\infty\omega_0^2[\omega_0^2 - \omega^2(1 - \varepsilon'/M)]}{[\omega_0^2 - \omega^2(1 - \varepsilon'/M)]^2 + (\omega^2\varepsilon''/M - \omega\Gamma)^2} \right\}^{\frac{1}{2}} \qquad (4)$$

(with $\varepsilon = \varepsilon' + i\varepsilon''$ the linear dielectric constant, and $M = m^* c^2/\hbar\omega_0$), we get also two solutions, χ_1 and χ_2. The absolute values $|\chi_1|$ and $|\chi_2|$ were determined for parameters similar to those of CuCl but for varying ratios of damping, Γ/ω_0, to coupling, M. For large damping, the lowest solution is the same as that produced in Fig. 11 near the sharp resonance. For smaller values of damping appropriate to the most likely value in CuCl the two solutions cross over. If one adopts a naive view of second harmonic generation in the presence of spatial dispersion, then two second harmonic waves are generated always and the total power produced is given simply by the sum of the contribution of both waves:

$$\text{Power generated} \propto \{|\chi_1|^2 |\varepsilon_1 - n_F^2|^{-2} + |\chi_2|^2 |\varepsilon_2 - n_F^2|^{-2}\} P_F^2 . \qquad (5)$$

Here n_F is the index at the incident photon energy. With this naive assumption, neglecting interference between the two generated waves no description of the experimental results can be obtained because the higher terms of $|\chi_1|$ and $|\chi_2|$ on either side of the resonance give larger contributions to the generated power than the lower terms which gave such good fits without spatial dispersion in Fig. 11.

Let me now return to the general topic, the connection between non-linear optics and high exciton densities. The next higher, second

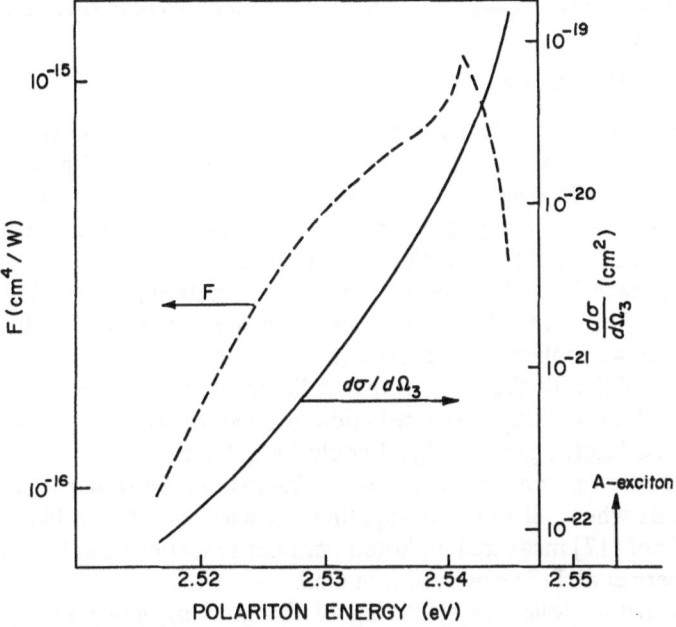

Fig. 13. Calculated elastic polariton-polariton scattering cross section $d\sigma/d\Omega$ as a function of polariton energy (solid line) with a value of πr_e^2 at the position of the A-exciton (r_e = exciton radius). Also shown (dashed curve) is the calculated bimolecular rate constant F of the scattering process. (After Andersen, Ref. [16])

order non-linear susceptibility $\chi^{(3)}$ is the one that involves polariton scattering. As shown in Fig. 12, $\chi^{(3)}$ can mediate a process in which two polaritons, say from a laserbeam, can decay in a solid into two others, of higher and lower energy, and with conservation of energy. Two such particular processes are possible. One is almost elastic and the two polaritons change very little in energy but do change momentum. This elastic process is the one that might contribute to the establishment of equilibrium in the bottleneck region and so an exploration of the cross section of this process as a function of incident photonenergy $\hbar\omega_p$ would be most helpful in evaluating the importance of elastic exciton-exciton scattering. Anderson [16] has recently calculated the cross section of this process in terms of the exciton-exciton cross section. Assuming a billiard-ball cross section πr_e^2 (with r_e = exciton radius) for exciton-exciton scattering in the bottle neck region the following values for the elastic scattering cross section as a function of photonenergy were obtained (Fig. 13). Although the differential elastic polariton-polariton cross section $d\sigma/d\Omega$ changes continuously from the A-exciton position to lower photon energies (solid line), the bimolecular rate constant F defined as:

$$\text{Power scattered} = FI_p^2 \tag{6}$$

where I_p is the incident (pump) intensity at ω_p, has a maximum near 2.54 eV, but does not change very much (dashed line). With an incident intensity of 10^6 W/cm^2 a scattered power per steradian of 10^{-3} Watt is predicted. Preliminary experiments of the 3M laboratories with very small incident light intensities were not successful.

We have looked at inelastic polariton scattering of the kind indicated in Fig. 8. In a first experiment with the 5145 Å line of the Argon ion laser we looked for scattered radiation near 5500–5600 Å but were unsuccessful with the CdS sample at liquid N_2 temperature. In a more recent attempt with pulsed radiation at 5300 Å (SH from a Q-switched Nd-glass laser) again no signal could be detected.

4-photon parametric effects were observed in the transparent region of solids where all four participating particles are photon-like. Eichler and Knof [17] measured 4-photon parametric scattering in Rutil (TiO$_2$), Grinberg et al. [17] observed it in CdS.

Polariton-polariton scattering of the kind imagined in Fig. 8 involves the creation of an exciton-like state ($n = 2, 3, \ldots \infty$). The theory of 3-photon parametric scattering, involving $\chi^{(2)}$, with one of the end products approaching a resonant state (in this case a phonon state rather than an exciton state) was recently summarized by Montgomery and Giallorenzi [18]. Their very careful theoretical analysis of the problem, aided by carefully controlled experiments, should now be carried over to the 4-particle polariton-polariton scattering of Fig. 8 involving the resonant behavior of $\chi^{(3)}$ and phase space arguments of the experimental set up.

In addition to the spontaneous decay into two other polaritons discussed so far there is also the possibility of stimulated decay. If in Fig. 7 a laserbean at ω_H would be simultaneously incident on the sample with the pump laser beam at $\omega_p(F)$, then greatly enhanced scattering should take place at ω_G and ω_F. Such an experiment was performed by Eichler et al. [19] in CdS with all four particles photon-like, i.e., in the transparent region of the crystal. As shown in Fig. 14, radiation intersects inside the sample at the correct angle for phase matching. No values were quoted for $\chi^{(3)}$ because no theory was as yet available to extract that information from experiment. For an incident pump power at 6943 Å of 2 MW/cm^2 and with a secondary laser beam at $\lambda = 7658$ Å of 50 kW/cm^2 a scattered power of 1 mW–1 W was measured at $\lambda = 6350$ Å. From older measurements of third harmonic generation Maker and Terhune [20] quoted $\chi^{(3)} \approx 1.6 \cdot 10^{12}$ cm^3 erg^{-1}.

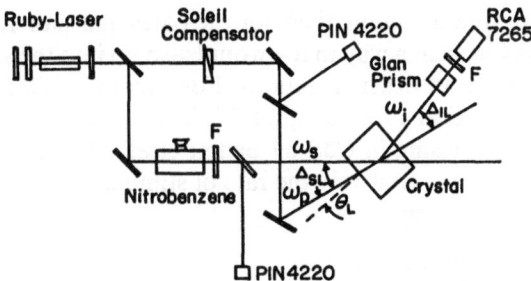

Fig. 14. Experimental set-up used to observe stimulated 4-photon parametric scattering in CdS. The pump radiation originates from a ruby laser; the stimulating radiation ω_L is produced by stimulated Raman scattering in Nitrobenzene. (After Eichler, Fery and Hermann, Ref. [19])

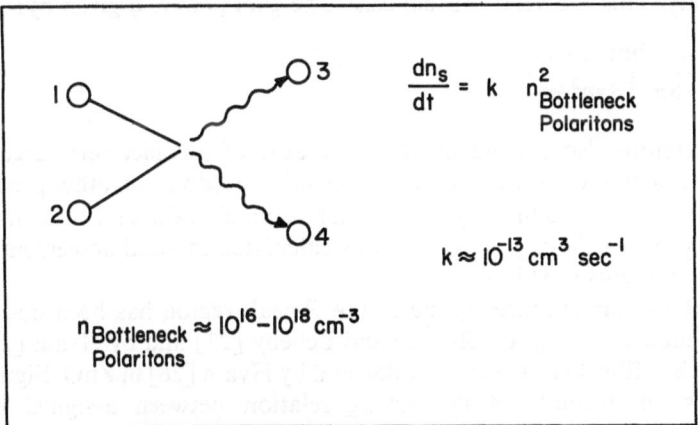

Fig. 15. Bottleneck exciton-exciton scattering

A similar experiment for exciton-like polaritons was recently proposed by Andersen [21]. Greatly enhanced scattering efficiencies were calculated for this elastic component of polariton-polariton scattering. For an incident pump beam and auxiliary beam of 10^5 Watts/cm^2 at $\hbar\omega_0 = 2.504$ eV and with a CdS sample at 20 K a scattered power of $4 \cdot 10^{-2}$ Watts is calculated.

The same process, stimulated 4-polariton interaction, may happen without externally incident radiation as in Fig. 8 at ω_4 by amplification of spontaneous emission produced inside the crystal. This process would then give rise to one-passage stimulated emission, often loosely called lasing. If in Fig. 15 a photon-like polariton of frequency ω_4 would come along into a given direction, then, for the particular pair 1, 2, the probability to decay in such a way that 4 would also go into the same direction is enhanced over all other directions. For exciton-exciton scattering in the bottleneck region a follow-up of this idea, which was originally proposed to explain lasing in CdS by Benoît et al. [9] was given by Mahr and Tang [22]. Quantitatively the rate of stimulated emission per mode and photon is given by:

$$R = \frac{kn^2 \text{ bottleneck-excitons}}{8\pi v^2 \varDelta v n_0^3/c^3} \tag{7}$$

where k is the bimolecular rate constant for exciton-exciton scattering, $\varDelta v$ the bandwidth of spontaneous emission in the P-peak and n_0 the index of refraction. The rate of stimulated emission is therefore proportional to n^2, where n is the average steady-state bottleneck polariton density. This is a very unusual case. The gain per cm g given by:

$$g = \frac{kn^2 \text{ bottleneck}}{8\pi v^2 \varDelta v n_0^2/c^2} \tag{8}$$

is therefore also dependent on n^2. Because of this fact very large gain can be achieved for high excitation densities, outdoing all other processes that might depend linearly on n. With reasonable values for the constants of Eq. (8) a gain of $g = 60 \text{ cm}^{-1}$ was calculated in good agreement with measurements [23] in CdS.

At He-temperatures lasing in the P-peak region has been observed by Benoît et al. [9]; by Shaklee and Leheny [23] and by Hvam [24] in CdS; by Hilinski et al. [25] in CdSe and by Hvam [26] in ZnO. Figure 16 shows an example of the strong relation between assigned $n = 2$, $n = 3 \rightarrow n = \infty$ peaks within P to features in the luminescence spectrum. Very interesting in the context of gain quadratic in exciton densities is a recent experiment with ZnO by Klingshirn [27]. Using two-photon excitation, i.e., volume excitation throughout a large sample with fairly good reflecting surfaces low gain lasing was observed at He-temperatures with impurity exciton emission and $A + LO$ emission, both processes that are linear in exciton concentration. Lasing occurs long before excitation densities are reached that allow observation of exciton-exciton scattering. It is only by excluding low gain processes that lasing on the exciton-exciton scattering becomes possible!

One should also not become carried away with exciton-exciton scattering. Lasing with high gain has been observed in many instances. There is no reason to believe, before substantial evidence is found, that in all cases exciton-exciton scattering should be responsible or that even a common explanation has to exist. Recent measurements [28] in GaAs

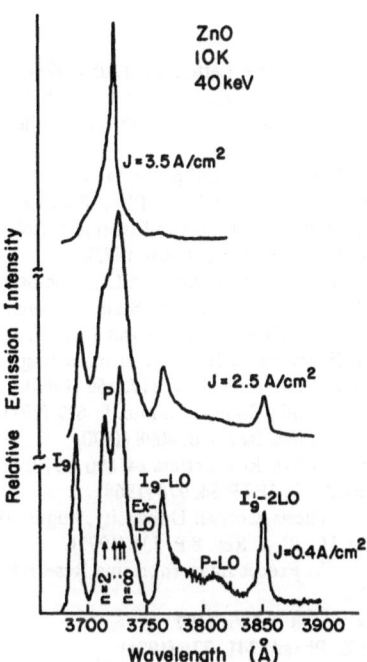

Fig. 16. Emission spectra of pure ZnO at 10 K excited with electron beams of varying current densities. Shown are the measured superlinear growth of emission in the P-band region and the calculated locations of P-band exciton-exciton scattering processes leading to occupation of $n = 2, 3, \infty$ final states. (After Hvam, Ref. [26])

show that lasing there, under certain curcumstances, is due to both band-to-band and single exciton-emission processes.

Finally there should be lasing with the polariton-polariton scattering experiment of Fig. 8. With a strong pump emitting high peak power short pulses, it should be possible to increase the scattering to the point where it would become stimulated. In this case a variable laser pump at $\hbar\omega_p$ would give rise to lasing at output frequencies ω_4 shifted farther and farther from the input. Such a light-frequency transducer might be useful in applications.

So far all considerations were concerned with the real part of $\chi^{(3)}$. The imaginary part of $\chi^{(3)}$ for elastic polariton-polariton scattering gives rise to two-photon absorption. Hanamura [29] suggested that truly giant two-photon absorption would take place at the position of the excitonic molecule (bi-exciton) if $\hbar\omega_p \approx E_M/2$ (E_M = molecular exciton position). Such a large value of $\mathrm{Im}\,\chi^{(3)}$ should also give rise to a correspondingly large value of $\mathrm{Re}\,\chi^{(3)}$. We predict therefore giant elastic 4-polariton scattering near $\hbar\omega_p \approx E_M/2$.

References

1. Heim, U., Wiesner, P.: Phys. Rev. Letters **30**, 1205 (1973)
2. Benoît à la Guillaume, C., Bonnot, A., Debever, J. M.: Phys. Rev. Letters **24**, 1235 (1970).
3. Gross, E., Permogorov, S., Travnikov, V., Selkin, A.: Solid State Commun. **10**, 1071 (1972).
4. Toyozawa, Y.: Progr. Theor. Phys., Suppl. **12**, 111 (1959).
5. Leheny, R. F., Nahory, R. E., Shaklee, K. L.: Phys. Rev. Letters **28**, 437 (1972).
6. Kuroda, Shionoya, S.: Technical Report of Institute of Solid State Physics, University of Tokyo, Japan, Series A No. 603 (August 1973).
7. Shionoya, S., Saito, H., Hanamura, E., Akinoto, O.: Solid State Commun. **12**, 223 (1973).
8. Saito, H., Shionoya, S., Hanamura, E.: Solid State Commun. **12**, 227 (1973).
9. Benoît à la Guillaume, C., Debever, J. M., Salvan, J.: In: Thomas, D. G. (Ed): Proceedings "International Conference on II–VI Semiconducting Compounds", 1967, p. 669. New York: A. Benjamin, Inc. Phys. Rev. **117**, 566 (1969).
10. Figueira, J. F., Mahr, H.: Solid State Commun. **9**, 679 (1971).
11. Magde, M. D., Mahr, H.: Phys. Rev. **2** B, 4098 (1970).
12. Magde, M. D., Mahr, H.: Phys. Rev. Letters **24**, 890 (1970).
13. Keldysh, L. V., Kozlov, A. N.: JETP **54**, 978 (1968).
14. Haueisen, D. C.: Ph. D. Thesis, Cornell University, August 1972 (unpublished).
    Haueisen, D. C., Mahr, H.: Phys. Rev. **8** B, 734 (1973).
15. Mahr, H., Haueisen, D. C.: Proceedings Taormina Research Conference, "Polaritons", Pergamon Press 1974.
16. Andersen, R. J.: Phys. Rev. **8** B, 3861 (1973).
17. Eichler, H., Knof, J.: Z. Physik **241**, 271 (1971).
    Grinberg, E. N., Ryvkin, S. M., Fishman, I. M., Yaroshetskii, E. D.: JETP Letters **7**, 253 (1968).
18. Montgomery, G. P., Jr., Giallorenzi, T. G.: Phys. Rev. **8** B, 808 (1973).
19. Eichler, H., Fery, H., Hermann, F.: Optics Commun. **6**, 152 (1972).
20. Maker, P. D., Terhune, R. W.: Phys. Rev. **137**, A 801 (1965).
21. Andersen, R. J.: Phys. Rev. (to be published).
22. Mahr, H., Tang, C. L.: J. Appl. Phys. **43**, 1818 (1972).
23. Shaklee, K. L., Leheny, R. F.: Appl. Phys. Letters **18**, 475 (1971).
24. Hvam, J. M.: Proceedings Tenth International Conference on the Physics of Semiconductors, Cambridge, Mass. (1970).
25. Filinski, I., Wojtowicz-Natanson, B., Hvam, J. M.: J. Phys. Chem. Solids **32**, 2193 (1971).
26. Hvam, J. M.: Solid State Commun. **12**, 95 (1973).
27. Klingshirn, C.: Solid State Commun. **13**, 297 (1973).
28. Göbel, E., Herzog, H., Pilkuhn, M. H., Zschauer, D. H.: Solid State Commun. **13**, 719 (1973).
29. Hanamura, E.: Solid State Commun. **12**, 951 (1973).

Dr. H. Mahr
Laboratory of Atomic and Solid State Physics
Cornell University
Ithaca, New York 14850
USA

Polaritons at High Light Intensities and in Bose Condensed Exciton Systems ⋆

H. Haken, J. Goll, and A. Schenzle

Contents

1. Introduction

The present paper deals with the resonant interaction of coherent light with excitons. We have found that a number of new effects may be expected at high excitations. In the following we will consider four different cases:

a) A spatially homogeneously distributed light wave interacts with excitons. While for small light intensities the usual polariton concept with the well known dispersion curve is obtained, at high light intensities the dispersion curve undergoes intensity dependent shifts [3].

b) If a short light pulse is impinging on the crystal the effect of self-induced transparency [4, 5] may be found for excitons [6] or, in other words, a new bound state between many excitons and photons arises which moves through the crystal in a soliton-like[1] manner [1, 2].

c) When the exciton system is in a Bose condensed state a new type of polariton occurs with a dispersion curve again shifted against the usual polariton curve but in a manner different from the dispersion curve mentioned in a). To measure the dispersion curve in that case small light intensities are sufficient in contrast to a).

⋆ This work was carried out within the frame of the agreement on French-German scientific cooperation between C.N.R.S., Paris, Université Louis Pasteur, Strasbourg, and Universität Stuttgart.

[1] "Soliton" is a name for the following phenomenon of wave propagation. While wave packets usually change their shape during propagation on account of dispersion effects, solitons are propagating pulses which keep their form due to nonlinear effects. For more details compare the literature [10].

d) A spatially homogeneous wave of high intensity creates excitons at a high density within a mode with the same wave vector k. We consider a situation in which all the excitons show Bose condensation into state k.

Before we describe the main steps of the mathematical treatment, we briefly indicate their physical meaning.

a) When the light intensity becomes very high an appreciable amount of excitons is created which may be described in a picture using localized atomic states in the following manner. At each site there is an appreciable depletion of the electrons in the groundstate and an appreciable creation of a local inversion. This high local inversion changes appreciably the polarisability of the crystal and thus leads ultimately to a new dispersion curve.

b) Selfinduced transparency. This effect had been predicted and discovered by McCall and Hahn [4, 5] for gas atoms. The basic idea is this: if a short coherent light pulse is impinging on an atom and if it is in resonance with the atomic optical transition the field produces a coherent oscillation of the electron bringing it to the upper state from where it falls down to the ground state emitting coherent light.

This process is only possible if there are no competing relaxation processes destroying the phase and the actual inversion. While this process may be easily understood for atoms it is not quite so obvious for excitons because excitons are delocalized excited states belonging to the whole crystal. By forming suitable wave packets, however, one may convince oneself that the impinging light creates locally coherently exciton wave packets which then travel jointly with the light through the crystal. The velocity of light propagation may be appreciably decreased by this process [5].

c) We consider here the situation in which a high density of excitons had already been created and the excitons had come to a quasi equilibrium in the form of a Bose condensed state. We then consider the bound states of an exciton and a photon in this system.

d) In the present case it is assumed that the high light intensity of a light wave with sharp k-vector stabilizes the exciton system at that k-state so that one may expect Bose condensation into that k-state. Again we derive the energy of the total system of excitons and photons which is analyzed in the framework of a normal mode analysis.

2. Polaritons at High Light Intensity
(Giant Polaritons and Selfinduced Transparency)

We use the formalism of second quantization and consider two bands: the valence band and the conduction band. We first use the localized description, distinguishing different lattice sites by the index l. The Hamiltonian consists of three parts, namely that of the electrons of the crystal, that of the free light field, and the interaction term. The Hamiltonian of the electrons has the following form [8]:

$$
\begin{aligned}
H_{el} = E_0 &+ \sum_{l_1,l_2} H^e_{l_1,l_2} a^+_{l_1} a_{l_2} - \sum_{l_1,l_2} H^h_{l_1,l_2} d^+_{l_1} d_{l_2} \\
&+ \tfrac{1}{2} \sum_{l_1,l_2} W^{e-e}(l_1 - l_2)\, a^+_{l_1} a^+_{l_2} a_{l_2} a_{l_1} \\
&+ \tfrac{1}{2} \sum_{l_1,l_2} W^{h-h}(l_1 - l_2)\, d^+_{l_1} d^+_{l_2} d_{l_2} d_{l_1} \\
&+ \sum_{l_1,l_2} W^{e-h}_{ex}(l_1 - l_2)\, a^+_{l_1} d^+_{l_1} d_{l_2} a_{l_2} \\
&- \sum_{l_1,l_2} W^{e-h}_c(l_1 - l_2)\, a^+_{l_1} d^+_{l_2} d_{l_2} a_{l_1}
\end{aligned}
\tag{2.1}
$$

a^+_l, a_l, d^+_l, d_l are the creation and annihilation operators of an electron and a hole, respectively, at lattice site l. The coefficients H^e, H^h are the corresponding transition matrix elements for the electron and hole motion from one lattice site to another one. W^{e-e}, W^{h-h} are the usual Coulomb interaction matrix elements between electrons localized at the corresponding sites. The last two terms in (2.1) describe the Coulomb exchange interaction and the Coulomb interaction between electrons and holes, respectively.

$$
H_{light} = (4\pi)^{-1} \int d^3x (-\omega_0^2 c^{-2} A(x,t) A^+(x,t) + \sum_j \nabla A_j(x,t) \nabla A^+_j(x,t)) \tag{2.2}
$$

is the usual Hamiltonian of the free light field.

ω_0 is the center frequency of the light field under consideration.

$$
H_{el-light} = i\hbar \sum_l a^+_l d^+_l A^+(l,t)\, g + \text{h.c.} \tag{2.3}
$$

is the interaction Hamiltonian describing the interaction between the field and electrons. g is the optical matrix element for the band-to-band transition. We have neglected antiresonant terms. Without going into all the details which can be found in our original publications [3, 7] we just describe the main steps of our procedure.

We consider quite generally Wannier excitons [9]. The wave function for an electron at site l_1 and the hole at lattice site l_2 having the center of gravity at site L and describing the internal state v of the exciton is denoted by $\varphi^{L,v}_{l_1,l_2}$. By means of these wave functions we define operators for localized excitons as follows:

$$
B^+_{L,v} = \exp(i(k_0 L - \omega_0 t)) \sum_{l_1,l_2} \varphi^{*L,v}_{l_1,l_2} a^+_{l_1} d^+_{l_2} \tag{2.4}
$$

where $k_0 = \omega_0/c$ and where c is the light velocity in the medium without exciton transition.

Assuming further that the effective masses of electron and hole are equal we define the number operator of electrons at site L by

$$N_{L,v}^e = \sum_{l_1,l_2} \varphi_{l_1,l_2}^{L,v} a_{l_1}^+ a_{l_2} \tag{2.5}$$

and that of holes by

$$N_{L,v}^h = \sum_{l_1,l_2} \varphi_{l_1,l_2}^{L,v} d_{l_1}^+ d_{l_2} . \tag{2.6}$$

In the following we will use also the abbreviation

$$N_{L,v} = \tfrac{1}{2}(N_{L,v}^e + N_{L,v}^h) . \tag{2.7}$$

As is well known from laser theory [1] it is now best to proceed to the Heisenberg equations of motion which for an arbitrary operator Ω may be found by the well known rule

$$\dot{\Omega} = (i/\hbar) [H, \Omega] . \tag{2.8}$$

One readily obtains the following equations of motion

$$\begin{aligned}
\dot{B}_L^+ = {}& i\omega_0 B_L^+ + (i/\hbar) \sum_{L'} E(L - L') B_{L'}^+ \\
& - (2i/\hbar) \sum_{L'} (W_c^{e-h}(L - L') - W^{e-e}(L - L')) B_L^+ N_{L'} \\
& - (2i/\hbar) \sum_{L'} W_{ex}^{e-h}(L - L') B_{L'}^+ N_L \\
& - g A(L, t) (2N_L - \varphi(0)),
\end{aligned} \tag{2.9}$$

$$\begin{aligned}
\dot{N}_L = {}& (i/\hbar) \sum_{L'} W_{ex}^{e-h}(L - L') (B_L^+ B_L - B_L^+ B_{L'}) \\
& + g(A^+(L, t) B_L^+ + A(L, t) B_L),
\end{aligned} \tag{2.10}$$

$$\partial A/\partial t + c\, \partial A/\partial L = -(\Lambda^{-2}/\varphi(0))\, g B_L^+ , \tag{2.11}$$

where

$$\Lambda^{-2} = 2\pi\hbar c^2/\omega_0 \Delta\Omega \tag{2.11a}$$

which can be easily interpreted as follows: (2.9) describes the temporal change of the exciton amplitude caused by the interaction with all other excitons and caused by the light field. (2.10) represents the change of exciton number on account of the motion of excitons as well as on account of the interaction with the lightfield. (2.11) finally describes the field propagation under the influence of the oscillating exciton amplitude. In order to solve the Eqs. (2.9–2.11) we now distinguish two cases:

a) Exciton amplitude and field oscillate harmonically with the same frequency and are both represented by running waves with the same wave vector. Then the Eqs. (2.9–2.11) reduce to algebraic equations immediately giving the dispersion curve. The dispersion curve is governed by the following two equations

$$\Omega = \tfrac{1}{2} \left(\omega_k + E(k) + 2 W_{ex}^{e-h}(k) \, N \right)$$
$$\pm \tfrac{1}{2} [(\omega_k - E(k) - 2 W_{ex}^{e-h}(k) \, N)^2 + (2g/\Lambda)^2 \, (\varphi(0) - 2N)]^{\tfrac{1}{2}}, \tag{2.12}$$

$$(\Lambda^4/g^2) \, \Omega^2 \, A^2 - N(\varphi(0) - N) = 0. \tag{2.13}$$

The definition of $E(k)$ and $W_{ex}^{e-h}(k)$ is given below in Eqs. (2.18), (2.19).

b) We now consider the question of selfinduced transparency. We anticipate that one may find pulse wave solutions which propagate through the crystal with velocity v. We thus introduce the new coordinate

$$\tau = t - L/v. \tag{2.14}^2$$

For the following analysis we expand the right hand side of Eq. (2.10) into a power series at L and confine the expansion to the first derivative. Under this assumption and using (2.14) as a new variable it is possible to find the following integrals of motion

$$(1 - v_0/v) \, B^+ B + N^2 = \lambda^2 \, \varphi(0) \, A^+ A, \tag{2.15}$$

$$N - (v_1/v) \, B^+ B = \lambda^2 \, \varphi(0) \, A^+ A \tag{2.16}$$

where we have used the abbreviations

$$\lambda^{-2} = -\Lambda^{-2} (1 - c/v)^{-1}, \tag{2.17}$$

$$E(k) = N^{-1/2} \sum_L e^{ikL} E(L), \tag{2.18}$$

$$W_{ex}^{e-h}(k) = N^{-1/2} \sum_L e^{ikL} W_{ex}^{e-h}(L), \tag{2.19}$$

$$v_0 = \hbar^{-1} \, \partial E(k)/\partial k, \tag{2.20}^2$$

$$v_1 = \hbar^{-1} \, \partial W_{ex}^{e-h}(k)/\partial k. \tag{2.21}^2$$

By means of the integrals of motion (2.15) and (2.16) it is not too difficult to find the exact solution (exact in the sense that in (2.9) and (2.10) we have kept in the expansion only terms up to first order). The solution has the following form

$$|A(\tau)|^2 = 2\lambda^{-2}(1 - v_0/v)^{-1} [\alpha \cosh 2T^{-1}(\tau - \tau_0) + \beta]^{-1}. \tag{2.22}$$

---

[2] Note that we now consider a one-dimensional case.

The coefficients α and β are defined as follows:

$$\alpha = \left[1 + (\tau/2)^2 \cdot \left(\frac{\omega_0 - E(k_0) - W_{ex}^{e-h}(k_0)}{1 - v_0/v}\right)^2\right]^{\frac{1}{4}}$$

$$\cdot \left[1 + (\tau/2)^2 \left(\frac{\omega_0 - E(k_0) + W_{ex}^{e-h}(k_0)}{1 - v_0/v}\right)^2\right]^{\frac{1}{4}},$$

$$\beta = \left[1 + (\tau/2)^2 \cdot \frac{(\omega_0 - E(k_0))^2 - (W_{ex}^{e-h}(k_0))^2}{(1 - v_0/v)^2}\right].$$

The phase of the lightfield Φ obeys the equation

$$\dot{\Phi}(\tau) = \tfrac{1}{2}(\omega_0 - E(k_0) - W_{ex}^{e-h}(k_0)) + \tfrac{1}{2}W_{ex}^{e-h}(k_0)\,\lambda^2(1 - v_0/v)\,|A(\tau)|^2 \quad (2.23)$$

and can immediately be found by quadratures. Finally the pulse width is explicitly given by

$$T^{-2} = (1 - v_0/v)^{-2}\left[(g/\lambda)^2 - \tfrac{1}{4}(\omega_0 - E(k_0) - W_{ex}^{e-h}(k_0))^2\right]. \quad (2.24)$$

Our solution shows the possibility to produce, at least in principle, soliton-like solutions consisting of excitons and photons or in other words to produce the phenomenon of selfinduced transparency. In conclusion of this chapter we discuss some necessary conditions for the observation. The pulse duration must be smaller than the characteristic relaxation time constants which are the recombination rate of excitons characterized by time $T_{||}$ and phase destroying processes, in particular by phonons which are characterized by $T_\perp$. In order to keep these times large it is obviously necessary to use crystals at low temperatures.

In the following estimates we assume that $T_{||}$ and $T_\perp$ are of the order of 10^{-11} to 10^{-12} sec. The just mentioned condition

$$T \ll T_{||}, T_\perp \quad (2.25)$$

gives an upper limit on the pulse duration of the impinging laser light. For a given pulse duration T we may calculate by means of (2.24) the pulse velocity v and thus using (2.22) the field amplitude or equivalently the number of created excitons. Assuming as diameter of an exciton 10^{-7} cm and as diameter of the crystal at its face perpendicular to the impinging light 10^{-2} cm^2, the required laser intensity is found to be 10 MW. Several processes competing with the process of selfinduced transparency should be mentioned. One of them is light scattering, another one two- or multiple-photon absorption. We have estimated that the second process should become important at light intensities 10^2 MW so that there seems to be a good chance to observe selfinduced transparency of excitons.

3. Polaritons in Bose Condensed Exciton Systems

We now proceed to consider light propagation in Bose condensed exciton systems. The Hamiltonian has again the structure

$$H = H_{el} + H_{light} + H_{el\text{-}light} \, . \tag{3.1}$$

It is, however, advantageous to use the Hamiltonian in k-space. Thus the Hamiltonian referring to the electrons reads

$$
\begin{aligned}
H_{el} = & \sum_k (E_g + \hbar^2 k^2/2m_e) \, a_k^+ \, a_k + \sum_k (\hbar^2 k^2/2m_h) \, d_k^+ \, d_k \\
& + \tfrac{1}{2} \sideset{}{'}\sum_{k,k',q} W_q a_{k+q}^+ a_{k'-q}^+ a_{k'} a_k \\
& + \tfrac{1}{2} \sideset{}{'}\sum_{k,k',q} W_q d_{k+q}^+ d_{k'-q}^+ d_{k'} d_k \\
& - \sideset{}{'}\sum_{k,k',q} W_q a_{k+q}^+ d_{k'-q}^+ d_{k'} a_k
\end{aligned}
\tag{3.2}
$$

where the indices k etc. refer to k-space. m_e, m_h are the effective masses of the electron and hole, while

$$W_q = 4\pi e^2/V \varepsilon q^2 \tag{3.3}$$

is the Fourier transform of the Coulomb matrix element. E_g is the band gap energy.

The Hamiltonian of the lightfield and of the interaction acquires the form

$$H_{light} = \sum_\lambda \hbar \omega_\lambda b_\lambda^+ b_\lambda \, , \tag{3.4}$$

$$H_{el\text{-}light} = \sum_{l,k} (b_k^+ + b_{-k}) \, d_{-l+k} a_l \tilde{g}_{-l+k,l} + \text{h.c.} \tag{3.5}$$

where $\tilde{g}$ is the optical matrix element for band-to-band transitions. Denoting the wave function of the internal motion of the exciton in the internal state v by $\phi_v(p)$ we define exciton operators by

$$Q_{K,v}^+ = \sum_p \phi_v(p) \, a_{p+(m_e/M)K}^+ \, d_{-p+(m_h/M)K}^+ \tag{3.6}$$

which have the inverse

$$a_{p+(m_e/M)K}^+ \, d_{-p+(m_h/M)K}^+ = \sum_\mu \phi_\mu^*(p) \, Q_{K,\mu}^+ \, . \tag{3.7}$$

We anticipate that all excitons are in the groundstate so that we drop in the following the index v. Since the exciton state with $K=0$ is macroscopically occupied we consider only the mutual interaction between the condensed excitons and the interaction between the non-condensed and condensed ones, but neglect the mutual interaction between the non-condensed excitons. This leads us to the following Heisenberg equations of motion

$$-i\hbar \, \partial Q_K^+/\partial t = [E_g + (K^2\hbar^2/2M) + \varepsilon^{(0)} + 2\varepsilon^{(1)} N_0/V] \, Q_K^+$$
$$+ (\varepsilon^{(1)} N_0/V) \, Q_{-K} + (G_K^{(0)} + G_K^{(1)} N_0/V)(b_K^+ + b_{-K}), \tag{3.8}$$

$$-i\hbar \, \partial b_K^+/\partial t = \hbar \omega_K b_K^+ + G_K^{(0)*} Q_K^+ - G_K^{(0)*} Q_{-K} \tag{3.9}$$

where

$$\varepsilon^{(0)} = \sum_k (k^2\hbar^2/2m) \, \phi^2(k) - \sum_{k,k'}' W_{k-k'} \phi(k) \, \phi(k'), \tag{3.10a}$$

$$\varepsilon^{(1)}/V = 2 \sum_{k,k'}' W_{k-k'} \phi^3(k) \, \phi(k') - 2 \sum_{k,k'}' W_{k-k'} \phi^2(k) \, \phi^2(k'), \tag{3.10b}$$

$$G_K^{(0)} = \sum_p \phi(p) \, \tilde{g}_{-p+(m_h/M)K, \, p+(m_e/M)K}, \tag{3.11a}$$

$$G_K^{(1)}/V = - \sum_p \phi(p) \, \tilde{g}_{-p+m_h K/M, \, p+m_e K/M} \tag{3.11b}$$
$$\cdot (\phi^2(p + m_e K/M) + \phi^2(-p + m_h K/M)) \, ;$$

m is the reduced mass. To linearize the Eqs. (3.8) and (3.9) we replaced the operators Q_0 by $\sqrt{N_0}$, where N_0 is the number of condensed excitons.

We seek the normal modes by introducing operators of the form

$$\alpha_K^+ = w_K Q_K^+ + x_K b_K^+ + y_K Q_{-K} + z_K b_{-K} \tag{3.12}$$

and find after a short straightforward calculation for the energy of the new type of polaritons

$$E_K^2 = \tfrac{1}{2}[\tilde{\varepsilon}_K^2 + (\hbar\omega_K)^2]$$
$$\pm \tfrac{1}{2}[(\tilde{\varepsilon}_K^2 - (\hbar\omega_K)^2)^2 + 16\hbar\omega_K G_K^{0*} G_K(E_g + K^2\hbar^2/2M + \varepsilon^{(0)} + 3\varepsilon^{(1)} N_0/V)]^{\frac{1}{2}} \tag{3.13}$$

where $G_K = G_K^{(0)} + G_K^{(1)} N_0/V$ and

$$\tilde{\varepsilon}_K^2 = (E_g + K^2\hbar^2/2M + \varepsilon^{(0)} + 2\varepsilon^{(1)} N_0/V)^2 - (\varepsilon^{(1)} N_0/V)^2. \tag{3.13a}$$

This is the required dispersion relation which is plotted in Fig. 1 for different concentrations.

Here we don't discuss the possibilities to check experimentally such a dispersion curve but we rather suggest an experiment which could prove the existence of the Bose condensate directly. Since in the Bose condensed state excitons have the k-vector $k=0$, they are unable to emit a light quantum which has a small but finite k-vector. However, it is possible that the condensate emits two polaritons with the same energy, but opposite momentum (Fig. 2). Because the decay rate is proportional to the number of condensed excitons the investigation of this anharmonic decay could be used to prove the existence of the condensate.

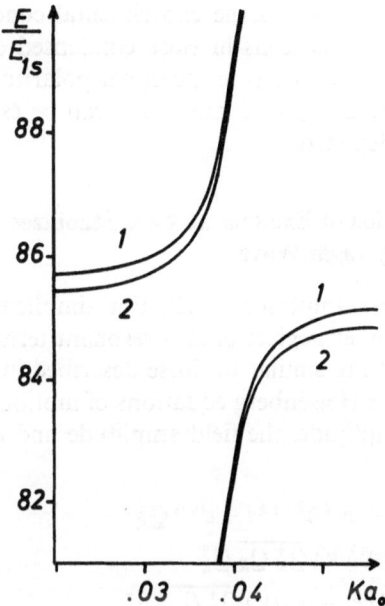

Fig. 1. Polariton dispersion. *1* in a Bose condensed system ($N_0 a_0^3/V = 0.005$); *2* at low excitation level (a_0: Bohr radius of exciton ground state)

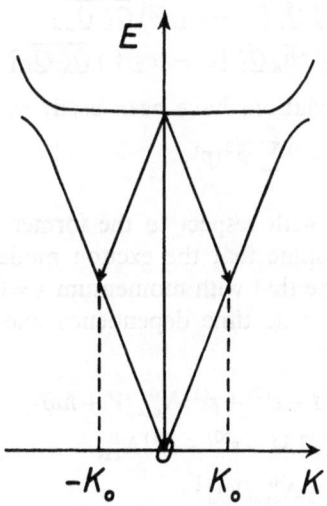

Fig. 2. Anharmonic decay

In conclusion we discuss the experimental conditions under which one might observe polaritons in Bose condensed exciton systems. We must require that the excitons in the upper polariton branch thermalize so quickly that the Bose condensed state can be established before the upper branch is depleted.

4. Bose Condensation of Excitons at $k \neq 0$ Stabilized by a High Intensity Light Wave

We again use the Hamiltonian (3.2). For simplicity we will represent results obtained under neglect of anti-resonant terms occurring in (3.5). The essential steps are similar to those described in Section 3.

Again we derive Heisenberg equations of motion, in the present case for the exciton amplitude, the field amplitude and the exciton number, respectively.

$$
\begin{aligned}
-i\hbar\, \partial Q_K^+/\partial t = &(E_g + K^2\hbar^2/2M + \varepsilon^{(0)})\, Q_K^+ \\
&+ (\varepsilon^{(1)}/V)\, \overline{Q_K^+ Q_K}\, Q_K^+ \\
&+ \tilde{g}[c_1 - (c_2/V)\, \overline{Q_K^+ Q_K}]\, b_K^+ ,
\end{aligned} \tag{4.1}
$$

$$
-i\hbar\, \partial b_K^+/\partial t = \hbar\omega_K b_K^+ + \tilde{g}^* c_1 Q_K^+ , \tag{4.2}
$$

$$
-i\hbar\, \partial \overline{N}_{\text{exc}}/\partial t = \tilde{g} c_1 \overline{b}_K^+ \overline{Q}_K - \tilde{g}^* c_1 \overline{b}_K \overline{Q}_K^+ . \tag{4.3}
$$

We further have to set up the equation of motion for $\overline{Q_K^+ Q_K}$

$$
\begin{aligned}
-i\hbar\, \partial \overline{Q_K^+ Q_K}/\partial t = &\tilde{g}\overline{b}_K^+ \overline{Q}_K [c_1 - (c_2/V)\, \overline{Q_K^+ Q_K}] \\
&- \tilde{g}^* \overline{b}_K \overline{Q}_K^+ [c_1 - (c_2/V)\, \overline{Q_K^+ Q_K}] .
\end{aligned} \tag{4.4}
$$

The following abbreviations have been used:

$$
c_1 = \sum_p \phi(p), \qquad c_2/V = 2 \sum_p \phi^3(p) .
$$

The main difference with respect to the former equations rests in the fact that we now assume that the exciton mode k is macroscopically occupied and no more that with momentum $k = 0$.

Assuming a harmonic time dependence one readily finds the following dispersion law

$$
\begin{aligned}
E_K = &\tfrac{1}{2}(E_g + K^2\hbar^2/2M + \varepsilon^{(0)} + \varepsilon^{(1)} N_{\text{exc}}^0/V + \hbar\omega_K) \\
&\pm \tfrac{1}{2}[(E_g + K^2\hbar^2/2M + \varepsilon^{(0)} + \varepsilon^{(1)} N_{\text{exc}}^0/V - \hbar\omega_K)^2 \\
&+ 4\,|\tilde{g}|^2 (c_1^2 - c_1 c_2 N_{\text{exc}}^0/V)]^{\frac{1}{2}}
\end{aligned} \tag{4.5}
$$

where $\varepsilon^{(0)}$ and $\varepsilon^{(1)}$ have been defined in (3.10a), (3.10b). $N_{\text{exc}}^0 \approx \overline{Q_k^+ Q_k}$ is the mean number of excitons in the k-mode under consideration.

Using Eq. (4.2) the exciton operators can be expressed by the photon operators

$$Q_K^+ = (\tilde{g}^* c_1)^{-1} (E_K - \hbar\omega_K) b_K^+ \tag{4.6}$$

Thus the ratio $\overline{Q_K^+ Q_K}/\overline{b_K^+ b_K}$ is given by

$$\overline{Q_K^+ Q_K}/\overline{b_K^+ b_K} = |\tilde{g}|^{-2} c_1^{-2} (E_K - \hbar\omega_K)^2 . \tag{4.7}$$

References

For a review see

1. Haken, H.: Handbuch der Physik, Bd. XXV/2c: Berlin-Heidelberg-New York: Springer 1970.
2. Courtens, E.: In: Arecchi, F. T., Schulz-Dubois, E. O. (Eds.): Laser Handbook, part E. Amsterdam: Nort Holland Publ. Comp. 1972.
3. Haken, H., Schenzle, A.: Phys. Letters **41** A, 405 (1972).
4. McCall, S. L., Hahn, E. L.: Phys. Rev. Letters **18**, 908 (1967).
5. McCall, S. L., Hahn, E. L.: Phys. Rev. **183**, 457 (1969).
6. Schenzle, A., Haken, H.: Opt. Commun. **6**, 96 (1972).
7. Haken, H., Schenzle, A.: Z. Physik **258**, 231 (1973).
8. Haken, H.: Quantenfeldtheorie des Festkörpers. Stuttgart: B. G. Teubner 1972.
9. Haken, H., Schenzle, A.: Proceedings of the First Taormina Research Conference, ed. E. Burstein, 1974.
10. See e.g. Hershkowitz, N., Romesser, T., Montgomery, D.: Phys. Rev. Letters **29**, 1586 (1972), where many further references are given.

Prof. Dr. H. Haken
Dipl. Phys. J. Goll
Dr. A. Schenzle
Institut für Theoretische Physik der Universität
7000 Stuttgart-Vaihingen, Pfaffenwaldring 57
Federal Republic of Germany

Subject Index

SPRINGER TRACTS
IN MODERN PHYSICS

Ergebnisse der exakten Naturwissenschaften

Atomic and Molecular Physics

Dettmann, K.: High Energy Treatment of Atomic Collisions (Vol. 58)

Donner, W., Süßmann, G.: Paramagnetische Felder am Kernort (Vol. 37)

Langbein, D.: Theory of Van der Waals Attraction (Vol. 72)

Racah, G.: Group Theory and Spectroscopy (Vol. 37)

Seiwert, R.: Unelastische Stöße zwischen angeregten und unangeregten Atomen (Vol. 47)

Zu Putlitz, G.: Determination of Nuclear Moments with Optical Double Resonance (Vol. 37)

Elementary Particle Physics

Current Algebra

Furlan, G., Paver, N., Verzegnassi, C.: Low Energy Theorems and Photo- and Electroproduction Near Threshold by Current Algebra (Vol. 62)

Gatto, R.: Cabibbo Angle and $SU_2 \times SU_2$ Breaking (Vol. 53)

Genz, H.: Local Properties of σ-Terms: A Review (Vol. 61)

Kleinert, H.: Baryon Current Solving SU (3) Charge-Current Algebra (Vol. 49)

Leutwyler, H.: Current Algebra and Lightlike Charges (Vol. 50)

Mendes, R. V., Ne'eman, Y.: Representations of the Local Current Algebra. A Constructional Approach (Vol. 60)

Müller, V. F.: Introduction to the Lagrangian Method (Vol. 50)

Pietschmann, H.: Introduction to the Method of Current Algebra (Vol. 50)

Pilkuhn, H.: Coupling Constants from PCAC (Vol. 55)

Pilkuhn, H.: S-Matrix Formulation of Current Algebra (Vol. 50)

Renner, B.: Current Algebra and Weak Interactions (Vol. 52)

Renner, B.: On the Problem of the Sigma Terms in Meson-Baryon Scattering. Comments on Recent Literature (Vol. 61)

Soloviev, L. D.: Symmetries and Current Algebras for Electromagnetic Interactions (Vol. 46)

Stech, B.: Nonleptonic Decays and Mass Differences of Hadrons (Vol. 50)

Stichel, P.: Current Algebra in the Framework of General Quantum Field Theory (Vol. 50)

Stichel, P.: Current Algebra and Renormalizable Field Theories (Vol. 50)

Stichel, P.: Introduction to Current Algebra (Vol. 50)

Verzegnassi, C.: Low Energy Photo and Electroproduction, Multipole Analysis by Current Algebra Commutators (Vol. 59)

Weinstein, M.: Chiral Symmetry. An Approach to the Study of the Strong Interactions (Vol. 60)

Electromagnetic Interactions

Deep Inelastic Lepton Scattering

Drees, J.: Deep Inelastic Electron-Nucleon Scattering (Vol. 60)

Landshoff, P. V.: Duality in Deep Inelastic Electroproduction (Vol. 62)

Llewellyn Smith, C. H.: Parton Models of Inelastic Lepton Scattering (Vol. 62)

Rittenberg, V.: Scaling in Deep Inelastic Scattering with Fixed Final States (Vol. 62)

Rubinstein, H. R.: Duality for Real and Virtual Photons (Vol. 62)

Rühl, W.: Application of Harmonic Analysis to Inelastic Electron-Proton Scattering (Vol. 57)

Experimental Techniques

Panofsky, W. K. H.: Experimental Techniques (Vol. 39)

Magnetism

Fischer, K.: Magnetic Impurities in Metals: the *s—d* exchange model (Vol. 54)

Schmid, D.: Nuclear Magnetic Double Resonance – Principles and Applications in Solid State Physics (Vol. 68)

Stierstadt, K.: Der Magnetische Barkhauseneffekt (Vol. 40)

Optical Properties of Crystals

Bäuerle, D.: Vibrational Spectra of Electron and Hydrogen Centers in Ionic Crystals (Vol. 68)

Daniels, J., v. Festenberg, C., Raether, H., Zeppenfeld, K.: Optical Constants of Solids by Electron Spectroscopy (Vol. 54)

Excitons in High Density, *Haken, H., Nikitine, S.* (Volume Editors). Contributors: *Bagaev, V. S., Biellmann, J., Bivas, A., Goll, J., Grosmann, M., Grun, J. B., Haken, H., Hanamura, E., Levy, R., Mahr, H., Nikitine, S., Novikov, B. V., Rashba, E. I., Rice, T. M., Rogachev, A. A., Schenzle, A., Shaklee, K. L.* (Vol. 73)

Godwin, R. P.: Synchrotron Radiation as a Light Source (Vol. 51)

Pick, H.: Struktur von Störstellen in Alkalihalogenidkristallen (Vol. 38)

Raether, H.: Solid State Excitations by Electrons (Vol. 38)

Quantum Statistics

Agarwal, G. S.: Quantum Statistical Theories of Spontaneous Emission and their Relation to Other Approaches (Vol. 70)

Graham, R.: Statistical Theory of Instabilities in Stationary Nonequilibrium Systems with Applications to Lasers and Nonlinear Optics (Vol. 66)

Haake, F.: Statistical Treatment of Open Systems by Generalized Master Equations (Vol. 66)

Semiconductors

Feitknecht, J.: Silicon Carbide as a Semiconductor (Vol. 58)

Grosse, P.: Die Festkörpereigenschaften von Tellur (Vol. 48)

Schnakenberg, J.: Electron-Phonon Interaction and Boltzmann Equation in Narrow Band Semiconductors (Vol. 51)

Superconductivity

Lüders, G., Usadel, K.-D.: The Method of the Correlation Function in Superconductivity Theory (Vol. 56)

X-Ray, Neutron-, Electron-Scattering

Steeb, S.: Evaluation of Atomic Distribution in Liquid Metals and Alloys by Means of X-Ray, Neutron and Electron Diffraction (Vol. 47)

Springer, T.: Quasi-Elastic Scattering of Neutrons for the Investigation of Diffusive Motions in Solids and Liquids (Vol. 64)

To Appear in Volume 74

Bauer, G.: Determination of Electron Temperatures and of Hot-Electron Distribution Functions in Semiconductors

Borstel, G., Falge, H. J., Otto, A.: Surface and Bulk Phonon-Polarisations Observed by Attenuated Total Reflection